Nucleic Acids and Molecular Biology

Volume 21

Series Editor

H.J. Gross, Institut für Biochemie, Biozentrum, Am Hubland,
97074 Würzburg, Germany, hj.gross@biozentrum.uni-wuerzburg.de

Holger Heine
Editor

Innate Immunity of Plants, Animals, and Humans

Springer

Dr. Holger Heine
Department of Immunology and Cell Biology
Research Center Borstel
Parkallee 22
23845 Borstel
Germany
hheine@fz-borstel.de

ISBN 978-3-540-73929-6   e-ISBN 978-3-540-73930-2

Nucleic Acids and Molecular Biology ISSN 0933-1891

Library of Congress Control Number: 2007935972

© 2008 Springer-Verlag Berlin Heidelberg

This work is subject to copyright. All rights are reserved, whether the whole or part of the material is concerned, specifically the rights of translation, reprinting, reuse of illustrations, recitation, roadcasting, reproduction on microfilm or in any other way, and storage in data banks. Duplication of this publication or parts thereof is permitted only under the provisions of the German Copyright Law of September 9, 1965, in its current version, and permission for use must always be obtained from Springer. Violations are liable to prosecution under the German Copyright Law.

The use of general descriptive names, registered names, trademarks, etc. in this publication does not imply, even in the absence of a specific statement, that such names are exempt from the relevant protective laws and regulations and therefore free for general use.

Cover design: WMXDesign GmbH, Heidelberg, Germany

Printed on acid-free paper

9 8 7 6 5 4 3 2 1

springer.com

Preface

All living organisms are in a constant battle against their environment. Since uncontained microorganisms would simply overgrow all higher animals, from the beginning of the evolution of multicellular organisms the need was clearly evident for adequate and efficient defense mechanisms to protect their own integrity and to ensure their own survival. Usually, the first encounter with pathogens occurs at epithelial interfaces, which present the first barrier against invading pathogens and already comprises a number of mechanical and chemical defense mechanisms. However, in addition to these passive mechanisms an arsenal of active weapons also evolved. As it turned out, some of them were so efficient that basically all organisms rely at least partly on them: there is no known species that does not produce antimicrobial peptides, which represent very likely the most ancient immune defense molecules and the most common effector molecules of the innate immune response.

Over recent decades, the appreciation of the innate immune system has vastly increased. A pivotal event and possibly the beginning of the modern era of innate immunity was Charles Janeway's opening lecture at the annual Cold Spring Harbor Symposium of Quantitative biology in 1989. He hypothesized that recognition of certain patterns or characteristics of infectious microorganisms through pattern recognition receptors whose specificity is "hard-wired" into the genome is vitally important for the immune response. However, it took about seven years before the involvement of the *Drosophila* Toll protein in the immune response was discovered by Jules Hoffmann's group in Strasbourg. One year later, the first human counterpart was discovered by Medzhitov and Janeway and the era of mammalian Toll-like receptors and the search for their ligands began.

Finally, people began to understand just how specific the so-called "unspecific" innate immune response really is. Since then, these receptors has been found and investigated in many species. It became clear that in higher animals the innate and the adaptive immune system is strongly intertwined and that the activation of the innate immune system is required for the activation of adaptive immune system. However, one has to remember that for many species the innate immune system is the sole active defense system and that it comprises many more mechanisms than only the detection of pathogen-associated molecular patterns through Toll-like receptors.

This book wants to give an overview of our current knowledge about the innate immune system of plants, animals and humans. In the first six chapters, the innate immune mechanisms and responses of so diverse organisms such as plants, Cnidaria, *Drosophila*, urochordates and zebrafish are presented and reviewed in great detail. Shunyuan Xiao presents an overview about the evolution of plant resistance genes, which evolved as a response to the recognition of pathogen effector proteins in plants. The next chapters cover organisms that are at critical places on the evolutionary tree. First, Thomas C.G. Bosch et al. provide fascinating information about one of the earliest multicellular species, the ancient group of Cnidaria, which diverged from the so-called Bilateria long before insects and worms evolved. Since the innate immune system of the fruitfly *Drosophila melanogaster* is among the best studied of all species, two chapters cover the field. Neal Silverman's group discusses the molecular mechanisms of pathogen recognition and signal transduction that leads to the elimination of invading microbes, whereas the group of Louisa Wu further elucidates two very important aspects of the cellular innate immunity: the encapsulation and phagocytosis of pathogens by *Drosophila* hemocytes. Next, Konstantin Khalturin et al. present an overview of the innate immune responses of the urochordates, which present the vertebrates closest relatives and thus provide insight into innate immune mechanisms just before the sudden appearance of adaptive immunity. Moving along the evolutionary tree, Con Sullivan and Carol H. Kim provide a review about innate immune responses of the zebrafish, *Danio reo*. In contrast to all species covered so far, the zebrafish is the first species that in addition to its innate immune defenses also contains an adaptive immune system.

The last four chapters deal with different aspects of the mammalian innate immune system: Andrei Medvedev and Stefanie Vogel provide detailed information about the human and mouse Toll-like receptor (TLR) family including their ligands and signal transduction. Besides the family of TLRs that all are expressed on cell or endosomal membranes, a new family of intracellular and cytosolic pattern recognition receptors has recently emerged. Named after the unifying expression of the nucleotide oligomerization domain (NOD) and with respect to the TLRs the members of this family are called NOD-like receptors. This family consists of 22 members and a number of mutations have been found in these proteins that are surprisingly often linked to inflammatory diseases.

Finally, two chapters present the major effector mechanisms of the innate immune system: Regine Gläser, Jürgen Harder, and Jens-Michael Schröder provide an up-to-date overview about human antimicrobial peptides; and Bob Sim et al. review the complement system.

Contents

1 Evolution of Resistance Genes in Plants 1
Shunyuan Xiao, Wenming Wang, and Xiaohua Yang

    1   Evolution of the Plant *R* Gene System 2
    2   Conservation and Diversity of Plant *R* Genes 3
    3   NBS, LRR and TIR – Domains of Defense 6
    4   Proliferation and Diversification of *NBS-LRR* Genes in Plants 9
    5   Mechanisms of R-Avr Recognition 10
        5.1   Direct R-Avr Interaction – the "Gene-For-Gene" Hypothesis . 10
        5.2   Indirect R-Avr Interaction – the "Guard" Hypothesis 11
    6   Patterns of *R-Avr* Coevolution 13
        6.1   Diversifying Selection Results from Direct R-Avr Recognition? 14
        6.2   Balancing Selection Results from Indirect Recognition? 15
        6.3   A General Model for Evolution of the Plant R Gene System .. 17
    7   New Perspectives 19
    References 20

2 The Path Less Explored: Innate Immune Reactions in Cnidarians.. 27
Thomas C.G. Bosch

    1   Cnidaria Are Among the Earliest Multicellular Animals
        in the Tree of Life 28
    2   Immune Reactions in Invertebrates 30
    3   Immune Reactions in Cnidaria 30
        3.1   How to Fight for a Space to Live? Intraspecies
               Competition in Sea Anemones 31
        3.2   How to Detect Approaching Allogeneic Cells
               as Foreign and to Eliminate Them? Allorecognition
               and Cell Lineage Competition in Colonial *Hydractinia* 32
        3.3   How to Detect and Disarm Microbial Attackers?
               Antimicrobial Defense Reactions in the Freshwater
               Polyp *Hydra* and the Jellyfish *Aurelia* 34

    3.4 How to Distinguish Between Friends and Foes:
         Symbiotic Relationships in Corals and *Hydra* 36
  4 How to Explore the Path They Went? Why Cnidarians Matter 38
  References 39

3 Bug Versus Bug: Humoral Immune Responses in *Drosophila melanogaster* ... 43
Deniz Ertürk-Hasdemir, Nicholas Paquette, Kamna Aggarwal, and Neal Silverman

  1 Introduction... 44
    1.1 A Brief History 44
    1.2 Overview of the *Drosophila* Immune Response 45
  2 Microbial Recognition – the Peptidoglycan Recognition Proteins .. 45
    2.1 Peptidoglycan 47
    2.2 NF-κB Proteins 48
  3 The Toll Pathway 49
  4 The IMD Pathway... 55
  5 Down-Regulation of the IMD Pathway by PGRP Amidases 60
  6 JAK/STAT Pathway 61
  7 Concluding Remarks 62
  References 63

4 Cellular Immune Responses in *Drosophila melanogaster* 73
Adrienne Ivory, Katherine Randle, and Louisa Wu

  1 Introduction... 74
  2 Encapsulation 75
    2.1 Recognition Centers on Membrane Differences 76
    2.2 Lamellocyte Proliferation: Necessary for Successful
         Encapsulation Response 78
    2.3 Adhesion Requires Integrins, Rac, and Rho 78
    2.4 Encapsulation Terminates with the Formation
         of Basement Membrane................................... 82
  3 Phagocytosis.. 83
    3.1 Proteins Opsonize Invading Bacteria and Fungi
         to Promote Phagocytosis 84
    3.2 Transmembrane and Circulating Peptidoglycan Recognition
         Proteins are Involved in the Recognition of Bacteria 85
    3.3 Receptors with Scavenger-Like Activity Recognize
         a Variety of Microbes 86
    3.4 Phagocytosis Requires Reorganization of the Actin
         Cytoskeleton 88
    3.5 Engulfed Pathogens are Degraded in Phagolysosomes....... 89
    3.6 Interactions Between Cellular and Humoral
         Immune Responses 90
  References 91

5 Immune Reactions in the Vertebrates' Closest Relatives, the Urochordates 99
Konstantin Khalturin, Ulrich Kürn, and Thomas C.G. Bosch

1 Introduction.. 100
2 Urochordates are at the Root of Vertebrate Evolution 100
3 Natural History and Ecology of Urochordates 101
4 Immunity in Urochordates 102
   4.1 Antimicrobial Peptides from Urochordates................. 103
   4.2 Allorecognition in Urochordates 104
   4.3 Complement in Urochordates 106
   4.4 Despite the Absence of MHC, Urochordate Blood Contains NK-Like Cells 107
5 Conclusion 108
References 108

6 Innate Immune System of the Zebrafish, *Danio rerio* 113
Con Sullivan and Carol H. Kim

1 Overview... 114
2 Components of Innate Immunity 115
   2.1 General Description..................................... 115
   2.2 *Drosophila* Toll: Identification and Recognition of a Dually Functioning Pathway 116
   2.3 TLRs and TIR-Bearing Adaptor Proteins 117
3 Zebrafish as a Model for Infectious Disease and Innate Immune Responses.. 118
   3.1 Overview 118
   3.2 Forward and Reverse Genetics 119
   3.3 An Infectious Disease and Innate Immunity Model 120
4 NK-Like Cells... 122
5 Additional Innate Immunity Receptors in Zebrafish 124
6 Zebrafish Phagocytes 126
7 Conclusion 126
References 127

7 Toll-Like Receptors in the Mammalian Innate Immune System 135
Andrei E. Medvedev and Stefanie N. Vogel

1 Introduction.. 136
2 TLRs as Primary Sensors of Pathogenic PAMPS and Endogenous "Danger" Molecules 137
3 TLR Signaling Pathways 141
   3.1 Interaction of TLRs with PAMPs and Co-Receptors Initiates Signaling 141
   3.2 Role of TIR-Containing Adapter Molecules in TLR Signaling..................................... 142

3.3　TLR Specificity for PAMPs in the Ectodomain
　　　　　and Adapters in the TIR Domain Underlie a Dual
　　　　　Recognition/Response System 146
　　　3.4　The IRAK Family: Key Regulators of TLR Signaling 147
　4　Mutations in TLRs and IRAK-4: Implications for Disease 149
　5　Conclusions 155
　References 156

8　NLRs: a Cytosolic Armory of Microbial Sensors Linked to Human Diseases 169
Mathias Chamaillard

　1　Introduction 170
　2　NLRs, a Conserved Cytosolic Arm of the Innate Immune System .. 171
　3　Physiological Role of NLRs in Innate and Adaptive Immunity:
　　　NLRs Join TLRs 174
　　　3.1　Host Sensing of Non-TLR PAMPs: Lessons from NOD1
　　　　　and NOD2 Studies 174
　　　3.2　NLRs Promote Maturation of TLR-Induced Il-1β
　　　　　and IL-18 Release 176
　4　What Can we Learn from NLRs Linked to Human Diseases?...... 177
　　　4.1　NOD1 and NOD2 Mutations Linked to Chronic
　　　　　Inflammatory Diseases 178
　　　4.2　Auto-Inflammatory Diseases 179
　　　4.3　Reproduction Diseases 179
　5　Concluding Remarks: Towards the Development
　　　of "Magic" Bullets 179
　References 180

9　Antimicrobial Peptides as First-Line Effector Molecules of the Human Innate Immune System 187
Regine Gläser, Jürgen Harder, and Jens-Michael Schröder

　1　Introduction... 188
　2　Epithelial Antimicrobial Peptides and Proteins 189
　　　2.1　Lysozyme 189
　　　2.2　Human Beta Defensins 189
　　　2.3　Human Alpha Defensins 194
　　　2.4　RNases 195
　　　2.5　S100 Proteins: S100 A7 (Psoriasin) 196
　　　2.6　Others 198
　3　Phagocyte Antimicrobial Peptides 200
　　　3.1　Human Alpha Defensins 200
　　　3.2　Cathelicidins 200
　　　3.3　S100 Proteins: S100 A8/9 (Calprotectin) and S100A12
　　　　　(Calgranulin C) 201
　　　3.4　Others... 202

	4	Putative Action of Antimicrobial Peptides in the Healthy Human .	202
	5	Antimicrobial Peptides and Diseases	205
		5.1 Skin Diseases	205
		5.2 Wound Healing.	206
		5.3 Diseases of the Airway Epithelia: Cystic Fibrosis	207
		5.4 Gastrointestinal Diseases: Inflammatory Bowel Diseases . . .	208
		5.5 Diseases Associated with Phagocyte Dysfunction	209
	6	General Conclusion and Future Aspects	210
	References		210

10 The Complement System in Innate Immunity **219**
K.R. Mayilyan, Y.H. Kang, A.W. Dodds, and R.B. Sim

	1	The Complement System in Mammals	220
		1.1 Classical Pathway	221
		1.2 The Lectin Pathway	223
		1.3 Alternative Pathway	225
		1.4 Regulation of the Complement System	228
		1.5 Complement Receptors	229
	2	The Structure of Complement Proteins	230
	3	Complement Across Species	232
	References		233

Index. **237**

Contributors

Shunyuan Xiao
Center for Biosystems Research, University of Maryland Biotechnology Institute, 9600 Gudelsky Drive, Rockville, MD 20850, USA, xiao@umbi.umd.edu

Thomas C.G. Bosch
Zoological Institute, Christian-Albrechts-University Kiel, Olshausenstrasse 40, 24098 Kiel, Germany, tbosch@zoologie.uni-kiel.de

Neal Silverman
Division of Infectious Diseases, Department of Medicine, University of Massachusetts Medical School, Worcester, MA 01605, USA, Neal.Silverman@umassmed.edu

Louisa Wu
Center for Biosystems Research, University of Maryland Biotechnology Institute, 5115 Plant Sciences Bldg, College Park, MD 20742, USA, wul@umbi.umd.edu

Konstantin Khalturin
Zoological Institute, Christian-Albrechts-University, Olshausenstrasse 40, 24098 Kiel, Germany, kkhalturin@zoologie.uni.kiel.de

Carol H. Kim
Department of Biochemistry, Microbiology, and Molecular Biology, University of Maine, Orono, ME 04469, USA, carolkim@maine.edu

Andrei E. Medvedev
Department of Microbiology and Immunology, School of Medicine, University of Maryland, Baltimore (UMB), 660 W. Redwood Street, Rm. HH 324, Baltimore, MD 21201, USA, amedvedev@som.umaryland.edu

Mathias Chamaillard
INSERM U795, Physiopathology of Inflammatory Bowel Disease, Swynghedauw Hospital, Rue A. Verhaeghe, 59037 Lille, France, m-chamaillard@chru-lille.fr

Regine Gläser
Department of Dermatology, Venerology and Allergology, Clinical Research Unit, University Hospital Schleswig-Holstein, Campus Kiel, Schittenhelmstrasse 7, 24105 Kiel, Germany, rglaeser@dermatology.uni-kiel.de

R.B. Sim
MRC Immunochemistry Unit, Department of Biochemistry, University of Oxford, South Parks Road, Oxford OX1 3QU, UK, bob.sim@bioch.ox.ac.uk

Chapter 1
Evolution of Resistance Genes in Plants

Shunyuan Xiao(✉), Wenming Wang, and Xiaohua Yang

1	Evolution of the Plant *R* Gene System..	2
2	Conservation and Diversity of Plant *R* Genes...	3
3	NBS, LRR and TIR – Domains of Defense...	6
4	Proliferation and Diversification of *NBS-LRR* Genes in Plants.......................................	9
5	Mechanisms of R-Avr Recognition	10
	5.1 Direct R-Avr Interaction – the "Gene-For-Gene" Hypothesis...............................	10
	5.2 Indirect R-Avr Interaction – the "Guard" Hypothesis	11
6	Patterns of *R-Avr* Coevolution	13
	6.1 Diversifying Selection Results from Direct R-Avr Recognition?.........................	14
	6.2 Balancing Selection Results from Indirect Recognition?......................................	15
	6.3 A General Model for Evolution of the Plant *R* Gene System	17
7	New Perspectives	19
References..		20

Abstract Potential pathogens deliver effector proteins into plant cells to suppress microbe-associated molecular pattern (MAMP)-triggered immunity in plants, resulting in host–pathogen coevolution. To counter pathogen suppression, plants evolved disease resistance (R) proteins to detect the presence of the pathogen effectors and trigger R-dependent defenses. Most isolated *R* genes encode proteins possessing a leucine-rich-repeat (LRR) domain, of which the majority also contain a nucleotide-binding site (NBS) domain. There is structural similarity and/or domain homology between plant R proteins and animal immunity proteins, suggesting a common origin or convergent evolution of the defense proteins. Two basic strategies have evolved for an R protein to recognize a pathogen effector (then called avirulence factor; Avr): direct physical interaction and indirect interaction via association with other host proteins targeted by the Avr factor. Direct R-Avr recognition leads to high genetic diversity at paired *R* and *Avr* loci due to diversifying selection, whereas indirect recognition leads to simple and stable polymorphism at the *R* and *Avr* loci due to balancing selection. Based on these two patterns of *R-Avr* coevolution, investigation of the sequence features at paired *R* and *Avr* may help infer the

---

Center for Biosystems Research, University of Maryland Biotechnology Institute,
9600 Gudelsky Drive, Rockville, MD 20850, USA, *xiao@umbi.umd.edu*

R-Avr interaction mechanisms, assess the role and strength of natural selection at the molecular level in host–pathogen interactions and predict the durability of *R* gene-triggered resistance.

Abbreviations *R*, resistance gene; *Avr*, avirulence gene; HR, hypersensitive response; MAMP, microbe associated molecular patterns; MTI, MAMP-triggered immunity; ETI, Effector-triggered immunity; TIR, toll and interleukin receptor; NBS, nucleotide binding site; LRR, leucine rich repeat; RLP, receptor-like protein; RLK, receptor-like kinase

1 Evolution of the Plant *R* Gene System

Plant innate immunity consists of preformed physical and chemical barriers (such as leaf hairs, rigid cell walls, pre-existing antimicrobial compounds) and induced defenses. Should an invading microbe successfully breach the preformed barriers, it may be recognized by the plant, resulting in the activation of cellular defense responses that stop or restrict further development of the invader (Nurnberger et al. 2004). Apart from virus-induced RNA silencing, an ancient, evolutionary conserved antiviral defense mechanism in both plants and animals (which is not discussed in this chapter), two evolutionarily interrelated mechanisms have evolved in plants for detection of the invading microbes. First, plants are able to recognize some conserved microbe-derived molecules which are collectively described as microbe-associated molecular pattern (MAMP) by cell-surface receptors and trigger immune response (Gomez-Gomez and Boller 2000; Zipfel et al. 2006). Evidence is accumulating that this so-called MAMP-triggered immunity (MTI) is evolutionarily ancient and may be a general feature of plant resistance against a broad-spectrum of potential pathogens (Nurnberger et al. 2004; He et al. 2006). This type of resistance occurs at or above the species level, and is often referred to as non-host resistance. It can be envisaged that microbes that successfully breached constitutive defensive barriers of plants but were restricted by MTI gradually evolved strategies to target and sabotage the MTI. Increasing evidence indicates that successful microbes suppressed MTI by sending effector proteins into the plant cell to interfere with the host defense system, resulting in the breakdown of non-host resistance and the establishment of a host–pathogen interaction. The "defeated" host then faced selection pressure imposed by the successful pathogen to evolve novel defense mechanism to survive. This led to the evolution of the second recognition mechanism for which plants evolved disease resistance (R) proteins to specifically detect the presence of the pathogen effectors [called avirulence factors (Avr) once recognized by R proteins] and subsequently trigger a much stronger defense response to counter the suppression of MTI by the pathogen (Chisholm et al. 2006). Thus, *R* gene-dependent, pathogen-*e*ffector-*t*riggered host *i*mmunity (ETI) most likely evolved on top of MTI to fortify the plant immune system. Recent publications strongly support this inference (Kim et al. 2005; He et al. 2006; Nomura et al. 2006). For example, He and colleagues recently found that HopM1, a conserved effector protein of *Pseudomonas syringae*, targets an immunity-associated protein, AtMIN7 in *Arabidopsis thaliana*. HopM1 mediates the destruction of AtMIN7 via the host proteasome (Nomura et al. 2006). Sheen and colleagues found that AvrPto and AvrPtoB, two effector proteins of the bacterial pathogen *P. syringae* suppress MTI at an early step upstream of MAPK signaling (He et al. 2006). Both AvrPto and AvrPtoB are recognized by the plant R protein Pto in tomato, thereby triggering Pto-dependent resistance (Kim et al. 2002).

Evolution of the ETI system in plants marks a higher level of plant–pathogen coevolution in which the major players are plant *R* and pathogen *Avr* genes. Unlike MTI, which is expressed in all plants of a given species, ETI is often expressed in some but not all genotypes within a plant species. This correlates to the phenomenon that there are often two likely outcomes from a given host–pathogen interaction: (a) compatible interaction in which the pathogen is able to suppress host defenses and colonize the plant; (b) incompatible interaction in which the pathogen is detected by the plant containing an *R* gene and the plant is resistant. Therefore, genetically defined *R* genes are polymorphic determinants of host resistance against specific pathogens.

MTI in plants resembles the innate immune system of animals in that structurally similar cell-surface receptors are deployed to recognize MAMPs such as flagellin and lipopolysaccharides and the induction of host defenses involves MAPK signaling cascades (Nurnberger et al. 2004). Thus, MTI seems to be a highly conserved defense mechanism evolved in both plants and animals. Interestingly, so far there is no clear evidence to indicate the existence of ETI in animals. Therefore, it appears that the evolution of an elaborative plant ETI system in which a large array of R proteins function as receptors to recognize pathogen-specific effectors constitutes an important distinction between the plant and animal innate immune systems (Ausubel 2005). This probably reflects the consequence of adaptive evolution: plants are sessile, lack a circulating system and live relatively longer than most invertebrate animals; thus evolution of a greater capacity in every single cell to respond and mount effective defenses against numerous microbial invaders seems to be a logical choice for plants.

In the following sections, we focus our review on the current understanding of evolution and maintenance of plant *R* genes within the context of concomitant evolution of pathogen *Avr* genes that interact with *R* genes. For detailed molecular mechanisms of *R* gene evolution, we strongly recommend several excellent earlier review articles (Michelmore and Meyers 1998; Bergelson et al. 2001; Holub 2001; Meyers et al. 2005).

2 Conservation and Diversity of Plant *R* Genes

Since the isolation of the first plant *R* gene, *Hm1* in maize in 1992 (Johal and Briggs 1992), over 60 plant *R* genes controlling resistance against pathogens ranging from viruses, bacteria, fungi to nematodes have been isolated from different plant species

(Xiao 2006). Most isolated *R* genes seem to activate common or overlapping sets of defense programs in local areas infected by pathogens. Those defense responses include transcriptional induction of pathogenesis-related (*PR*) genes, production of reactive oxygen species, fortification of the cell wall, synthesis of antimicrobial compounds and, in many cases, a hypersensitive response (HR) which is a form of plant programmed cell death analogous to animal apoptosis (Hammond-Kosack and Jones 1997; Dangl and Jones 2001; Nurnberger et al. 2004). The primary local resistance triggered by *R* genes may also lead to activation of a secondary defense termed systemic acquired resistance in the uninfected tissues, which is a more long-lasting immune response throughout the whole plant against a broad range of pathogens (Durrant and Dong 2004).

Based on features of the deduced domain structures and/or biochemical functions, R proteins can be divided into three classes (Table 1). The largest class contains a nucleotide-binding site (NBS) and leucine-rich-repeat (LRR) motifs (Hammond-Kosack and Jones 1997; Dangl and Jones 2001). These R proteins confer resistance to various pathogens and can be further subdivided into two groups, based on their N-terminal features. The first group contain an N-terminal domain resembling the cytoplasmic signaling domain of the *Drosophila t*oll and human *i*nterleukin-1 *r*eceptors (TIR) and are called TIR-NBS-LRRs (Whitham et al. 1994; Lawrence et al. 1995). The second group contain (in most cases) a coiled-coil (CC) domain and thus often are referred to as CC-NBS-LRRs (Bent et al. 1994; Grant et al. 1995). An exceptional case in the TIR-NBS-LRR group is the *Arabidopsis* RRS1-R protein that has a WRKY domain attached to the LRR at the C-terminus (Deslandes et al. 2002). The WRKY domain is found in a group of transcription factors implicated in the signal transduction of *R* genes (Eulgem 2005). The structural feature of RRS1-R implies a direct link between Avr-recognition and the transcriptional activation of defense genes (Deslandes et al. 2003).

The second class of R proteins comprise cell surface receptor-like transmembrane proteins (RLP) and receptor-like kinases (RLK) (Table 1). The common feature of these proteins is that they possess an extracellular LRR (eLRR) domain. Representatives of RLP R proteins are tomato Cf proteins conferring resistance to the tomato fungal pathogen *Cladosporium fulvum* (Jones et al. 1994; Hammond-Kosack and Jones 1997) and *Arabidopsis RPP27* conferring resistance to the oomycete *Hyaloperonospora parasitica* (Tor et al. 2004). RLK R proteins are represented by rice Xa21 and Xa26, both of which confer resistance to multiple strains of *Xanthomonas oryzae* pv. *oryzae* (Song et al. 1995; Sun et al. 2004).

The remaining *R* genes encode proteins that either resemble the overall structure or a domain of the above two classes with some degree of structural variation, or have a novel protein structure that does not show significant homology to any other R proteins (Table 1). Therefore, they are atypical *R* genes in comparison with the LRR-encoding *R* genes. For example, tomato *Pto* and *Arabidopsis PBS1* encode members of a conserved protein kinase family (Martin et al. 1993; Swiderski and Innes 2001) that resemble the cytoplasmic protein kinase domain of RLK R proteins. The broad-spectrum powdery mildew *R* gene *RPW8* from *Arabidopsis* encodes a small protein containing an N-terminal transmembrane domain and a CC

Table 1 Conservation and diversity of plant R proteins. *TIR* Toll and interleukin-1 receptor, *NBS* nucleotide binding site, *(e)LRR* (extracellular) leucine rich repeats, *CC* coiled coil, *Kin* kinase, *TM* transmembrane helix predicted by TMpred and TMHMM

R protein class	Schematic domain structure	Predicted function	Examples	References
NBS-LRR		Receptor	N, L	Whitham et al. (1994), Lawrence et al. (1995)
		Receptor	RPM1, RPS2	Bent et al. (1994), Grant et al. (1995)
		Receptor	RRS1-R	Deslandes et al. (2002)
eLRR		Receptor	Cf9; RPP27	Jones et al. (1994), Tor et al. (2004)
		Receptor	Xa21, Xa26	Song et al. (1995), Sun et al. (2004)
Atypical		Host target?	Pto, PBS1	Martin et al. (1993), Swiderski and Innes (2001)
		?	RPW8	Xiao et al. (2001)
		?	Xa27	Gu et al. (2005)
		Fertility	Xa13	Chu et al. (2006)
		Negative regulator of PCD	MLO	Buschges et al. (1997)

(Xiao et al. 2001) that shows homology to the CC domain of NRG1 (Peart et al. 2005) and At5g66910 (Meyers et al. 2003), both of which are NBS-LRR proteins. The recessive *R* gene *mlo* from barley confers non-race-specific resistance against powdery mildew (*Blumeria graminis* f. sp. *hordei*) and the MLO wild-type protein contains seven transmembrane helices (Buschges et al. 1997). Two recently cloned rice *R* genes *Xa27* and *xa13*, each conferring resistance in rice to certain strains of the bacterial pathogen *X. oryzae* pv. *oryzae*, encode novel proteins that do not show homology to any other R proteins (Gu et al. 2005; Chu et al. 2006). Notably, *xa13* is also recessive and the deduced Xa13 protein contains eight transmembrane helices, which is reminiscent of *mlo*/MLO (Table 1).

Several points can be highlighted from the findings regarding R proteins:

1. The majority of cloned *R* genes encode NBS-LRR proteins, suggesting a prominent role of NBS-LRRs in host resistance (i.e. ETI).
2. The most common protein domain in R proteins is LRR, which is present in both NBS-LRR and eLRR R proteins, implying a role in pathogen detection for this domain.
3. Except for the kinase *R* genes, the rare/unique structure of some individual R proteins coincides with its unique representation as an *R* gene in a specific plant species, implying that evolution of these *R* genes may have involved different mechanisms.

From an evolutionary point of view, there are several challenging questions to ask: Why are the structures of NBS-LRR and eLRR receptors the choice for most R proteins? Is there any evolutionary link between plant R proteins and animal immunity proteins? Why are *NBS-LRR* genes so abundant in the plant genomes? How did a specific *R* gene originate and coevolve with the cognate pathogen *Avr* gene and how have *R* and *Avr* genes been maintained in the plant and pathogen populations respectively? Why and how did rare *R* genes evolve? Recent progress in this field has made it possible to at least partially address these questions.

3 NBS, LRR and TIR – Domains of Defense

Perhaps among the most exciting discoveries in the biology of host immunity is that plant R proteins possess domains that share homology to animal proteins involved in innate immunity and/or apoptosis (Hammond-Kosack and Jones 1997; Dangl and Jones 2001; Ausubel 2005). First, the TIR domain in the TIR-NBS-LRR R proteins shows homology to the Toll and IL-1 receptor. TIR domain-containing animal proteins such as *Drosophila* Toll and mammalian Toll-like receptors (such as TLR2, 4, 5, 9) are involved in innate immunity in the animal system that resembles plant MTI (Roeder et al. 2004; Takeda and Akira 2004). Second, the overall structure of the cytoplasmic NBS-LRR proteins is similar to that of mammalian proteins such as Nod1, Nod2 and NALP3, which function as intracellular receptors for bacterial MAMPs and play a key role in innate immunity (Staskawicz et al.

2001; Inohara and Nunez 2003). Last, an eLRR domain is present in both RLP and RLK R proteins and in Toll and TLR receptors, as well as in FLS2 and ERF, the two recently characterized plant MAMP-receptors (Gomez-Gomez and Boller 2000; Zipfel et al. 2006).

The structural similarity and/or domain homology of R proteins to animal immunity proteins, plus the similarity in the overall signaling structure of MTI in both plants and animals (Dangl and Jones 2001; Nurnberger et al. 2004) provoked speculation that the domains of these defense protein might have evolved in an ancient unicellular eukaryote pre-dating the separation of the plant and animal kingdoms and that the occurrence of plant- and animal-specific receptors consisting of these domains could be a result of divergent evolution (Fluhr and Kaplan-Levy 2002; Nurnberger and Brunner 2002). After careful comparative examination of the overall mechanisms of recognition and signaling for animal immunity proteins and plant R proteins, Ausubel (2005) recently proposed that these seemingly analogous regulatory modules used in plant and animal innate immunity evolved independently by convergent evolution and reflect inherent constraints on how an innate immune system can be constructed.

Regardless of the antiquity of these analogous defense protein domains, a fundamental question to be asked is why these characteristic domains, i.e. TIR, NBS and (e)LRR, were recruited for the construction of plant and animal immunity proteins.

In animals, the TIR domain is found only in immune-related proteins such as Toll/TLRs and in downstream adaptor proteins such as MyD88 and Mal. The function of the TIR is thought to mediate protein–protein interaction for signal transduction from the receptors to downstream components in the immune system (O'Neill et al. 2003; Yamamoto et al. 2004). In plants, the TIR domain is found not only in TIR-NBS-LRR proteins with a likely role in signaling and Avr recognition (Ellis et al. 1999; Luck et al. 2000), but also in putative TIR-NBS and TIR-X proteins (X is variable) in *Arabidopsis* (Meyers et al. 2002). It is likely that the *TIR-NBS* and *TIR-X* genes may also be involved in plant defense (Meyers et al. 2002). Interestingly, while *CC-NBS-LRR* genes are present in both dicot and monocot plant genomes, *TIR-NBS-LRR* genes are not found in rice and other monocot genomes (Bai et al. 2002; Cannon et al. 2002; Zhou et al. 2004), indicating that the function played by TIR in immunity may be dispensable and/or can be complemented by other domains, at least in monocot plants. However, it is noted that homologs of *Arabidopsis* signaling components also exist in rice, such as *EDS1* and *PAD4* that are required for the function of many *TIR-NBS-LRR* genes but not for *CC-NBS-LRR* genes (Aarts et al. 1998; Goff et al. 2002), implying that these genes may have a more general function in defense signaling. Indeed, *EDS1* and *PAD4* have been shown to play an important role in basal resistance against biotrophic pathogens in the absence of specific *R-Avr* recognition (Wiermer et al. 2005), implying that TIR-NBS-LRR-specified immunity may be mechanistically superimposed onto a more evolutionarily ancient basal resistance mechanism.

The NBS of intracellular plant R proteins and animal Nod1/Nod2 proteins is part of a larger domain called NB-ARC, because it is shared between R proteins and both human apoptotic protease-activating factor 1 (APAF-1) and its *Caenorhabditis elegans* homolog CED-4 (Van der Biezen and Jones 1998b). The NBS domain is proposed to work as an NTP-hydrolyzing (ATPase or GTPase, etc) switch, regulating signal transduction by conformational changes (Leipe et al. 2004). In vitro binding studies on the I-2 protein (a CC-NBS-LRR) suggest that the NBS domain provides a molecular switch by alternating between two different states: the resting ADP-bound state and the active ATP-bound state (Takken et al. 2006; Tameling et al. 2006). The NBS domain is highly conserved among R proteins (McHale et al. 2006), making it possible to design degenerate primers against this domain to amplify *R* gene analogs from various plant genomes without prior genomic information (Cannon et al. 2002).

The LRR domain is a common motif found in numerous proteins from viruses to eukaryotes that function in diverse processes from development to disease resistance. The predicted biochemical function of the LRR domain is to mediate protein–protein interaction. This domain consists of up to 47 duplicated LRRs (Baumgarten et al. 2003). Each typical LRR of R proteins comprises a core of about 26 amino acids containing the L-xx-L-xx-Lx-L-xx motif (where x is any amino acid), which forms a β-sheet (McHale et al. 2006). Sequences encoding putative solvent-exposed residues (i.e. the x residues) are hypervariable among different R proteins and show significantly elevated ratios of non-synonymous to synonymous substitutions, indicating that the LRR domain is subject to positive selection for amino acid diversification (Parniske et al. 1997; McDowell et al. 1998; Michelmore and Meyers 1998). It has been hypothesized and experimentally confirmed that the LRR domain of R proteins is involved in the specific recognition of pathogen effectors (Hammond-Kosack and Jones 1997; Ellis et al. 2000; Van der Hoorn et al. 2001).

How the individual domains of the R proteins function coordinately to perceive signals from pathogens and activate downstream signaling cascades is not well understood. It is assumed that R proteins form protein complexes with other host proteins via intermolecular interaction to implement their function (Belkhadir et al. 2004). For example, SGT1, a host protein required for the function of many but not all R proteins (Muskett and Parker 2003) appears to associate with Bs2, a CC-NBS-LRR protein from pepper when transiently coexpressed in *Nicotiana benthamiana* (Leister et al. 2005). Exciting recent studies indicate that there is also intramolecular interaction between different domains of the NBS-LRR proteins (Moffett et al. 2002; Leister et al. 2005; Rairdan and Moffett 2006). Moffett and colleagues found that the CC domain and LRR domain of Rx (a CC-NBS-LRR conferring resistance to potato virus x) could function *in trans* and the LRR domain interacted physically *in planta* with the CC-NBS, as did CC with NBS-LRR (Moffett et al. 2002). More recently, they further established that the ARC1 motif within the NBS domain is required for the binding between the CC-NBS and the LRR of the Rx protein, and the ARC2 motif of the NBS domain is required for switching on/off the Rx molecule through its interaction with the LRR domain (Rairdan and Moffett 2006), which is consistent with the inference on the NBS function from another recent study (Tameling et al. 2006). Their results suggest that the ARC region of the NBS domain, through its interaction with the LRR, translates Avr elicitor-induced modulations of the LRR into a signal initiation event (Rairdan and Moffett 2006). This constitutes an important step towards our understanding how NBS-LRR proteins work.

4 Proliferation and Diversification of *NBS-LRR* Genes in Plants

Revealing the whole-genome sequences of *Arabidopsis* and rice enabled thorough sequence mining and analyses for *NBS-LRR* genes at the genome scale. It was found that the *Arabidopsis thaliana* genome contains 55 non-*TIR-NBS-LRR* and 94 *TIR-NBS-LRR* genes, which represents ~0.4% of the total predicted genes (Meyers et al. 2003). A similar analysis with the rice genome identified 480 non-*TIR NBS-LRR* genes in the current version of the *Oryza sativa* L. (var. Nipponbare) genome sequence. This represents about 1% of all the predicted genes in this species (Zhou et al. 2004). Surprisingly, as mentioned earlier, while *TIR-NBS-LRR* gene are abundant in dicot genomes, none is found in the rice genome or in other monocot genomes (Bai et al. 2002; Meyers et al. 2002). However, *TIR-NBS* and *TIR-X* genes are found in the rice genome (Meyers et al. 2002), suggesting that *TIR-NBS-LRR* genes might have evolved before separation of the dicot and monocot genomes, and they were lost during monocot evolution (Meyers et al. 2002). Phylogenetic analyses indicated that separation of the TIR and non-TIR subfamilies represent an ancient division that may date back to the common ancestor of angiosperms and gymnosperms around 300 million years ago (Meyers et al. 1999; Young 2000) and that members of the TIR subfamily are relatively more conserved, whereas members of the non-TIR subfamily are more divergent (Cannon et al. 2002; Meyers et al. 2003).

Studies indicated that the presence of a large number of *NBS-LRR* genes in contemporary plant genomes may be attributed to both initial duplications of large segmental genomic sequences that happened during early stages of plant genome evolution (Richly et al. 2002; Leister 2004) and small-scale gene duplications that occurred locally to form many *R* gene clusters (Michelmore and Meyers 1998; Meyers et al. 2003; Zhou et al. 2004). Duplicated *NBS-LRR* genes would have provided the potential templates for the evolution of divergent members of the superfamily through various molecular mechanisms, including inter- and intra-genic recombinations, gene conversion, unequal crossing-over, deletion, insertion and point mutations (Michelmore and Meyers 1998; Baumgarten et al. 2003; Meyers et al. 2005). The concomitant evolution and maintenance of so many *NBS-LRR* genes in plant genomes reflect the great capacity that plants have evolved to fight against diverse pathogens and the high selective pressure imposed on plants by pathogens.

If the proliferation and diversification of *R* genes at the genome level represents the macroevolutionary aspect of plant *R* genes, the evolution and maintenance of genetic polymorphisms at specific *R* loci is the microevolutionary aspect of plant *R* genes. Based on the rate of sequence diversification at two complex *R* clusters,

RGC2 in lettuce and *R1* in *S. demissum*, Kuang and coauthors recently hypothesized that there are two types of *NBS-LRR* genes, denoted by type I and type II (Kuang et al. 2004, 2005). Type I comprise fast-evolving genes, characterized by chimeric structures resulting from frequent sequence exchange among group members and consequently lacking a clear allelic/orthologous relationship between different genotypes. Unlike Type I, Type II *NBS-LRR* genes comprise slow-evolving genes, exhibiting infrequent sequence exchange between paralogous sequences and obvious allelic/orthologous relationships among different genotypes (Kuang et al. 2004, 2005). In general, this is in agreement with the findings that some *R* loci exhibit very high levels of sequence polymorphism (Ellis et al. 1999; Noel et al. 1999; Rose et al. 2004) while others show simple, lower levels of sequence polymorphism (Stahl et al. 1999; Tian et al. 2002). A challenging question is: what evolutionary forces are behind the formation of the two contrasting patterns of *R* gene sequences?

5 Mechanisms of R-Avr Recognition

Different mechanisms for sequence evolution have been documented for *R* genes (Michelmore and Meyers 1998; Meyers et al. 2005). Until recently, relatively little is known about the selective forces driving the evolution of specific *R* genes at the molecular level. It is clear that the ultimate driving force for *R* gene evolution comes from disease pressure imposed on plants by pathogens; but the type and strength of selection may vary, depending on the mechanisms by which plants recognize pathogens and the levels of pathogen virulence and host resistance. It is conceivable that mechanisms of R-Avr recognition may profoundly influence the patterns of *R-Avr* coevolution. Recent advances in understanding the molecular mechanisms of R-Avr recognition have provided exciting opportunities to investigate this question.

5.1 Direct R-Avr Interaction – the "Gene-For-Gene" Hypothesis

How a plant R protein recognizes a specific Avr factor to establish the specificity of plant–pathogen interaction has been a focus of intensive research. Over half a century ago, by studying the interaction specificity between different flax genotypes and different strains of flax rust pathogen, *Melampsora lini*, Flor proposed a "gene-for-gene" relationship to interpret the plant–pathogen interaction specificity at the genetic level (Flor 1956). The molecular extrapolation from this gene-for-gene theory is that the product of a dominant plant *R* gene functions as a receptor for a pathogen-derived ligand encoded by the *Avr* gene, implying a direct R (receptor)–Avr (ligand) interaction relationship. In the past decade, over 60 *R* genes have been characterized from different plant species (Xiao 2006) and many *Avr* genes (some of which are recognized by the cloned *R* genes) have also been isolated from bacterial, oomycete and fungal pathogens (Mudgett 2005; Chisholm et al. 2006; Ellis et al. 2006). In contrast to R proteins, Avr proteins have diverse deduced structures, presumably performing a wide range of functions inside the host cell to enhance pathogen virulence (Mudgett 2005). Isolation of the *R* and the cognate *Avr* genes in several cases (Chisholm et al. 2006; Table 2) provided opportunities to test whether the "gene-for-gene" model for R-Avr interaction holds true at the protein level.

Among the available R-Avr pairs, three pairs, i.e. Pita-AvrPita, RRS1-R-PopP-2 and L-AvrL567, show direct physical interaction when the corresponding components are co-expressed in the yeast-2-hybrid system or in vitro (Jia et al. 2000; Deslandes et al. 2003; Dodds et al. 2006). The L-AvrL567 interaction is particularly interesting as the interaction specificity of the *R* alleles and *Avr* alleles largely matches the recognition specificity *in planta*, demonstrating that Flor's gene-for-gene model holds true at the protein level (Dangl and McDowell 2006; Dodds et al. 2006).

5.2 Indirect R-Avr Interaction – the "Guard" Hypothesis

However, the majority of the available R-Avr pairs failed to show any physical interaction despite repeated experiments. This, together with the reasoning that, even though *R* genes are abundant in plants, there are still far fewer *R* genes in plant genomes than potential genes encoding effector proteins in pathogens, led to the indirect recognition model, i.e. the "guard" hypothesis (Van der Biezen and Jones 1998a; Dangl and Jones 2001). This model assumes that Avr proteins are pathogen effectors that are delivered into the host cell by pathogens to perform virulence functions by targeting relevant host proteins. The R proteins monitor the status change of the host proteins targeted by the Avr factors and initiate defense signaling. The "guarded" host targets may be signaling components of the plant defense system or components in other metabolic pathways. This indirect R-Avr recognition provides an explanation for mechanisms by which a plant R protein could perceive multiple effectors that might converge on the same target and that multiple R proteins might sense alterations of the same host protein mediated by different effectors (Chisholm et al. 2006; Dangl and McDowell 2006).

Compelling evidence for the guard hypothesis came from the indirect recognition between the *Arabidopsis* CC-NBS-LRR protein RPS5 and the effector protein AvrPphB of *P. syringae*. A genetic screen identified *PBS1*, encoding a protein kinase, as a host protein required for *RPS5* function (Warren et al. 1999; Swiderski and Innes 2001). It was later found that AvrPphB, a cysteine protease of *P. syringae* delivered into the host cell via the type III secretion system, targets and cleaves PBS1, and the cleaved product of PBS1 activates RPS5-triggered resistance (Shao et al. 2003). These findings indicate that AvrPphB is detected by RPS5 indirectly via its enzymatic activity on the host target PBS1, conforming to the "guard" hypothesis. A similar scenario is demonstrated in Cf2 (a RLP type R protein from tomato) and Avr2 (a cysteine-rich protein secreted by *C. fulvum*) indirect recognition

Table 2 Patterns of R-Avr Evolution. *NA* Not available

R/Avr locus	Plant/pathogen	R/Avr genetic diversity[a]	R-Avr recognition	R-Avr coevolution	References
L/AvrL567	Flax/Melampsora lini	High (13a)/high (12p)	Direct	Diversifying selection	Dodds et al. (2004, 2006)
RPP1/ATR1	Arabidopsis/Hyaloperonospora parasitica	High (4p)/high (6p)	?	Diversifying selection	Botella et al. (1998), Rehmany et al. (2005)
RPP13/ATR13	Arabidopsis/Hyaloperonospora parasitica	High(19a)/high	?	Diversifying selection	Allen et al. (2004), Rose et al. (2004)
Mla/AvrMla	Barley/Blumeria graminis	High (31p)/NA	?	Diversifying selection?	Wei et al. (2002), Halterman and Wise (2004)
Pm3/AvrPm3	Wheat/Blumeria graminis	High (10p)/NA	?	Diversifying selection?	Yahiaoui et al. (2006)
RRS1-R/PopP2	Arabidopsis/Ralstonia solanacearum	NA/NA	Direct	?	Deslandes et al. (2003)
Pita/AvrPita	Rice/Magnaporthe grisea	NA/NA	Direct	?	Jia et al. (2000)
RPM1/AvrRpm1	Arabidopsis/Pseudomonas syringae	Low (2a)/NA	Indirect	Balancing selection	Stahl et al. (1999), Mackey et al. (2002)
RPS2/AvrRpt2	Arabidopsis/Pseudomonas syringae	Low/NA	Indirect	Balancing selection	Mackey et al. (2003), Mauricio et al. (2003)
RPS5/AvrPphB	Arabidopsis/Pseudomonas syringae	Low (2a)/NA	Indirect	Balancing selection	Tian et al. (2002), Shao et al. (2003)
Cf2/Avr2	Tomato/Cladosporium fulvum	Low/NA	Indirect	Balancing selection?	Luderer et al. (2002), Rooney et al. (2005)
Pto/AvrPto	Tomato/Pseudomonas syringae	Low/NA	Direct	Balancing selection?	Tang et al. (1996), Rose et al. (2005)

[a] There are two types of R/Avr loci: (a) complex loci that contain highly homologous genes and (b) single-gene loci that either comprise multiple functionally distinct alleles or simple presence/absence alleles. Numbers in parentheses indicate the paralogous genes ("p") or alleles ("a")

through the host target protein Rcr3 (Rooney et al. 2005). The most complex R-Avr indirect recognition scenario so far characterized involves the host target protein RIN4. It has been shown that RIN4 is targeted by at least three different Avr proteins, AvrRpm1, AvrB and AvrRpt2, secreted by *P. syringae* and is guarded by at least two R proteins, RPM1 and RPS2 (Mackey et al. 2002, 2003; Axtell and Staskawicz 2003; Kim et al. 2005), both of which are CC-NBS-LRR proteins. RPM1 recognizes both AvrRpm1 and AvrB, two unrelated effector proteins, via perception of the modification (i.e. likely phosphorylation) of RIN4 by the two Avr proteins, while RPS2 recognizes AvrRpt2 through detection of RIN4 elimination mediated by AvrRpt2. From an evolutionary point of view, a likely time-sequence of evolution for this scenario has been envisaged (Chisholm et al. 2006). Perhaps the pathogen first delivered AvrB and AvrRpm1 into the host cell to suppress MTI by modifying RIN4 via phosphorylation. The plant then evolved RPM1 to detect this modification of RIN4, inducing RPM1-dependent resistance. The pathogen in turn evolved AvrRpt2 to interfere with RPM1 function by elimination of RIN4. In response to AvrRpt2's interference, the plant further evolved RPS2 to detect the disappearance of RIN4, inducing the RPS2-dependent defense response (Chisholm et al. 2006).

A special case of R-Avr direct interaction is between Pto of tomato and AvrPto of *P. syringae*, where the R protein Pto is a protein kinase (non-LRR type) and does not seem to be a receptor for AvrPto. Intriguingly Prf, a CC-NBS-LRR protein, is required for Pto function (Salmeron et al. 1996). Attempts to explain how Pto functions to recognize AvrPto and then trigger Prf-dependent resistance actually inspired the initial idea of the "guard" hypothesis in which Pto was thought to be a host target of AvrPto and guarded by Prf (Van der Biezen and Jones 1998a). A few years later, a similar but more convincing case involving RPS5 (the CC-NBS-LRR keeping guard), PBS1 (the protein kinase being guarded) and AvrPphB (the ligand; Shao et al. 2003) was characterized, providing direct evidence for the "guard" hypothesis. It should be noted that, in the case of Prf-Pto-AvrPto, it is the kinase gene rather than the CC-NBS-LRR gene that is genetically defined as the polymorphic determinant of disease resistance (i.e. the *R* gene). Molecular evidence for Pto being a host protein guarded by Prf still remains to be seen. If confirmed, this would suggest that a host protein targeted by a pathogen effector may be under selection and evolve into an *R* gene in some cases. This may represent a potential mechanism for the origin of other atypical *R* genes.

6 Patterns of *R-Avr* Coevolution

As indicated in Section 4, evolutionary analyses revealed two types of *R* genes with contrasting sequence features. One comprises *R* genes either located in complex loci with a number of highly homologous genes or in single-gene loci with multiple divergent alleles in the population, which is indicative of diversifying selection. The other type comprises *R* genes in single-gene loci with simple, low-level (including presence/absence) polymorphisms, which is indicative of balancing selection. A few recent studies indicated that there may be similar contrasting sequence patterns at *Avr* loci in pathogens that match those in the corresponding *R* loci in the hosts (Allen et al. 2004; Rehmany et al. 2005; Araki et al. 2006). An emerging theme is that these two patterns of *R* or *Avr* gene coevolution appear to be determined by the two modes of R-Avr recognition mechanisms described in the previous section.

6.1 Diversifying Selection Results from Direct R-Avr Recognition?

If an R protein detects an Avr protein by direct interaction and triggers host resistance, selection pressure should be imposed on the pathogen carrying the *Avr* gene to evade recognition. This could be achieved by structural changes in the Avr protein through mutations without necessarily affecting its virulence function. In this case, the Avr recognition by the R gene may be unrelated to its virulence function, so mutations that abolish recognition may retain function with little or no fitness penalty to the pathogen (Dodds et al. 2006). Once the host resistance is overcome, the selection pressure would be on the plant to generate a new *R* gene to recognize the mutated *Avr*. This *R-Avr* coevolutionary "arms-race" may have two outcomes. One is that single *R* (*Avr*) alleles of the corresponding genes sweep through the host (pathogen) populations, due to directional selection for advantageous alleles (Bergelson et al. 2001). However, this situation has rarely been documented in natural plant populations and a genome-wide survey of *R* gene polymorphism in *Arabidopsis* failed to find convincing evidence for directional selection (Bakker et al. 2006). The other outcome is the generation and maintenance of high sequence diversity at both the *R* and *Avr* loci in the corresponding populations, respectively, for plants to better achieve and the pathogen to better evade detection, through diversifying selection.

Among those characterized *R* loci that have been subject to comparative sequence analysis either at the interspecific or the intraspecific level (Table 2), the *Arabidopsis R* gene *RPP13* showed extremely high allelic polymorphism at the amino acid level (Rose et al. 2004), indicating diversifying selection. The *RPP1* locus, which comprises several tightly linked homologues genes, also exhibited a high level of sequence polymorphism (Botella et al. 1998). The corresponding *Avr* genes *ATR13* and *ATR1* were recently isolated and also shown to have high allelic diversity in the oomycete pathogen *H. parasitica* (Allen et al. 2004; Rehmany et al. 2005). This parallel high amino acid sequence diversity at the paired *R* and *Avr* loci suggests that the diversifying selections on the cognate *R* and *Avr* genes are potentially driven by the pressure imposed by pathogen virulence and plant resistance, respectively, in an interactive way. The most enlightening discovery in this aspect comes from recent studies by Ellis and coworkers on the flax *L* locus and the cognate *AvrL567* locus in the flax rust pathogen, *M. lini*. The *L* locus encodes at least 11 *R* alleles (including *L5*, *L6*, *L7*) capable of recognizing distinct *Avr* genes belonging to different loci, including *AvrL567* in the pathogen (Ellis et al. 1999). The *AvrL567* locus also contains multiple *Avr* genes that are recognized by the *R* alleles at the *L* locus (Dodds et al. 2004). More significantly, they demonstrated that the R proteins L5, L6 and L7 physically interact with the corresponding Avr proteins in the yeast-two-hybrid system in a specific manner that matches the specificity of the genetic interaction observed (Flor 1956; Dodds et al. 2006). These results strongly suggest that diversifying selection at the *R* and the *Avr* loci for high levels of amino acid sequence polymorphism is a consequence of the *R-Avr* "arms-race" between plants and pathogens. However, a generalization that direct *R-Avr* recognition leads to high levels of amino acid polymorphism at engaged *R* and *Avr* loci requires more evidence. Therefore, it will be of great interest to see whether the matching high genetic diversity at paired *R-Avr* loci, such as *RPP13-ATR13* and *RPP1-ATR1*, also correlates with direct-recognition between these R and cognate Avr proteins (Table 2). Conversely, it will be interesting to see whether the *R* and *Avr* genes involved in *Pita-AvrPita* and *RRS1-R-PopP2* direct interaction (Jia et al. 2000; Deslandes et al. 2003) belong to *R* and *Avr* loci that possess high levels of allelic diversity or consist of highly homologous genes.

Given that microbes reproduce much faster than plants, the pathogen would always have an upper hand in this "arms-race". Yet, *R* genes engaged in this "arms-race" have been amplified and maintained in the contemporary plant genomes, indicating that plants may somehow compensate for the slower pace in evolving new *R* genes. This compensation is perhaps reflected by the structural features of *R* proteins. Whereas Avr proteins from bacterial, oomycete and fungal pathogens are structurally diverse (Mudgett 2005; Chisholm et al. 2006; Ellis et al. 2006), presumably performing very different virulence functions in the host, most R proteins contain the hyper-variable LRR domain for achieving new recognition specificity and the highly conserved NBS domain for activating a central defense signaling network. This means a new *Avr* allele that evades *R* recognition may only evolve from the existing alleles through mutations or allelic recombination; however, a new *R* gene that regains recognition of a new *Avr* allele could evolve not only from the old (defeated) alleles (Holub 2001; Yahiaoui et al. 2006), but also from alleles that recognize other unrelated *Avr* genes and from highly homologous *R* genes at the same locus, if there are any, through point mutations, intra- and inter-genic recombination, gene conversion and unequal crossing over (McDowell et al. 1998; Michelmore and Meyers 1998; Kuang et al. 2004; Meyers et al. 2005).

6.2 Balancing Selection Results from Indirect Recognition?

R proteins that are engaged in indirect recognition of Avr proteins are predicted to function in protein complexes as guards of virulence targets of the pathogen Avr proteins. Compared with direct detection of pathogens via structural R-Avr recognition, it is conceivable that there may be selective advantages for indirect detection by monitoring the virulence function of Avr proteins, because structural modification of an Avr protein by mutations cannot escape R recognition unless the virulence function is affected. A likely outcome is the deletion of the Avr gene in the pathogen to avoid *R*-dependent resistance. However, if that virulence activity is important for the pathogen, loss of the *Avr* could incur a penalty in fitness for the pathogen. In this case, natural selection would balance the benefit (enhanced virulence) and cost (elicited resistance) of keeping this *Avr* gene in the pathogen population, resulting in a balanced polymorphism at the *Avr* locus. Likewise, while expression of the *R* gene has an obvious benefit for the host in the presence of the cognate *Avr*-expressing pathogen, it may incur a cost of resistance in the absence of the pathogen (Tian et al. 2003). Thus natural selection would also favor a balanced polymorphism at the *R* locus (Van der Hoorn et al. 2002). The most likely outcome of this *R-Avr* coevolution is maintenance of the *R* and cognate *Avr* genes at temporally and spatially variable frequencies in the plant and pathogen populations, respectively, for a very long time, through balancing selection (also referred to as frequency-dependent selection).

Compelling evidence for balancing selection on *R* genes comes from evolutionary analyses at several well studied *R* loci. *Arabidopsis* RPM1 and RPS5, both of which appear to detect their Avr proteins by guarding the host targets of the Avr proteins (Mackey et al. 2002; Shao et al. 2003), have been shown to be subject to balancing selection: there is simple but stable presence/absence polymorphism at both *R* loci; and the *R-Avr* recognition in both cases appears to be of ancient origin and to have been maintained for millions of years (Stahl et al. 1999; Tian et al. 2002). Relatively low genetic diversity with simple resistance/susceptibility allelism has been found at the *Arabidopsis RPS2* locus (Caicedo et al. 1999; Mauricio et al. 2003), correlating to the indirect recognition of AvrRpt2 by RPS2 (Axtell and Staskawicz 2003; Mackey et al. 2003). It is conceivable that, under situations where disease pressure is intermediate, to better balance the cost and benefit associated with an *R* gene, the plant may evolve and maintain partially functional *R* variants. Therefore, it is possible that some of the *RPS2* divergent alleles may be partially functional.

On the pathogen side, it is well observed that bacterial type III effectors can be found as presence/absence alleles among different pathovars (Rohmer et al. 2004; Pitman et al. 2005). However, relatively little is known about the sequence polymorphism in pathogen populations at genetic loci that encode *Avr* effectors indirectly recognized by *R* proteins. A recent report by Bergelson and coworkers suggested that balancing selection is also responsible for the maintenance of two pathogenicity islands (PAIs) that contain clusters of genes encoding for pathogen effectors in *Pseudomonas viridiflava*, a prevalent bacterial pathogen of *Arabidopsis* (Araki et al. 2006). This presence/absence sequence feature is strikingly similar to the stable balanced polymorphism for resistance and susceptibility alleles at the *R* loci described above.

The indirect R-Avr association renders it possible for independent evolution of *R* genes in plants to detect virulence function of the same *Avr* genes. For example, it has been shown that RPM1 from *Arabidopsis* and Rpg1 from soybean appear to have evolved independently to indirectly recognize AvrB of *P. syringae*, a common pathogen causing disease of many plant species (Ashfield et al. 2004). Another possibility is that, while maintaining the ability to bind to a host target, an Avr may evolve to alter conformational changes in the target that are caused by binding, thus evading detection by guarding R proteins (Birch et al. 2006). As a result, the host target may be under selection pressure to evolve a new version that can trigger the guarding R protein-dependent resistance. Although it is the same plant defense program that is elicited by the *Avr* gene, in this case the polymorphic determinant (i.e. genetically defined *R* gene) is being guarded rather than being the guard. A likely example for this scenario is the *Pto-AvrPto* triggered *Prf*-dependent resistance in tomato against *P. syringae* (Van der Biezen and Jones 1998a; see earlier text). Whether other non-LRR *R* genes are "mutated" host targets remains an open interesting question.

The diversifying selection and balancing selection on *R* and *Avr* genes described above represent probably just two simplified, typical scenarios of *R-Avr* coevolution. In nature, the actual situation is far more complicated due to the coexistence of different types of pathogens with opposite modes of parasitism, alternate host species and temporally and spatially variable environmental conditions that may favor or restrict plant or pathogen growth. Also, the strength of the diversifying selection or balancing selection may vary depending on the level of the cost associated with expression of the *R* gene and the fitness penalty associated with loss of the *Avr* gene.

Similar or entirely different evolutionary mechanisms may be associated with some non-LRR types of *R* gene. For example, the *Arabidopsis* broad-spectrum *R* gene *RPW8* is located in a complex locus (Xiao et al. 2001). Intra- and inter-specific sequence analyses indicated that *RPW8* originated from its likely progenitor gene via diversifying selection; and the maintenance of *RPW8* in the *Arabidopsis* population seems to be a consequence of balancing selection (Xiao et al. 2004; Orgil et al. 2007). In the case of two rice *R* genes, *Xa27* and *Xa13* that confer resistance to bacterial blight, it is the transcription of the genes rather the protein sequences that specifies resistance, suggesting that selection acts on promoters (Gu et al. 2005; Chu et al. 2006). Resistance specified by *Xa13* is associated with low expression levels of this gene, conforming to the recessive nature of the resistance. Interestingly, a high expression level of *Xa13*, which is required for establishing disease by the cognate *X. oryzae* pv. *oryzae* strains, is also required for rice fertility. Indeed, sequence polymorphism is found in the promoters of the *Xa13* alleles from different rice haplotypes, indicating that balancing selection has acted on the promoter sequence, presumably to balance the benefit (fertility) and cost (disease) of expression of this gene, which is unusual among *R* genes (Chu et al. 2006).

6.3 *A General Model for Evolution of the Plant* **R** *Gene System*

In summary, a general model for evolution of the plant *R* gene system is proposed (Fig. 1). The main points for this model are as follows:

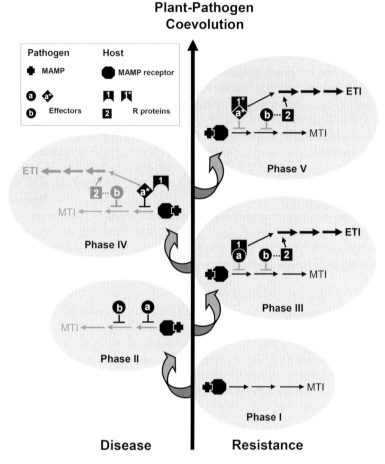

Fig. 1 A general model for plant–pathogen coevolution. Plants possess a multilayered defense system to ward off potential pathogens. MAMP-triggered immunity (MTI) constituted an important part of non-host resistance (*Phase I*). The pathogen delivered effector proteins "*a*" and "*b*" into the host cell to suppress the MTI and eventually overcame non-host resistance (*Phase II*). In response to the pathogen-imposed pressure, the plant evolved specific R proteins "*1*" and "*2*" to recognize the effector protein (referred to as Avr factors) "*a*" directly in a receptor–ligand interactive manner and "*b*" indirectly by detecting its virulence activity, respectively, thereby inducing a stronger defense response (*Phase III*). In response to the plant-imposed pressure, the pathogen accumulated mutations that change the structure of effector "*a*" to escape recognition by R protein "*1*", resulting in suppression of R-resistance. In the other scenario, effector gene "*b*" might have been deleted in the pathogen population if this incurred little fitness penalty in virulence, or "*b*" became rare in the pathogen population as a consequence of *R*-resistance, which in turn relaxed the selection for *R* gene "*2*", resulting in a low frequency of *R* gene "*2*" in the plant population as a result of balancing selection for the benefit and cost of *R* gene expression (*Phase IV*). Next, the plant faced pressure to evolve a new *R* gene "*1**" (either from the defeated or other related *R* gene) to recognize "*a**" to activate defense; or as the frequency of the effector gene "*b*" increased, the frequency of *R* gene "*2*" also increased due to its selective advantage in the presence of "*b*" (*Phase V*). The first *R*-*Avr* (i.e. "*1*"–"*a*") coevolution scenario leads to high genetic diversity at both the *R* and *Avr* loci due to diversifying selection; and the second scenario (i.e. "*2*"–"*b*") leads to simple, stable polymorphism at the *R* and *Avr* loci due to balancing selection

1. In the early stages of plant–microbe interaction, the invading microbes evolved effectors to suppress evolutionarily ancient MAMP-triggered plant immunity, beginning the host–pathogen coevolutionary process.
2. The plant R gene system evolved to counter pathogen suppression of MTI by detecting the effector proteins and triggering a stronger defense response.
3. Recognition of effectors (i.e. Avr factors) by R proteins involves either direct physical interaction between the R and the cognate Avr or indirect interaction via association with a host protein targeted by the Avr.
4. Diversifying selection acted on both the R and the Avr genes engaged in direct interaction to achieve and evade detection, respectively, resulting in high genetic diversity at the R and the Avr loci. Balancing selection acted on both the R and the Avr genes, involving indirect recognition to achieve a balance between benefit and cost of resistance/virulence in the plant and pathogen, respectively, leading to simple, stable polymorphisms at the corresponding R and Avr loci.

7 New Perspectives

Tremendous progress has been made in understanding the evolution and molecular mechanism of plant R gene-based resistance (Hammond-Kosack and Parker 2003; Chisholm et al. 2006; this chapter). Many important questions regarding R gene evolution, R-Avr recognition and defense signaling remain to be addressed. These include when and how the earliest plant *NBS-LRR* genes originated, whether *NBS-LRR* genes function in MAMP recognition or other cellular processes, how MTI and ETI are connected at the molecular level, what other components are in the R recognition complex and why *TIR-NBS-LRRs* are absent in monocots. Also not known is whether atypical R genes represent host targets and activate *NBS-LRR*-dependent defenses or involve different defense mechanisms through novel yet unidentified signaling pathways.

What can we learn from our current knowledge about the plant R gene system at the molecular level to improve disease control for agricultural crops? First, based on the R-Avr coevolution, R genes recognizing Avr genes via direct R-Avr interaction may be easily overcome by mutations in cognate Avr genes; thus monoculture with crops of such genotypes may be subject to high risk of epidemics caused by virulent pathogens carrying mutated versions of Avr genes. Given that genetic diversity at the R gene loci is the consequence of R-Avr coevolution in natural plant populations and is important for the generation of new R genes or alleles, it would be a better strategy to grow crop cultivars containing distinct R genes or R alleles in a mixed structure (polyculture) which resembles the natural plant populations, thereby preventing disease epidemics and reducing disease incidence (Dangl and Jones 2001). The potential of this strategy has been demonstrated in rice to control bacterial blight (Zhu et al. 2000).

Second, identification and deployment of durable R genes is a primary target of crop breeding programs. However, there is no easy way to assess and predict the durability of R genes. A high fitness penalty of *Avr* genes in the pathogen could be used for predicting the durability of the cognate *R* genes (Leach et al. 2001), as a high fitness penalty would set constraints for *Avr* genes to escape from *R* recognition. But it may not be accurate in situations where *Avr* genes directly recognized by *R* genes may evade recognition by mutations that do not necessarily affect its virulence function (see Section 6.1). Also it may be practically difficult to assess the fitness penalty of *Avr* genes. One of the most exciting inferences from *R-Avr* coevolution is that R-Avr indirect recognition leads to stable balanced polymorphisms at the *R* and *Avr* loci (Table 2), which implies that *R* genes of this category are likely to be durable. Even though this rule needs to be vigorously tested with more *R* and *Avr* pairs, it gives a promising molecular tool to identify and predict durable *R* genes that guard critical virulence targets based on the feature of the sequence polymorphism in the natural plant populations (Dangl and McDowell 2006). Comparative genomic analyses have identified two types of *NBS-LRR* genes based on the rate of sequence diversification: fast evolving and slow evolving (Kuang et al. 2004, 2005). A key question is: would the existence of these two types of *NBS-LRR* genes reflect the two paths of *R* gene evolution largely determined by the two different R-Avr recognition mechanisms in general? If true, prediction of *R* recognition mechanism and resistance durability by coupling data from genetic mapping and sequence polymorphism could be applied at the whole genome scale.

Third, the natural evolution of *R* genes that recognize new or mutated *Avr* genes is after all a slow process. If such *R* gene resources are unavailable from natural plant populations, a viable strategy is to generate *R* genes artificially in laboratories by in vitro mutagenesis or sequence shuffling with relevant existing genes/alleles and to screen for DNA clones that can induce target *Avr*-dependent HR through transient coexpression with cognate *Avr* genes in a suitable host. Such synthetic genes may function as new *R* genes in the native host to recognize the pathogens carrying newly mutated versions of *Avr* genes and can be introduced to desirable plant cultivars through genetic transformation.

Acknowledgements We would like to thank Donald Nuss and Dacheng Tian for critical review and valuable comments on the manuscript. Work in S.X.'s laboratory is supported by the National Research Initiative of the USDA Cooperative State Research, Education and Extension Service, Grants 2005-35319-15656 and 2006-35301-16883.

References

Aarts N, Metz M, Holub E, Staskawicz BJ, Daniels MJ, et al (1998) Different requirements for EDS1 and NDR1 by disease resistance genes define at least two R gene-mediated signaling pathways in *Arabidopsis*. Proc Natl Acad Sci USA 95:10306–10311

Allen RL, Bittner-Eddy PD, Grenville-Briggs LJ, Meitz JC, Rehmany AP, et al (2004) Host–parasite coevolutionary conflict between *Arabidopsis* and downy mildew. Science 306:1957–1960

Araki H, Tian D, Goss EM, Jakob K, Halldorsdottir SS, et al (2006) Presence/absence polymorphism for alternative pathogenicity islands in *Pseudomonas viridiflava*, a pathogen of *Arabidopsis*. Proc Natl Acad Sci USA 103:5887–5892

Ashfield T, Ong LE, Nobuta K, Schneider CM, Innes RW (2004) Convergent evolution of disease resistance gene specificity in two flowering plant families. Plant Cell 16:309–318

Ausubel FM (2005) Are innate immune signaling pathways in plants and animals conserved? Nat Immunol 6:973–979

Axtell MJ, Staskawicz BJ (2003) Initiation of RPS2-specified disease resistance in *Arabidopsis* is coupled to the AvrRpt2-directed elimination of RIN4. Cell 112:369–377

Bai J, Pennill LA, Ning J, Lee SW, Ramalingam J, et al (2002) Diversity in nucleotide binding site-leucine-rich repeat genes in cereals. Genome Res 12:1871–1884

Bakker EG, Toomajian C, Kreitman M, Bergelson J (2006) A genome-wide survey of r gene polymorphisms in *Arabidopsis*. Plant Cell 18:1803–1818

Baumgarten A, Cannon S, Spangler R, May G (2003) Genome-level evolution of resistance genes in *Arabidopsis thaliana*. Genetics 165:309–319

Belkhadir Y, Subramaniam R, Dangl JL (2004) Plant disease resistance protein signaling: NBS-LRR proteins and their partners. Curr Opin Plant Biol 7:391–399

Bent AF, Kunkel BN, Dahlbeck D, Brown KL, Schmidt R, et al (1994) RPS2 of *Arabidopsis thaliana*: a leucine-rich repeat class of plant disease resistance genes. Science 265:1856–1860

Bergelson J, Kreitman M, Stahl EA, Tian D (2001) Evolutionary dynamics of plant R-genes. Science 292:2281–2285

Birch PR, Rehmany AP, Pritchard L, Kamoun S, Beynon JL (2006) Trafficking arms: oomycete effectors enter host plant cells. Trends Microbiol 14:8–11

Botella MA, Parker JE, Frost LN, Bittner-Eddy PD, Beynon JL, et al (1998) Three genes of the *Arabidopsis* RPP1 complex resistance locus recognize distinct *Peronospora parasitica* avirulence determinants. Plant Cell 10:1847–1860

Buschges R, Hollricher K, Panstruga R, Simons G, Wolter M, et al (1997) The barley Mlo gene: a novel control element of plant pathogen resistance. Cell 88:695–705

Caicedo AL, Schaal BA, Kunkel BN (1999) Diversity and molecular evolution of the RPS2 resistance gene in *Arabidopsis thaliana*. Proc Natl Acad Sci USA 96:302–306

Cannon SB, Zhu H, Baumgarten AM, Spangler R, May G, et al (2002) Diversity, distribution, and ancient taxonomic relationships within the TIR and non-TIR NBS-LRR resistance gene subfamilies. J Mol Evol 54:548–562

Chisholm ST, Coaker G, Day B, Staskawicz BJ (2006) Host–microbe interactions: shaping the evolution of the plant immune response. Cell 124:803–814

Chu Z, Yuan M, Yao J, Ge X, Yuan B, et al (2006) Promoter mutations of an essential gene for pollen development result in disease resistance in rice. Genes Dev 20:1250–1255

Dangl JL, Jones JD (2001) Plant pathogens and integrated defence responses to infection. Nature 411:826–833

Dangl JL, McDowell JM (2006) Two modes of pathogen recognition by plants. Proc Natl Acad Sci USA 103:8575–8576

Deslandes L, Olivier J, Theulieres F, Hirsch J, Feng DX, et al (2002) Resistance to Ralstonia solanacearum in *Arabidopsis thaliana* is conferred by the recessive RRS1-R gene, a member of a novel family of resistance genes. Proc Natl Acad Sci USA 99:2404–2409

Deslandes L, Olivier J, Peeters N, Feng DX, Khounlotham M, et al (2003) Physical interaction between RRS1-R, a protein conferring resistance to bacterial wilt, and PopP2, a type III effector targeted to the plant nucleus. Proc Natl Acad Sci USA 100:8024–8029

Dodds PN, Lawrence GJ, Catanzariti AM, Ayliffe MA, Ellis JG (2004) The *Melampsora lini* AvrL567 avirulence genes are expressed in haustoria and their products are recognized inside plant cells. Plant Cell 16:755–768

Dodds PN, Lawrence GJ, Catanzariti AM, Teh T, Wang CI, et al (2006) Direct protein interaction underlies gene-for-gene specificity and coevolution of the flax resistance genes and flax rust avirulence genes. Proc Natl Acad Sci USA 103:8888–8893

Durrant WE, Dong X (2004) Systemic acquired resistance. Annu Rev Phytopathol 42:185–209

Ellis JG, Lawrence GJ, Luck JE, Dodds PN (1999) Identification of regions in alleles of the flax rust resistance gene L that determine differences in gene-for-gene specificity. Plant Cell 11:495–506

Ellis J, Dodds P, Pryor T (2000) Structure, function and evolution of plant disease resistance genes. Curr Opin Plant Biol 3:278–284

Ellis J, Catanzariti AM, Dodds P (2006) The problem of how fungal and oomycete avirulence proteins enter plant cells. Trends Plant Sci 11:61–63

Eulgem T (2005) Regulation of the *Arabidopsis* defense transcriptome. Trends Plant Sci 10:71–78

Flor HH (1956) The complementary genic systems in flax and flax rust. Adv Genet Mol Genet Med 8:29–54

Fluhr R, Kaplan-Levy RN (2002) Plant disease resistance: commonality and novelty in multicellular innate immunity. Curr Top Microbiol Immunol 270:23–46

Goff SA, Ricke D, Lan TH, Presting G, Wang R, et al (2002) A draft sequence of the rice genome (*Oryza sativa* L. ssp. *japonica*). Science 296:92–100

Gomez-Gomez L, Boller T (2000) FLS2: an LRR receptor-like kinase involved in the perception of the bacterial elicitor flagellin in *Arabidopsis*. Mol Cell 5:1003–1011

Grant MR, Godiard L, Straube E, Ashfield T, Lewald J, et al (1995) Structure of the *Arabidopsis* RPM1 gene enabling dual specificity disease resistance. Science 269:843–846

Gu K, Yang B, Tian D, Wu L, Wang D, et al (2005) R gene expression induced by a type-III effector triggers disease resistance in rice. Nature 435:1122–1125

Halterman DA, Wise RP (2004) A single-amino acid substitution in the sixth leucine-rich repeat of barley MLA6 and MLA13 alleviates dependence on RAR1 for disease resistance signaling. Plant J 38:215–226

Hammond-Kosack KE, Jones JD (1997) Plant disease resistance genes. Annu Rev Plant Physiol Plant Mol Biol 48:575–607

Hammond-Kosack KE, Parker JE (2003) Deciphering plant–pathogen communication: fresh perspectives for molecular resistance breeding. Curr Opin Biotechnol 14:177–193

He P, Shan L, Lin NC, Martin GB, Kemmerling B, et al (2006) Specific bacterial suppressors of MAMP signaling upstream of MAPKKK in *Arabidopsis* innate immunity. Cell 125:563–575

Holub EB (2001) The arms race is ancient history in *Arabidopsis*, the wildflower. Nat Rev Genet 2:516–527

Inohara N, Nunez G (2003) NODs: intracellular proteins involved in inflammation and apoptosis. Nat Rev Immunol 3:371–382

Jia Y, McAdams SA, Bryan GT, Hershey HP, Valent B (2000) Direct interaction of resistance gene and avirulence gene products confers rice blast resistance. EMBO J 19:4004–4014

Johal GS, Briggs SP (1992) Reductase activity encoded by the HM1 disease resistance gene in maize. Science 258:985–987

Jones DA, Thomas CM, Hammond-Kosack KE, Balint-Kurti PJ, Jones JD (1994) Isolation of the tomato Cf-9 gene for resistance to *Cladosporium fulvum* by transposon tagging. Science 266:789–793

Kim MG, da Cunha L, McFall AJ, Belkhadir Y, DebRoy S, et al (2005) Two *Pseudomonas syringae* type III effectors inhibit RIN4-regulated basal defense in *Arabidopsis*. Cell 121:749–759

Kim YJ, Lin NC, Martin GB (2002) Two distinct *Pseudomonas* effector proteins interact with the Pto kinase and activate plant immunity. Cell 109:589–598

Kuang H, Woo SS, Meyers BC, Nevo E, Michelmore RW (2004) Multiple genetic processes result in heterogeneous rates of evolution within the major cluster disease resistance genes in lettuce. Plant Cell 16:2870–2894

Kuang H, Wei F, Marano MR, Wirtz U, Wang X, et al (2005) The R1 resistance gene cluster contains three groups of independently evolving, type I R1 homologues and shows substantial structural variation among haplotypes of *Solanum demissum*. Plant J 44:37–51

Lawrence GJ, Finnegan EJ, Ayliffe MA, Ellis JG (1995) The L6 gene for flax rust resistance is related to the *Arabidopsis* bacterial resistance gene RPS2 and the tobacco viral resistance gene N. Plant Cell 7:1195–1206

Leach JE, Vera Cruz CM, Bai J, Leung H (2001) Pathogen fitness penalty as a predictor of durability of disease resistance genes. Annu Rev Phytopathol 39:187–224

Leipe DD, Koonin EV, Aravind L (2004) STAND, a class of P-loop NTPases including animal and plant regulators of programmed cell death: multiple, complex domain architectures, unusual phyletic patterns, and evolution by horizontal gene transfer. J Mol Biol 343:1–28

Leister D (2004) Tandem and segmental gene duplication and recombination in the evolution of plant disease resistance gene. Trends Genet 20:116–122

Leister RT, Dahlbeck D, Day B, Li Y, Chesnokova O, et al (2005) Molecular genetic evidence for the role of SGT1 in the intramolecular complementation of Bs2 protein activity in *Nicotiana benthamiana*. Plant Cell 17:1268–1278

Luck JE, Lawrence GJ, Dodds PN, Shepherd KW, Ellis JG (2000) Regions outside of the leucine-rich repeats of flax rust resistance proteins play a role in specificity determination. Plant Cell 12:1367–1377

Luderer R, Takken FL, de Wit PJ, Joosten MH (2002) *Cladosporium fulvum* overcomes Cf-2-mediated resistance by producing truncated AVR2 elicitor proteins. Mol Microbiol 45:875–884

Mackey D, Holt BF 3rd, Wiig A, Dangl JL (2002) RIN4 interacts with *Pseudomonas syringae* type III effector molecules and is required for RPM1-mediated resistance in *Arabidopsis*. Cell 108:743–754

Mackey D, Belkhadir Y, Alonso JM, Ecker JR, Dangl JL (2003) *Arabidopsis* RIN4 is a target of the type III virulence effector AvrRpt2 and modulates RPS2-mediated resistance. Cell 112:379–389

Martin GB, Brommonschenkel SH, Chunwongse J, Frary A, Ganal MW, et al (1993) Map-based cloning of a protein kinase gene conferring disease resistance in tomato. Science 262: 1432–1436

Mauricio R, Stahl EA, Korves T, Tian D, Kreitman M, et al (2003) Natural selection for polymorphism in the disease resistance gene Rps2 of *Arabidopsis thaliana*. Genetics 163:735–746

McDowell JM, Dhandaydham M, Long TA, Aarts MG, Goff S, et al (1998) Intragenic recombination and diversifying selection contribute to the evolution of downy mildew resistance at the RPP8 locus of *Arabidopsis*. Plant Cell 10:1861–1874

McHale L, Tan X, Koehl P, Michelmore RW (2006) Plant NBS-LRR proteins: adaptable guards. Genome Biol 7:212

Meyers BC, Dickerman AW, Michelmore RW, Sivaramakrishnan S, Sobral BW, et al (1999) Plant disease resistance genes encode members of an ancient and diverse protein family within the nucleotide-binding superfamily. Plant J 20:317–332

Meyers BC, Morgante M, Michelmore RW (2002) TIR-X and TIR-NBS proteins: two new families related to disease resistance TIR-NBS-LRR proteins encoded in *Arabidopsis* and other plant genomes. Plant J 32:77–92

Meyers BC, Kozik A, Griego A, Kuang H, Michelmore RW (2003) Genome-wide analysis of NBS-LRR-encoding genes in *Arabidopsis*. Plant Cell 15:809–834

Meyers BC, Kaushik S, Nandety RS (2005) Evolving disease resistance genes. Curr Opin Plant Biol 8:129–134

Michelmore RW, Meyers BC (1998) Clusters of resistance genes in plants evolve by divergent selection and a birth-and-death process. Genome Res 8:1113–1130

Moffett P, Farnham G, Peart J, Baulcombe DC (2002) Interaction between domains of a plant NBS-LRR protein in disease resistance-related cell death. EMBO J 21:4511–4519

Mudgett MB (2005) New insights to the function of phytopathogenic bacterial type III effectors in plants. Annu Rev Plant Biol 56:509–531

Muskett P, Parker J (2003) Role of SGT1 in the regulation of plant R gene signalling. Microbes Infect 5:969–976

Noel L, Moores TL, Van der Biezen EA, Parniske M, Daniels MJ, et al (1999) Pronounced intraspecific haplotype divergence at the RPP5 complex disease resistance locus of *Arabidopsis*. Plant Cell 11:2099–2112

Nomura K, Debroy S, Lee YH, Pumplin N, Jones J, et al (2006) A bacterial virulence protein suppresses host innate immunity to cause plant disease. Science 313:220–223

Nurnberger T, Brunner F (2002) Innate immunity in plants and animals: emerging parallels between the recognition of general elicitors and pathogen-associated molecular patterns. Curr Opin Plant Biol 5:318–324

Nurnberger T, Brunner F, Kemmerling B, Piater L (2004) Innate immunity in plants and animals: striking similarities and obvious differences. Immunol Rev 198:249–266

O'Neill LA, Fitzgerald KA, Bowie AG (2003) The Toll-IL-1 receptor adaptor family grows to five members. Trends Immunol 24:286–290

Orgil U, Araki H, Tangchaiburana S, Berkey R, and Xiao S (2007) Intraspecific Genetic Variations, Fitness Cost and Benefit of RPW8, A Disease Resistance Locus in *Arabidopsis thaliana*. Genetics 176:2317–2333

Parniske M, Hammond-Kosack KE, Golstein C, Thomas CM, Jones DA, et al (1997) Novel disease resistance specificities result from sequence exchange between tandemly repeated genes at the Cf-4/9 locus of tomato. Cell 91:821–832

Peart JR, Mestre P, Lu R, Malcuit I, Baulcombe DC (2005) NRG1, a CC-NB-LRR protein, together with N, a TIR-NB-LRR protein, mediates resistance against tobacco mosaic virus. Curr Biol 15:968–973

Pitman AR, Jackson RW, Mansfield JW, Kaitell V, Thwaites R, et al (2005) Exposure to host resistance mechanisms drives evolution of bacterial virulence in plants. Curr Biol 15:2230–2235

Rairdan GJ, Moffett P (2006) Distinct domains in the ARC region of the potato resistance protein Rx mediate LRR binding and inhibition of activation. Plant Cell 18:2082–2093

Rehmany AP, Gordon A, Rose LE, Allen RL, Armstrong MR, et al (2005) Differential recognition of highly divergent downy mildew avirulence gene alleles by RPP1 resistance genes from two *Arabidopsis* lines. Plant Cell 17:1839–1850

Richly E, Kurth J, Leister D (2002) Mode of amplification and reorganization of resistance genes during recent *Arabidopsis thaliana* evolution. Mol Biol Evol 19:76–84

Roeder A, Kirschning CJ, Rupec RA, Schaller M, Weindl G, et al (2004) Toll-like receptors as key mediators in innate antifungal immunity. Med Mycol 42:485–498

Rohmer L, Guttman DS, Dangl JL (2004) Diverse evolutionary mechanisms shape the type III effector virulence factor repertoire in the plant pathogen *Pseudomonas syringae*. Genetics 167:1341–1360

Rooney HC, Van't Klooster JW, van der Hoorn RA, Joosten MH, Jones JD, et al (2005) *Cladosporium* Avr2 inhibits tomato Rcr3 protease required for Cf-2-dependent disease resistance. Science 308:1783–1786

Rose LE, Bittner-Eddy PD, Langley CH, Holub EB, Michelmore RW, et al (2004) The maintenance of extreme amino acid diversity at the disease resistance gene, RPP13, in *Arabidopsis thaliana*. Genetics 166:1517–1527

Rose LE, Langley CH, Bernal AJ, Michelmore RW (2005) Natural variation in the Pto pathogen resistance gene within species of wild tomato (*Lycopersicon*). I. Functional analysis of Pto alleles. Genetics 171:345–357

Salmeron JM, Oldroyd GE, Rommens CM, Scofield SR, Kim HS, et al (1996) Tomato Prf is a member of the leucine-rich repeat class of plant disease resistance genes and lies embedded within the Pto kinase gene cluster. Cell 86:123–133

Shao F, Golstein C, Ade J, Stoutemyer M, Dixon JE, et al (2003) Cleavage of *Arabidopsis* PBS1 by a bacterial type III effector. Science 301:1230–1233

Song WY, Wang GL, Chen LL, Kim HS, Pi LY, et al (1995) A receptor kinase-like protein encoded by the rice disease resistance gene, Xa21. Science 270:1804–1806

Stahl EA, Dwyer G, Mauricio R, Kreitman M, Bergelson J (1999) Dynamics of disease resistance polymorphism at the Rpm1 locus of *Arabidopsis*. Nature 400:667–671

Staskawicz BJ, Mudgett MB, Dangl JL, Galan JE (2001) Common and contrasting themes of plant and animal diseases. Science 292:2285–2289

Sun X, Cao Y, Yang Z, Xu C, Li X, et al (2004) Xa26, a gene conferring resistance to *Xanthomonas oryzae* pv. *oryzae* in rice, encodes an LRR receptor kinase-like protein. Plant J 37:517–527

Swiderski MR, Innes RW (2001) The *Arabidopsis* PBS1 resistance gene encodes a member of a novel protein kinase subfamily. Plant J 26:101–112

Takeda K, Akira S (2004) Microbial recognition by Toll-like receptors. J Dermatol Sci 34:73–82

Takken FL, Albrecht M, Tameling WI (2006) Resistance proteins: molecular switches of plant defence. Curr Opin Plant Biol 9:383–390

Tameling WI, Vossen JH, Albrecht M, Lengauer T, Berden JA, et al (2006) Mutations in the NB-ARC domain of I-2 that impair ATP hydrolysis cause autoactivation. Plant Physiol 140:1233–1245

Tang X, Frederick RD, Zhou J, Halterman DA, Jia Y, et al (1996) Initiation of plant disease resistance by physical interaction of AvrPto and Pto kinase. Science 274:2060–2063

Tian D, Araki H, Stahl E, Bergelson J, Kreitman M (2002) Signature of balancing selection in *Arabidopsis*. Proc Natl Acad Sci USA 99:11525–11530

Tian D, Traw MB, Chen JQ, Kreitman M, Bergelson J (2003) Fitness costs of R-gene-mediated resistance in *Arabidopsis thaliana*. Nature 423:74–77

Tor M, Brown D, Cooper A, Woods-Tor A, Sjolander K, et al (2004) *Arabidopsis* downy mildew resistance gene RPP27 encodes a receptor-like protein similar to CLAVATA2 and tomato Cf-9. Plant Physiol 135:1100–1112

Van der Biezen EA, Jones JD (1998a) Plant disease-resistance proteins and the gene-for-gene concept. Trends Biochem Sci 23:454–456

Van der Biezen EA, Jones JD (1998b) The NB-ARC domain: a novel signalling motif shared by plant resistance gene products and regulators of cell death in animals. Curr Biol 8:R226–227

Van der Hoorn RA, Roth R, De Wit PJ (2001) Identification of distinct specificity determinants in resistance protein Cf-4 allows construction of a Cf-9 mutant that confers recognition of avirulence protein Avr4. Plant Cell 13:273–285

Van der Hoorn RA, De Wit PJ, Joosten MH (2002) Balancing selection favors guarding resistance proteins. Trends Plant Sci 7:67–71

Warren RF, Merritt PM, Holub E, Innes RW (1999) Identification of three putative signal transduction genes involved in R gene-specified disease resistance in *Arabidopsis*. Genetics 152:401–412

Wei F, Wing RA, Wise RP (2002) Genome dynamics and evolution of the Mla (powdery mildew) resistance locus in barley. Plant Cell 14:1903–1917

Whitham S, Dinesh-Kumar SP, Choi D, Hehl R, Corr C, et al (1994) The product of the tobacco mosaic virus resistance gene N: similarity to toll and the interleukin-1 receptor. Cell 78:1101–1115

Wiermer M, Feys BJ, Parker JE (2005) Plant immunity: the EDS1 regulatory node. Curr Opin Plant Biol 8:383–389

Xiao S (2006) Current perspectives on molecular mechanisms of plant disease resistance. In: Teixeira da Silva JA (ed) Floriculture, ornamental and plant biotechnology: advances and topical issues, vol 3. Global Science Books, London, pp 317–333

Xiao S, Ellwood S, Calis O, Patrick E, Li T, et al (2001) Broad-spectrum mildew resistance in *Arabidopsis thaliana* mediated by RPW8. Science 291:118–120

Xiao S, Emerson B, Ratanasut K, Patrick E, O'Neill C, et al (2004) Origin and maintenance of a broad-spectrum disease resistance locus in *Arabidopsis*. Mol Biol Evol 21:1661–1672

Yahiaoui N, Brunner S, Keller B (2006) Rapid generation of new powdery mildew resistance genes after wheat domestication. Plant J 47:85–98

Yamamoto M, Takeda K, Akira S (2004) TIR domain-containing adaptors define the specificity of TLR signaling. Mol Immunol 40:861–868

Young ND (2000) The genetic architecture of resistance. Curr Opin Plant Biol 3:285–290

Zhou T, Wang Y, Chen JQ, Araki H, Jing Z, et al (2004) Genome-wide identification of NBS genes in japonica rice reveals significant expansion of divergent non-TIR NBS-LRR genes. Mol Genet Genomics 271:402–415

Zhu Y, Chen H, Fan J, Wang Y, Li Y, et al (2000) Genetic diversity and disease control in rice. Nature 406:718–722

Zipfel C, Kunze G, Chinchilla D, Caniard A, Jones JD, et al (2006) Perception of the bacterial PAMP EF-Tu by the receptor EFR restricts *Agrobacterium*-mediated transformation. Cell 125:749–760

Chapter 2
The Path Less Explored: Innate Immune Reactions in Cnidarians

Thomas C.G. Bosch

1	Cnidaria Are Among the Earliest Multicellular Animals in the Tree of Life	28
2	Immune Reactions in Invertebrates	30
3	Immune Reactions in Cnidaria	30
	3.1 How to Fight for a Space to Live? Intraspecies Competition in Sea Anemones	31
	3.2 How to Detect Approaching Allogeneic Cells as Foreign and to Eliminate Them? Allorecognition and Cell Lineage Competition in Colonial *Hydractinia*	32
	3.3 How to Detect and Disarm Microbial Attackers? Antimicrobial Defense Reactions in the Freshwater Polyp *Hydra* and the Jellyfish *Aurelia*	34
	3.4 How to Distinguish Between Friends and Foes: Symbiotic Relationships in Corals and *Hydra*	36
4	How to Explore the Path They Went? Why Cnidarians Matter	38
References		39

Abstract The phylum Cnidaria is one of the earliest branches in the animal tree of life. Cnidarians possess most of the gene families found in bilaterians and have retained many ancestral genes that have been lost in *Drosophila* and *Caenorhabditis elegans*. Characterization of the innate immune repertoire of extant cnidarians is, therefore, of both fundamental and applied interest – it not only provides insights into the basic immunological "tool kit" of the common ancestor of all animals, but is also likely to be important in understanding human barrier disorders by describing ancient mechanisms of host/microbial interactions and the resulting evolutionary selection processes. The chapter summarizes four aspects of immunity which can be studied particularly well within cnidarians – and which may be of interest from a comparative point of view to all immunologists: intraspecies competition in sea anemones, allorecognition and cell lineage competition in the marine hydrozoan *Hydractinia*, antimicrobial defense reactions in *Hydra* and jellyfish, and symbiotic

Zoological Institute, Christian-Albrechts-University Kiel, Olshausenstrasse 40,
24098 Kiel, Germany, *tbosch@zoologie.uni-kiel.de*

H. Heine (ed.), *Innate Immunity of Plants, Animals, and Humans.*
Nucleic Acids and Molecular Biology 21.
© Springer-Verlag Berlin Heidelberg 2008

relationships in both corals and *Hydra*. Studies in cnidarians reveal that there is no problem in innate immunity these basal metazoans did not attempt to solve. Thus, whatever we experience with our own innate immune system, whatever we hope to learn, we will see that the cnidarians have been there before us.

1 Cnidaria Are Among the Earliest Multicellular Animals in the Tree of Life

The phylum Cnidaria is one of the earliest branches in the animal tree of life (Fig. 1). To understand cnidarian immunity we must first consider their anatomy and life history. Cnidaria are the first in evolution that have a defined body plan including an axis, a nervous system, and a tissue layer construction. Cnidarians are diploblastic, consisting of two epithelia, the ectoderm and the endoderm surrounding a gastric cavity. In the cnidarian *Hydra* (Fig. 2A) there are about 20 cell types distributed among three cell lineages (for a review, see Bosch 2007a, b). Each of the epithelial layers is made up of a cell lineage, while the remaining cells are part of the interstitial cell lineage which reside among the epithelial cells of both layers. Multipotent interstitial stem cells give rise to neurons, secretory cells, and gametes in a position-dependent manner (Bosch and David 1987). In sharp contrast to the morphological simplicity, cnidarians have attained remarkable diversity through modifications largely in colonial organization and life

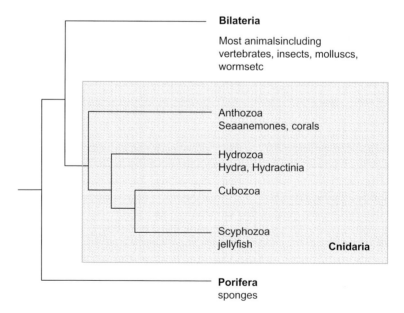

Fig. 1 Phylogeny of the "lower" Metazoa. Within the Cnidaria, the Anthozoa (to which the coral Acropora and the sea anemones Nematostella belong) are basal

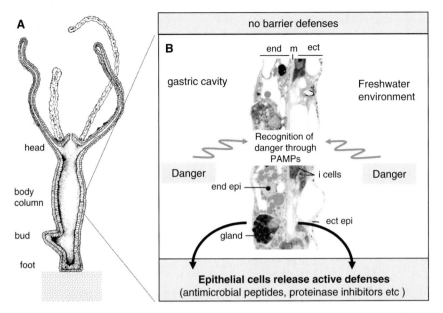

Fig. 2 The freshwater polyp *Hydra* relies entirely on defense mechanisms in epithelial cells. In the absence of physical barrier defenses, immune responses to external antigens are induced in the presence of danger. Danger is recognized through pattern-recognition receptors, with the resultant release of active defenses, such as antimicrobials or antiproteinases. **A** Schematic longitudinal cross-section indicating the simple epithelial organization. **B** Photograph (courtesy of Dr. F. Anton-Erxleben) of a section of part of the epithelial lining of the body column, showing the diploblastic organization. *End* Endoderm, *ect* ectoderm, *m* mesoglea

histories. The diverse phylum includes hydroids (Hydrozoa), sea anemones and corals (Anthozoa), box jellies (Cubozoa), and the true jellyfish (Scyphozoa). Anthozoa is considered the ancestral group (Bridge et al. 1995; Medina et al. 2001; Collins et al. 2006). Since cnidarians possess most of the gene families found in bilaterians (Kusserow et al. 2005; Miller et al. 2005) and have retained many ancestral genes that have been lost in *Drosophila* and *Caenorhabditis elegans* (Miller et al. 2005; Technau et al. 2005), they provide insight into the contents of the "genetic tool kit" present in the cnidarian–bilaterian ancestor.

The diversity in cnidarian life histories (solitary vs colonial, sessile vs pelagic) and habitats (marine vs freshwater) raises several important issues relating to immunity. First, to maintain tissue integrity, colonial forms have to rely on their capacity of self/nonself discrimination to rapidly detect approaching allogeneic cells as foreign and to eliminate them. Second, since a solid substratum is rare in most of the cnidarians habitats, solitary cnidarians immediately after larval settlement have to defend themselves against other settlers and fight for the substratum. Third, in the absence of protective layers, cnidarians must have effective mechanisms to defend against microbial pathogens. And fourth, for cnidarians such as corals, successful growth means to be able to distinguish between friends and foes, i.e., to allow symbiotic algae to live within the endodermal epithelial cells and to close the doors for all other intruders. And finally, cnidarians such as sea anemones and corals have extremely long life spans – and must have some very effective immune systems in order to assue longevity. Thus, how do cnidarians manage to interact so successfully for more than 550 million years with their environment?

2 Immune Reactions in Invertebrates

The earliest study of immune function in invertebrates was done by Durham (1888) involving phagocytosis in starfish (reviewed by Leclerc 1996). Focusing on phagocytosis in various animals, Metchnikoff in 1892 described cnidarian (anthozoan) phagocytes that "responded to foreign bodies or their own cells by phagocytosing them" (cited by Bigger and Hildemann 1982). Although these observations must be taken *cum grane salis*, within the past 20 years the search for predecessors of the innate immune system in urochordates (for reviews, see Fujita 2002; Khalturin et al. 2004; see also Khalturin et al. in this volume: *Immune Reactions in the Vertebrates' Closest Relatives, the Urochordates*), insects (Cherry and Silverman 2006), and worms (Gravato-Nobre and Hodgkin 2005; Kim and Ausubel 2005) has revealed that invertebrates share many components of innate immunity with vertebrates. This ancient type of innate pathogen defense system is based on receptor-mediated recognition of abundant surface components of the pathogen that are not present in the host (Beutler 2004). Pathogen-associated molecular patterns (PAMPs) are of diverse structure and usually of essential importance for the pathogen. PAMPs bind to host receptors that share conserved functional domains, such as leucine-rich repeat (LRR) motifs and Toll/IL-1 receptor (TIR) domains. Upon ligand binding these receptors initiate signal transduction networks that activate a multifaceted defense response. In different species the downstream effector molecules, in contrast to the conserved signalling cascade, share little commonalities beside a protein or peptide structure which allows the destruction of bacterial cell membranes.

3 Immune Reactions in Cnidaria

Insects, worms, and vertebrates all belong to the "triploblast" or "Bilateria" clade of metazoans (Fig. 1). Since several animal phyla diverged, however, before the origin of this clade, the discovery of shared molecules tells us little about their origin and original roles, until we have comparative data from more basal animals. The aim of this chapter is to review the experimental evidence for innate immune reactions in Cnidaria, an ancient group of animals which diverged long before insects and nematodes were around (Fig. 1). Although cnidarians have a long history as model systems in comparative immunology (Campbell and Bibb 1970; Du Pasquier 1974, 2001), the underlying molecular mechanisms are poorly understood. Here

I summarize four aspects of immunity which can be studied particularly well within cnidarians – and which may be of interest from a comparative point of view to all immunologists: Intraspecies competition in sea anemones, allorecognition and cell lineage competition in the marine hydrozoan *Hydractinia*, antimicrobial defense reactions in Hydra and jellyfish, and symbiotic relationships in both corals and *Hydra*.

3.1 How to Fight for a Space to Live? Intraspecies Competition in Sea Anemones

Space on which to live is the most limiting resource in marine hard-substratum environments. Many hard surfaces in the sea are dominated by encrusting colonial or solitary cnidarians that must compete for space. Although "on the reef, it is important to know who you are, and who is rubbing shoulders with you" (David Miller, personal communication), little is known about the cellular and molecular mechanisms involved in mediating spatial competition. The effector mechanisms range from contact avoidance involving remote sensing, to barrier formation, stolon overgrowth, or deployment of fighting tentacles (reviewed by Raftos 1996). For example, the sea anemone *Anthopleura xanthogrammmica* "tolerates" adjacent clonal individuals, but attempts to "reject" heterogenic clones with which they come into contact (Sebens 1984). In the anthozoans *Stylophora pistillata* and *Montipora verrucosa* branches within one colony can easily fuse, while branches of genetically different individuals never undergo fusion (Hildemann et al. 1980; Müller et al. 1984; Chadwick-Furman and Rinkevich 1994). Fusion of two conspecific individuals is occasionally referred to as "natural transplantation". Observations in sea anemones (*Anthopeura elegantissima*, *Phymactis clematis*) and gorgonians (*Eunicella stricta*) indicate that individual colonies possess unique sets of histocompatibility elements, which are recognized as nonself by all other conspecific colonies (Lubbock 1980; Meinardi et al. 1995). In Hydrozoa, the same phenomenon was reported for *Millepora dichotoma* (Frank and Rinkevich 1994) and studied in great detail in the colonial marine hydroid *Hydractinia echinata* (see Section 3.2).

Many cnidarians use nematocysts as effector mechanisms in spatial competition. Nematocysts can discharge neurotoxic, myotoxic, hemolytic, and necrotic factors to damage the tissue of the competitor (Raftos 1996). The sea anemone *Anthopleura elegantissima*, for example, lives in large colonies of genetically identical clones on boulders around the tide line. Where two colonies meet they form a distinct boundary zone. Anemones that contact an animal from another colony will fight, hitting each other with special fighting tentacles (acrorhagi), catch tentacles, and sweeper tentacles, that leave patches of stinging cells stuck to their opponent. Besides specialized tentacles, the other structure used to defend against territorial invasion is a modified element of the mesentery, the mesenteric filament (for reviews, see Bigger 1988; Kass-Simon and Scappaticci 2002). When anemones make physical contact with one another, the acrorhagi tentacles expand and are repeatedly applied to the target organism. Then nematocysts are discharged, and the acrorhagial ectoderm adheres to the target organism. As a result of continued nematocyst discharge into the victim, the tissue beneath the acrorhagial peel becomes necrotic and dies. The specific molecules inititating this behavior remain to be characterized (Kass-Simon and Scappaticci 2002). Catch tentacles, which have been described for some sea anemones, are also specialized tentacles that are distinguished from the more slender feeding tentacles by their opacity and their blunt, wide form (for a review, see Bigger 1988). Catch tentacles develop from feeding tentacles that undergo a morphological change when an anemone comes into contact with appropriate conspecifics or other sea anemone species (reviewed by Kass-Simon and Scappaticci 2002). Another modified tentacle used in defense is the sweeper tentacle of scleractinian corals, whose tips, referred to as acrospheres, contain a large number of holotrichous isorhizas (den Hartog 1977; Richardson et al. 1979; Wellington 1980; Bigger 1988; Sebens and Miles 1988). Like catch tentacles, sweeper tentacles differentiate in response to contact with corals of another species (for references, see Kass-Simon and Scappaticci 2002). Sweeper tentacles extend at night and, as their name implies, flail or undulate (Bigger 1988). They can reach 5–10 times the length of feeding tentacles. Corals also use the digestive mesenterial filament as a weapon of aggression and defense by extruding mesenterial filaments through the body wall or oral cavity onto adjacent corals (Lang 1973). Although digestion by the filaments appears to be the major source of injury to the target organism, the great numbers of nematocysts (of unknown type) they possess no doubt contribute to the filaments' effectiveness as a weapon (Bigger 1988). These studies demonstrate that cnidarians clearly are able to recognize "self" in the use of nematocysts. In an interesting behavioral study, David Ayre from the University of Wollongong, Australia, and Rick Grosberg from the University of California at Davis recently showed (Ayre and Grosberg 2005) that clashing colonies of sea anemones fight as organized armies with distinct castes of warriors, scouts, reproductives, and other types. The study shows that very complex, sophisticated, and even coordinated behaviors can emerge in cnidarians at the level of the group, even when the group members are very simple organisms with nothing resembling a central nervous system.

3.2 How to Detect Approaching Allogeneic Cells as Foreign and to Eliminate Them? Allorecognition and Cell Lineage Competition in Colonial **Hydractinia**

Hydractinia is a colonial marine cnidarian composed of a limited number of repeating structural units, polyps, and stolons. The polyps may be specialized for feeding, predator defense, or reproduction (for a review, see Cadavid 2004). *Hydractinia* maintain self-perpetuating stem cell lineages throughout their life history, with the interstitial stem cells giving rise to the germ line and several other cell types. Polyps extend across the substratum as the colony grows and maintain gastrovascular connections between the polyps. After metamorphosis on a variety of hard

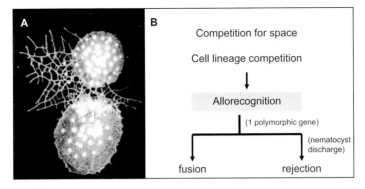

Fig. 3 **A** Aggressive rejection in *Hydractinia symbiolongicarpus* (modified from Shenk and Buss 1991). **B** *Hydractinia* colonies use allorecognition in intraspecific competition for space and to prevent somatic cell parasitism. The mechanisms, molecules and genes involved are currently under intensive investigation

substrata, including rocks, living snails, and empty shells occupied by hermit crabs, the planula larva forms the first polyp of a new *Hydractinia* colony. Whenever two or more planulae recruit to the same shell, the colonies may grow into contact and allorecognition interactions start. Allogeneic contacts have two major classes of outcome: rejection or fusion (Fig. 3). Within each major class there are two additional classes of outcomes: rejections may be aggressive or passive (Buss and Grosberg 1990); fusions may be permanent or metachronous (i.e., transitory; Hauenschild 1954; Buss and Shenk 1990). Lange et al. (1989) showed that all contacts follow a similar sequence of events. When stolons come into contact a large number of nematocysts are transported to the regions in contact. The nematocysts orient themselves in firing position in the contact zone. Once a threshold number of nematocysts have accumulated, they either disperse in fusion interactions or fire and damage the allogeneic tissue in rejection interactions (Lange et al. 1989).

These allorecognition responses play a fundamental role in maintaining the genetic and physiological integrity of the colony (Yund et al. 1987; Buss and Yund 1988; Yund and Parker 1989) because the germ line is not sequestered and because interstitial cells migrate within *Hydractinia* colonies. Thus, since allogeneic colonies which have the appropriate haplotype will fuse despite the fact that the rest of their genomes are different, in permanent fusions there is competition at the level of cell lineages. This may be as significant as selective death in rejections since, depending on the outcome of allorecognition reactions, the original members of the chimera risk losing access to the germ line, and thus risk reduced fitness, through "somatic cell parasitism" (Buss 1982; Buss and Shenk 1990).

Such allorecognition systems have long been of interest to geneticists by virtue of the substantial allotypic diversity they display (Cadavid 2004). Immunologists have long maintained that these phenomena lie at the root of vertebrate immunity. *Hydractinia* allorecognition is therefore of immediate interest in exploring the evolution of the immune system. *Hydractinia* in fact was among the first invertebrates shown to display a genetically based system of intolerance against allogeneic tissue: for more than 50 years it has been known that allorecognition and the ability to fuse between stolons of different colonies is under the control of one polymorphic gene (Hauenschild 1954, 1956; Ivker 1972). Isogeneic contacts result in fusion of the contacting tissues into a single physiological unit; stolonal contacts between clonal replicates of the same genotype are equivalent to the anastomosis of stolons which occurs during normal growth within a single colony.

More than five decades after Hauenschild's pioneering experiments, a robust model explaining the transmission genetics of allorecognition in *Hydractinia* is yet to be postulated (Cadavid 2004). Since allorecognition responses are controlled by highly variable genetic systems, and are considered a complex trait (Grosberg et al. 1997), an inbreeding program was initiated to allow the identification of an individual allorecognition locus in *Hydractinia* (Mokady and Buss 1996). Inbred lines were generated by brother–sister matings of fusible offspring for several generations. Colonies from an inbred line fused to one another and rejected those of the other line (Cadavid 2005). Current efforts to elucidate the genetic and molecular mechanisms controlling allorecognition in *Hydractinia* are based on positional or map-based cloning of allorecognition genes (Cadavid 2005). Preliminary data from defined genetic lines of *Hydractinia symbiolongicarpus* confirm that allorecognition is under the control of one polymorphic gene and show that the single chromosomal region contains at least two loci (Cadavid et al. 2004). Sequencing of the *Hydractinia* allorecognition complex will give population geneticists the tools necessary to explore the high diversity of this system and will reveal whether that diversity arises by an unusual genomic architecture. The results will also detect the differences and commonalities between allorecognition in *Hydractinia*, colonial urochordates such as *Botryllus* (see Khalturin et al. in this volume: *Immune Reactions in the Vertebrates' Closest Relatives, the Urochordates*), and vertebrates.

3.3 How to Detect and Disarm Microbial Attackers? Antimicrobial Defense Reactions in the Freshwater Polyp Hydra *and the Jellyfish* Aurelia

Cnidaria are soft-bodied animals lacking migratory phagocytic cells, hemolymph, and impermeable barriers, such as a cuticle or an exoskeleton, resulting in seemingly high vulnerability to pathogens. The animals live in habitats containing myriads of microbes and are constantly exposed to them. Scattered amidst the microbes are potential pathogens – bacteria, viruses, or protists – capable of tissue destruction and functional impairment.

Recently (Bosch et al., in preparation) we showed that, in the absence of migratory phagocytic cells and protective layers, the epithelium of the freshwater polyp *Hydra magnipapillata* is remarkably well equipped to survive in an environment teeming with potential pathogens and to prevent infectious agents from entering the body. To find out how the host perceives infection, we used a combined biochemical and transcriptome analysis approach to identify proteins and genes involved in epithelial defense (Bosch et al., in preparation). We could show that, in *Hydra*, all innate immune responses are mediated by the epithelial cells. Following pathogen invasion, there is an activation of an inducible defense system marked by an increased expression of genes encoding antimicrobial peptides. One of them is Hydramacin-1, a basic eight cysteines containing a cationic 60-aa peptide with a calculated molecular mass of 7009 Da (Bosch et al., in preparation). Hydramacin-1 is expressed exclusively in the endodermal epithelium, inducible by microbial products, and an antimicrobial peptide with extraordinary high activity against *Bacillus megaterium* (Bosch et al., in preparation). In the absence of conventional Toll-like receptors (TLRs; Miller et al. 2007), a leucine-rich repeat (LRR) lacking putative transmembrane receptor with a highly conserved Toll/IL-1 receptor (TIR) domain responds to microbial signals and cell death-associated molecules and is required for antimicrobial peptide induction (Bosch et al., in preparation). Interestingly, the expression of host-defense genes is affected by nerve cells (Kasahara and Bosch 2003; Bosch et al., in preparation). The results revealed several novel facts concerning the evolution of innate immune reactions: (a) the ancestral system of host defense is the inducible expression of antimicrobial peptides, (b) the epithelium represents the ancient system of host defense, and (c) even at the base of eumetazoan evolution the expression of immune effector genes is affected by nerve cells, suggesting that the neuro-immuno connection may be as old as the nervous system. In other animal groups communication and reciprocal regulation between the nervous and immune systems have been proposed to be essential for the stability of the organism. So far, however, studies have focused either on the evolution of the immune system or on the evolution the nervous system (Brogden et al. 2005). Our study (Kasahara and Bosch 2003) showed for the first time that both systems evolved in close relation to each other together and that *Hydra* will provide insight in a phylogenetically old, intriguing system that has developed to cope effectively with infections of various types.

Most recently, a novel antimicrobial peptide was purified from the mesoglea of the jellyfish *Aurelia aurita*, one of the most common and widely recognized types of jellyfish found near the coasts in the Atlantic, Arctic, and Pacific Oceans (Ovchinnikova et al. 2006). Aurelin, a 40-residue antimicrobial peptide with a molecular mass of 4296.95 Da was shown to exhibit activity against Gram-positive (*Listeria monocytogenes*, strain EGD) and Gram-negative (*Escherichia coli*, strain ML-35p) bacteria. Its primary structure, including six cysteines forming three disulfide bonds, as well as the primary structure of its molecular precursor, consisting of a canonical signal peptide, anionic propiece, and a mature cationic part, resembles the common structural features of animal defensins (Ovchinnikova et al. 2006). However, the distribution of cysteine residues makes it also similar to the K^+ channel-blocking toxins of sea anemones. Although aligning the aurelin and sea anemone toxin sequences shows rather moderate homology, Ovchinnikova et al. (2006) suggests that aurelin could be functionally related to the ShK-like toxins. Both the expression pattern as well as the in vivo function of aurelin remain to be clarified.

Similar to *Hydra* and jellyfish, corals also lack a physical barrier such as a hard exoskeleton. Although molecular and biochemical studies have not yet been published, there is evidence that corals depend on their anti-microbial mucus to remove and lyse bacterial invaders (Phillips 1963; Bigger and Hildemann 1982).

3.4 How to Distinguish Between Friends and Foes: Symbiotic Relationships in Corals and **Hydra**

Cnidaria are the phylogenetically oldest Eumetazoa phylum known to form symbiotic relationships with unicellular algae. There is increasing evidence from other mutualistic endosymbioses that the inter-partner signaling pathways involved during the onset of symbiosis are homologous to those driving animal host/pathogenic microbe interactions. In both types of relationships there are common principles and besides released signals, pathogen/symbiont surface molecules most likely are major determinants for an interaction with the host. Reduced host defense responses are necessary for the success of a symbiont when getting in contact with the host and are an essential requirement for the set-up of long-term symbiotic interactions or of persistent infections. Here I describe two examples of research in cnidarians focused towards an understanding of how signals and cell surface molecules allow the successful establishment of a sustained interaction of a potential symbiont with a cnidarian host.

One effort is investigating the role of host innate immunity in the recognition process during the onset of symbiosis in corals. Most shallow-water corals and anemones form mutualistic symbioses with photosynthetic microalgae. The microalgae are dinoflagellates, usually of the genus *Symbiodinium*, and are known colloquially as zooxanthellae. Cnidarian–dinoflagellate intracellular symbioses are one of the most important mutualisms in the marine environment. The coral–zooxanthella symbiosis is susceptible to abiotic stresses. When exposed to elevated temperature and irradiance, corals bleach, usually as a result of the expulsion of their zooxanthellae. Bleaching is a growing global environmental problem and can cause the destruction of the entire reef ecosystems. The cellular and molecular interactions underlying this interaction with particular emphasis on the establishment, maintenance, and breakdown of these cooperative partnerships are currently investigated in several coral species, including the Hawaiian stony coral, *Fungia scutaria*, the tropical sea anemone *Aiptasia pallida*, a temperate sea anemone found on the Oregon coast, *Anthopleura elegantissima*, and a Red Sea soft coral, *Heteroxenia fuscescens*. Most species must acquire symbionts anew with each generation and therefore must engage in a complex recognition and specificity process that results in the establishment of a stable symbiosis. To identify genes that initiate, regulate, and maintain this host/symbiont interaction, the Weis laboratory at Oregon State University recently conducted a comparative transcriptome analysis in the host sea anemone *A. elegantissima* using a cDNA microarray platform (Rodriguez-Lanetty et al. 2006). Although statistically significant differences in host gene expression profiles were detected between *A. elegantissima* in a symbiotic and nonsymbiotic state, the group of genes, whose expression is altered, is diverse, suggesting that the molecular regulation of the symbiosis is governed by changes in multiple cellular processes including lipid metabolism, cell adhesion, cell proliferation, apoptosis, and oxidative stress (Rodriguez-Lanetty et al. 2006). To search for symbiosis-specific proteins during the natural onset of symbiosis in early host ontogeny, Barneah et al. (2006) used two-dimensional polyacrylamide gel electrophoresis and compared patterns of proteins synthesized in symbiotic and aposymbiotic primary polyps of the Red Sea soft coral *Heteroxenia fuscescens* in the initiation phase, in which the partners interact for the first time. Surprisingly there were no changes detectable in the host proteome as a function of symbiotic state (Barneah et al. 2006). Since coral bleaching is caused by the breakdown of the polyp/algal symbiont association, another study at University of Newcastle upon Tyne investigated the frequency of apoptosis-like and necrosis-like cell death in the symbiotic sea anemone *Aiptasia* sp. subjected to environmentally relevant doses of heat stress (Dunn et al. 2004). The observations indicate that apoptosis and necrosis occur simultaneously in both host tissues and zooxanthellae (Dunn et al. 2004) and, therefore, must be considered as part of the cellular machinery involved in stress-induced bleaching. Taken together, both the transcriptom (Rodriguez-Lanetty et al. 2006) and proteomics study (Barneah et al. 2006) do not support the existence of symbiosis-specific genes involved in controlling and regulating the symbiosis. Instead, it appears that symbiosis is maintained by altering expression of existing genes involved in vital cellular processes (Rodriguez-Lanetty et al. 2006). Together with observations on apoptosis and necrosis in symbiotic *Aiptasia* sp., these first molecular examinations of a coral–dinoflagellate association seem to indicate that a suppression of apoptosis, together with a deregulation of the host cell cycle, is all what is necessary for symbiont and/or symbiont-containing host cell survival (Rodriguez-Lanetty et al. 2006).

Another model organism for studying the set-up of long-term symbiotic interactions is the freshwater polyp *Hydra*. *H. viridis* forms a stable symbiosis with intracellular green algae of the *Chlorella* group (Lenhoff and Muscatine 1963). The symbionts are located in endodermal epithelial cells. Each alga is enclosed by an individual vacuolar membrane (O'Brien 1982) resembling a plastid of eukaryotic origin at an evolutionary early stage of symbiogenesis. Proliferation of symbiont and host is tightly correlated. The photosynthetic symbionts provide nutrients to the polyps enabling *Hydra* to survive extended periods of starvation (Lenhoff and Muscatine 1963; Thorington and Margulis 1981). Symbiotic *Chlorella* is unable to grow outside the host, indicating a loss of autonomy during establishment of the intimate symbiotic interactions with *Hydra* (Habetha et al. 2003). During sexual reproduction of the host, *Chlorella* algae are translocated into the oocyte, giving rise to a new symbiont population in the hatching embryo. In an attempt to get insight in the underlying genetic machinery we screened *H. viridis* for symbiosis related genes using an unbiased approach based on cDNA representational difference analysis (RDA; Habetha and Bosch 2005). One of the characterized genes, HvAPX1, encoded an ascorbate peroxidase which is expressed exclusively during oogenesis. Sequence comparison showed that the gene is most closely related to plant peroxidases. Since the HvAPX1 gene in contrast to orthologous genes in plants does not contain introns, most likely during metazoan evolution it was translocated from a plant symbiont to the *Hydra* genome (Habetha and Bosch 2005). Five other *H. viridis* genes differentially expressed in presence of *Chlorella* algae are involved in the development of *Hydra* ovaries (Lange, Habetha, and Bosch, unpublished data). This is in agreement with the observation (Habetha et al. 2003) that *H. viridis* is hardly able to form ovaries in the absence of symbionts. Current efforts are directed on sequencing and the characterizing the RDA cDNA library as further steps in determining the molecular changes associated with symbiosis in *Hydra*.

4 How to Explore the Path They Went? Why Cnidarians Matter

A few years ago, outside the vertebrates the molecular nature of immune reactions could be approached only in a very few "model" species. Now the tide has turned: in silico approaches allow us to mine practically any species of choice. The question is no longer can we molecularly study this species or not. The question now is does this species tell us anything new about the history of the immune system which we did not know before and which we can learn from hardly anywhere else. In this respect, comparative immunology entered an era of radically widening horizons. As a sister group to the bilateria, Cnidaria are an important phylum, potentially providing key insight to the ancestry and evolution of immune reactions. As outlined above and elsewhere (Hemmrich et al. 2007; Miller et al. 2007), much effort in cnidarian research has recently been directed towards cloning evolutionarily conserved genes known to play critical roles in innate immunity in bilaterians. Data from several cnidarian taxa indicate convincingly that successful strategies for the detection and elimination of pathogens are present at that level of animal evolution. In the absence of an adaptive immune system, cnidarians employ an elaborate innate immune system to detect and eliminate nonself and to disarm microbial attackers. Beside an impressive accumulation of gene sequences, several novel tools and the development of genomic resources including the availability of transgenic Hydra (Wittlieb et al. 2006) have brought a new perspective on innate immunity in cnidarians and now pave the way for many important scientific and technological applications. Cnidaria emerge as kind of opener of discussions, an invitation to think about the structures and mechanisms of immune defense and self/nonself recognition.

To understand why Cnidaria matter, we must appreciate their flexibility and fathomless versatility of defense reactions that have helped them to survive for so long. As outlined above, there is no problem in innate immunity – intraspecies competition, cell lineage competition, host/pathogen and host/symbiont interactions – the cnidarians did not attempt to solve. Thus, whatever we experience with our own innate immune system, whatever we hope to learn, we see that the cnidarians have been there before us. There is a second reason why innate immune reactions in cnidarians matter: Cnidaria may emerge as attractive model systems to understand human barrier disorders by describing ancient mechanisms of host/microbial interactions and the resulting evolutionary selection processes or advantages. The identification of genes responsible for human diseases affecting biological barriers (e.g., skin or intestinal mucosa) often does not in itself provide a clue to etiopathogenesis or therapeutic targets, as the interaction of a suite of genes in a complex system such as the human is difficult to understand. Likewise the involved pathways that ultimately lead to the development of the disease phenotype are unclear. Searching for the evolutionary origin of the disease-causing genes and characterizing the variation in such genes under known evolutionary pressures may provide insights into the development of diseases in humans and identify new targets for therapy or prevention. Cnidaria may allow to unravel the complex interplay of host/pathogen signaling cascades that are also relevant to human barrier organs and its microbiota. Finally, in human medicine, the increasing prevalence of antibiotic-resistant microbes requires the development of new antimicrobials. Antimicrobial peptides of animal origin may be an effective alternative or additive of conventional antibiotics for therapeutic use. The recent identification of highly active antimicrobial peptides in hydra and jellyfish show that antimicrobial peptides from marine and freshwater cnidarians may represent a largely unexploited resource that can afford the design of new antibiotics with broad-spectrum antimicrobial activity.

Acknowledgement Supported by grants from the Deutsche Forschungsgemeinschaft (SFB 617).

References

Ayre DJ, Grosberg RK (2005) Behind anemone lines: factors affecting division of labour in the social cnidarian *Anthopleura elegantissima*. Anim Behav 70:97–110
Barneah O, Benayahu Y Weis VM (2006) Comparative proteomics of symbiotic and aposymbiotic juvenile soft corals. Mar Biotechnol 8:11–16
Beutler B (2004) Innate immunity: an overview. Mol Immunol 40:845–859
Bigger CH (1988) The role of nematocysts in anthozoan aggression. In: Hessinger DA, Lenhoff HM (eds) The biology of nematocysts. Academic Press, San Diego, pp 295–308
Bigger CH, Hildemann WH (1982) Cellular defense systems of the coelenterate. In: Cohen A, Sigel S (eds) Phylogeny and ontogeny. (The reticuloendothelial system: a comprehensive treatise, vol 3) Plenum, New York, pp 59–87
Bosch TCG (2007a) Symmetry breaking in stem cells of the basal metazoan *Hydra*. Prog Mol Subcell Biol 45:61–78
Bosch TCG (2007b) Why polyps regenerate and we don't: towards a cellular and molecular framework for *Hydra* regeneration. Dev Biol 303:421–433
Bosch TCG, David CN (1987) Stem cells of *Hydra magnipapillata* can differentiate into somatic cells and germ line cells. Dev Biol 121:182–191
Bridge D, et al (1995) Class-level relationships in the phylum Cnidaria: molecular and morphological evidence. Mol Biol Evol 12:679–689

Brogden KA, Guthmiller JM, Salzet M, Zasloff M (2005)The nervous system and innate immunity: the neuropeptide connection. Nat Immunol 6:558–564

Buss LW (1982) Somatic cell parasitism and the evolution of somatic tissue compatibility. Proc Natl Acad Sci USA 79:5337–5341

Buss LW, Grosberg RK (1990) Morphogenetic basis for phenotypic differences in hydroid competitive behavior. Nature 343:63–66

Buss LW, Shenk MA (1990) Hydroid allorecognition regulates competition at both the level of the colony and at the level of the cell lineage. In: Marchalonis JJ, Reinisch C (eds) Defense molecules. Liss, New York, pp 85–105

Buss LW, Yund PO (1988) A comparison of modern and historical populations of the colonial hydroid *Hydractinia*. Ecology 69:646–654

Cadavid LF (2004) Self-discrimination in colonial invertebrates: genetic control of allorecognition in the hydroid *Hydractinia*. Dev Comp Immunol 28:871–879

Cadavid LF (2005) Self/non-self discrimination in basal metazoa: genetics of allorecognition in the hydroid *Hydractinia*. Integr Comp Biol 45:623–630

Campbell RD, Bibb C (1970) Transplantation in coelenterates. Transplant Proc 2:202–211

Chadwick-Furman N, Rinkevich B (1994) A complex allorecognition system in a reef-building coral: delayed responses, reversals and nontransitive hierarchies. Coral Reefs 13:57–63

Cherry S, Silverman N (2006) Host–pathogen interactions in *Drosophila*: new tricks from an old friend. Nat Immunol 7:911–917

Collins AG, et al (2006) Medusozoan phylogeny and character evolution clarified by new large and small subunit rDNA data and an assessment of the utility of phylogenetic mixture models. Syst Biol55:97–115

Du Pasquier L (1974) The genetic control of histocompatibility reactions: phylogenetic aspects. Arch Biol 85:91–103

Du Pasquier L (2001) The immune system of invertebrates and vertebrates (review). Comp Biochem Physiol B Biochem Mol Biol 129:1–15

Dunn SR, Thomason JC, Le Tissier MD, Bythell JC (2004) Heat stress induces different forms of cell death in sea anemones and their endosymbiotic algae depending on temperature and duration. Cell Death Differ 11:1213–1222

Durham F (1888) On the emigration of ameboid corpuscles in the starfish. Proc R Soc Lond B 43:328–330

Frank U, Rinkevich B (1994) Nontransitive patterns of historecognition phenomena in the Red Sea hydrocoral *Millepora dichotoma*. Mar Biol 118:723–729

Fujita T (2002) Evolution of the lectin-complement pathway and its role in innate immunity (review). Nat Rev Immunol 2:346–353

Gravato-Nobre MJ, Hodgkin J (2005) *Caenorhabditis elegans* as a model for innate immunity to pathogens (review). Cell Microbiol 7:741–751

Grosberg RK, Hartt MW, Levitan DR (1997) Is allorecognition specificity in *Hydractinia symbiolongicarpus* controlled by a single gene? Genetics 145:857–860

Habetha M, Bosch TCG (2005) Symbiotic *Hydra* express a plant-like peroxidase gene during oogenesis. J Exp Biol 208:2157–2165

Habetha M, Anton-Erxleben F, Neumann K, Bosch TCG (2003) The *Hydra viridis/Chlorella symbiosis*. Growth and sexual differentiation in polyps without symbionts. Zoology 106:101–108

Hartog JC den (1977) The marginal tentacles of Rhodactissanctithomae (Corallimorphia) and the sweeper tentacles of *Monrastrea cavernosa* (Scleractinia): their cnidom and possible function. Proc Int Coral Reef Symp 3:463–469

Hauenschild VC (1954) Genetische und entwichlungphysiologische Untersuchungen ueber Intersexualitaet und Gewebevertraeglichkeit bei *Hydractinia echinata* Flem. Wilhelm Roux Arch Entwicklungsmech Org 147:1–41

Hauenschild VC (1956) Uber die Vererbung einer Gewebevertraeglichkeitseigenschaft bei dem Hydroidpolypen *Hydractinia echinata*. Z Naturforsch 1956:132–138

Hemmrich G, Miller DJ, Bosch TCG (2007) The evolution of immunity – a low life perspective. Trends Immunol (in press)

Hildemann WH, Jokiel PL, Bigger CH, Johnston IS (1980) Allogeneic polymorphism and alloimmune memory in the coral, *Montipora verrucosa*.Transplantation 30:297–301

Ivker FB (1972) A hierarchy of histo-compatibility in *Hydractinia echinata*. Biol Bull 143:162–174

Kasahara S, Bosch TCG (2003) Enhanced antibacterial activity in *Hydra* polyps lacking nerve cells. Dev Comp Immunol 27:79–166

Kass-Simon AA, Scappaticci AA (2002) The behavioral and developmental physiology of nematocysts. Can J Zool 80:1772–1794

Khalturin K, Bosch TCG (2006) Self/nonself discrimination at the basis of chordate evolution: limits on molecular conservation. Curr Opin Immunol 19:4–9

Khalturin K, Panzer Z, Cooper MD, Bosch TCG (2004) Recognition strategies in the innate immune system of ancestral chordates. Mol Immunol 41:1077–1087

Kim DH, Ausubel FM (2005) Evolutionary perspectives on innate immunity from the study of *Caenorhabditis elegans* (review). Curr Opin Immunol 17:4–10

Kusserow A, et al (2005) Unexpected complexity of the Wnt gene family in a sea anemone. Nature 433:156–160

Lang J (1973) Interspecific aggression by scleractinian corals. 2. Whythe race is not only to the swift. Bull Mar Sci 23:260–279

Lange R, Plickert G, Miller WA (1989) Histocompatibility in a low invertebrate, *Hydractinia echinata*: analysis of the mechanism of rejection. J Exp Zool 249:284–292

Leclerc M (1996) Humoral factors in marine invertebrate. In: Rinkevich B, Müller WEG (eds) Progress in molecular and subcellular biology: invertebrate immunology. Springer, Berlin Heidelberg New York, pp 1–9

Lenhoff HM, Muscatine L (1963) Symbiosis: on the role of algae symbiotic with hydra Science 142:956–958

Lubbock R (1980) Clone-specific cellular recognition in a sea anemone. Proc Natl Acad Sci USA 77:6667–6669

Medina M, et al (2001) Evaluating hypotheses of basal animal phylogeny using complete sequences of large and small subunit rRNA. Proc Natl Acad Sci USA 98:9707–9712

Meinardi E, Florin-Christensen M, Paratcha G, Azcurra JM, Florin-Christensen J (1995) The molecular basis of the self/nonself selectivity of a coelenterate toxin. Biochem Biophys Res Commun. 216:348–354

Miller DJ, Ball EE, Technau U (2005) Cnidarians and ancestral genetic complexity in the animal kingdom. Trends Genet 21:536–539

Miller DJ, Hemmrich G, Ball EE, Hayward DC, Khalturin K, Funayama N, Agata K, Bosch TCG (2007) The innate immune repertoire in cnidaria – ancestral complexity and stochastic gene loss. Genome Biol 8:R59

Mokady O, Buss LW (1996) Transmission genetics of allorecognition in *Hydractinia symbiolongicarpus* (Cnidaria: Hydrozoa). Genetics 143:823–827

Müller WEG, Müller I, Zahn RK, Maidhof A (1984) Intraspecific recognition system in scleractinian corals: morphological and cytochemical description of the autolysis mechanism. J Histochem Cytochem 32:285–288

O'Brien TL (1982) Inhibition of vacuolar membrane fusion by intracellular symbiotic algae in *Hydra viridis* (Florida strain). J Exp Zool 223:211–218

Ovchinnikova TV, Balandin SV, Aleshina GM, Tagaev AA, Leonova YF, Krasnodembsky ED, Men'shenin AV, Kokryakov VN (2006) Aurelin, a novel antimicrobial peptide from jellyfish *Aurelia aurita* with structural features of defensins and channel-blocking toxins. Biochem Biophys Res Commun S348:514–523

Phillips JH (1963) Immune mechanisms in the phylum Coelenterata. In: Dougherty EC (ed) The lower metazoan: comparative biology and phylogeny. University of California Press, Berkeley, pp 425–431

Raftos DA (1996) Histocompatibility reactions in invertebrates.In: Cooper EL (ed) Invertebrate immune responses: cell activities and the environment. (Advances in comparative and environmental physiology, vol 24) Springer, Berlin Heidelberg New York, pp 77–121

Richardson CA, Dustan P, Lang J (1979) Maintenance of living space by sweeper tentacles of *Montastrea cavernosa*, a Caribbean reef coral. Mar Biol 55:181–186

Rodriguez-Lanetty M, Phillips W, Weis VM (2006) Transcriptome analysis of a cnidarian–dinoflagellate mutualism reveals complex modulation of host gene expression. BMC Genomics 7:23

Sebens KP (1984) Agonistic behavior in the intertidal sea anemone *Arthropleura xanthogrammica*. Biol Bull (Woods Hole, Mass) 166:457–472

Sebens KP, Miles JS (1988) Sweeper tentacles in a gorgoniaoctocoral: morphological modifications for interference competition. Biol Bull (Woods Hole, Mass) 175:378–387

Shenk MA, Buss LW (1991) Ontogenetic changes in fusibility in the colonial hydroid *Hydractinia symbiolongicarpus*. J Exp Zool 257:80–86

Technau U, et al (2005) Maintenance of ancestral complexity and non-metazoan genes in two basal cnidarians. Trends Genet 21:633–639

Thorington G, Margulis L (1981) *Hydra viridis*: transfer of metabolites between *Hydra* and symbiotic algae. Biol Bull 160:175–188

Wellington GM (1980) Reversal of digestive interactions between Pacific reef corals: mediation by sweeper tentacles. Oecologia 47:340–343

Wittlieb J, Khalturin K, Lohmann J, Anton-Erxleben F, Bosch TCG (2006) Transgenic *Hydra* allow in vivo tracking of individual stem cells during morphogenesis. Proc Natl Acad Sci USA 103:6208–6211

Yund PO, Cunningham CW, Buss LW (1987) Recruitment and post-recruitment interactions in a colonial hydroid. Ecology 68:971–982

Yund PO, Parker HM (1989) Population structure of the colonial hydroid *Hydractinia* sp. nov. C in the Gulf of Maine. J Exp Mar Biol Ecol 125:63–82

Chapter 3
Bug Versus Bug: Humoral Immune Responses in *Drosophila melanogaster*

Deniz Ertürk-Hasdemir, Nicholas Paquette, Kamna Aggarwal, and Neal Silverman(✉)

1   Introduction..   44
   1.1   A Brief History   44
   1.2   Overview of the *Drosophila* Immune Response..   45
2   Microbial Recognition – the Peptidoglycan Recognition Proteins................................   45
   2.1   Peptidoglycan..   47
   2.2   NF-κB Proteins   48
3   The Toll Pathway   49
4   The IMD Pathway...   55
5   Down-Regulation of the IMD Pathway by PGRP Amidases   60
6   JAK/STAT Pathway   61
7   Concluding Remarks...   62
References...   63

Abstract Insects mount a robust innate immune response against a wide array of microbial pathogens. For example, the fruit fly *Drosphila melanogaster* uses both cellular and humoral innate immune responses to combat pathogens. The hallmark of the *Drosophila* humoral immune response is the rapid induction of antimicrobial peptide genes in the fat body, the homolog of the mammalian liver. Expression of these antimicrobial peptide genes is rapidly induced by two immune signaling pathways, which respond to distinct microorganisms. The Toll pathway is activated by fungal and Gram-positive bacterial infections, whereas the IMD pathway responds to Gram-negative bacteria. In this chapter, we discuss recent advances in understanding the mechanisms involved in microbial recognition, signal transduction, and immune protection mediated by these pathways, highlighting similarities and differences between *Drosophila* immune responses and mammalian innate immunity.

Division of Infectious Diseases, Department of Medicine, University of Massachusetts Medical School, Worcester, MA 01605, USA, *Neal.Silverman@umassmed.edu*

H. Heine (ed.), *Innate Immunity of Plants, Animals, and Humans.*
Nucleic Acids and Molecular Biology 21.
© Springer-Verlag Berlin Heidelberg 2008

1 Introduction

Insects are exposed to a multitude of pathogens in their natural environment. Therefore, they have developed sophisticated mechanisms to recognize and respond to infectious microorganisms. Even without adaptive immunity, insects have very effective immune responses to a wide range of pathogens (both microbial and larger). Moreover, the insect immune response has proven to be a useful and highly conserved model system for the study of innate immunity in general. In particular, the genetic, genomic, and molecular tools available for studying the immune response in the fruit fly *Drosophila melanogaster* make this a favored model system (Brennan and Anderson 2004; Cherry and Silverman 2006; Hultmark 2003; Lemaitre and Hoffmann 2007). *Drosophila* relies on several distinct effector mechanisms for immune protection, including clotting, melanization and encapsulation, cell-based phagocytosis, and the inducible production of a battery of antimicrobial peptides. This antimicrobial peptide response is critical for protection from experimental and natural infections. In this chapter, we focus on the inducible and systemic production of antimicrobial peptides. In another chapter, Louisa Wu and colleagues highlight the cellular immune response of *Drosophila*.

1.1 A Brief History

In 1865 Louis Pasteur discovered that microsporidia cause pébrine disease in silkworms, a condition characterized by melanization in the silkworm, and a significant economic concern (Brey 1998). In the following decades Carlos Finlay showed that mosquitoes were vectors for yellow fever (Chaves-Carballo 2005). This led to a great deal of interest in the early twentieth century on characterizing the microbial flora associated with insects (Steinhaus 1940).

By the end of 1960s, it was already known that pathogens such as fungi, protozoa, viruses, and bacteria could infect insects. In response to these infections, insects activated cellular and humoral immune defenses, including phagocytosis and the production of antimicrobial substances (Heimpel and Harshbarger 1965). A milestone in the insect immunity field was the study by Hans Boman and colleagues on the inducible antibacterial defense mechanisms of *Drosophila* (Boman et al. 1972). In subsequent years, a number of studies were done to characterize specific antimicrobial peptides (AMP) and the genes encoding them in various insects, including *Drosophila* (Hultmark et al. 1983; Kylsten et al. 1990; Samakovlis et al. 1990; Steiner et al. 1981; Sun et al. 1991). These antimicrobial peptides are small, cationic molecules that are effective against specific classes of pathogens. For example, in *Drosophila,* defensin acts against Gram-positive bacteria (Dimarcq et al. 1994), and Diptericin, Drosocin and Attacin are effective against Gram-negative bacteria (Asling et al. 1995; Michaut et al. 1996; Wicker et al. 1990), whereas drosomycin shows anti-fungal activity. Metchnikowin and Cecropin are both antibacterial and antifungal (Ekengren and Hultmark 1999; Levashina et al. 1995;

Samakovlis et al. 1992). These antimicrobial peptides are critical for resistance to infection, such that transgenic expression of a single antimicrobial peptide can protect immunodeficient flies (Tzou et al. 2002).

Since then, two major questions have shaped the field of insect immunology. How are microbes recognized? And, how is antimicrobial peptide gene expression regulated? Having the most powerful genetic and molecular tools available, *Drosophila* became the preferred experimental system to address these issues.

1.2 Overview of the Drosophila *Immune Response*

Drosophila has a multi-layered system for host defense. The chitin-based exoskeleton and chitinous internal structures form a physical barrier. If a pathogen breaches these barriers, several immune effector mechanism respond, including cellular responses (i.e. phagocytosis, encapsulation, melanization) and humoral responses (i.e. antimicrobial peptides). Antimicrobial peptides are found both locally, at the site of infections, and systemically in the insect sera, or hemolymph. In terms of the systemic humoral response, the fat body is the major site of antimicrobial peptide production, although other tissues also contribute, including the malphigian tubules and circulating blood cells, known as hemocytes. The local response induces antimicrobial peptide gene expression in epithelial tissues, like the trachea and the gut (Ferrandon et al. 1998; Liehl et al. 2006; Tzou et al. 2000). Recognition is the first step in a cascade of events that leads to these immune responses. Microbial products, often cell wall components, are detected by recognition receptors, which in turn stimulate signaling pathways that culminate in the induction of antimicrobial peptide gene expression.

At least two different pathways regulate the expression of antimicrobial peptide genes in *Drosophila*. The Toll pathway is stimulated by fungal and many Gram-positive bacterial pathogens. In contrast, the immune deficiency (IMD) pathway is triggered by Gram-negative bacteria. These two pathways are the focus of this chapter. In addition, we also discuss the *Drosophila* Janus kinase/signal transducers and activators of transcription (JAK/STAT) pathway, which is implicated in immunity, although its role is less clear.

2 Microbial Recognition – the Peptidoglycan Recognition Proteins

Peptidoglycan recognition protein (PGRP) was first identified as a peptidoglycan (PGN) binding factor from the hemolymph of the silkworm *Bombyx mori*, involved in activating the melanization cascade in vitro (Yoshida et al. 1996). PGRPs were subsequently cloned from both *Bombyx mori* and the moth *Trichoplusia ni* (Kang et al. 1998; Ochiai and Ashida 1999). *Drosophila* encodes 13 PGRP genes that are spliced into at least 17 PGRP proteins (Werner et al. 2000). All PGRP proteins contain a domain homologous to bacteriophage type-2 amidases, enzymes that cleave the amide bond connecting the stem-peptide to the carbohydrate backbone of peptidoglycan (Mellroth et al. 2003). Several PGRPs are known to have amidase activity (PGRP-SC1b and -LB) and some (PGRP-SB1, -SB2, -SC1 and -SC2) are predicted amidases (Kim et al. 2003; Mellroth and Steiner 2006; Mellroth et al. 2003). The non-catalytic PGRPs (PGRP-SA, -SD, -LA, -LC, -LE, -LF), which lack a cysteine residue that is critical for enzymatic activity, function as receptors for innate immune recognition of bacteria.

The *Drosophila* PGRPs can also be classified as either short PGRP proteins (seven different genes, seven proteins) or long PGRP proteins, with extended N-termini (ten genes, 13 proteins; Werner et al. 2000). Most short PGRP proteins have a signal sequence, lack a transmembrane domain, and are likely to be secreted (SA, SB1, SB2, SC1a, SC1b, SC2, SD). Some long PGRP proteins have a single-pass transmembrane domain and are likely transmembrane proteins (LAa, LAb, LCa, LCx, LCy, LD, LF). However, some long PGRP proteins lack a signal peptide and a transmembrane domain (LAc, LB, LE), and are likely intracellular proteins, or they could be secreted by a non-canonical mechanism (Takehana et al. 2002).

Mammals encode four PGRPs, termed PGRP-S, PGRP-L, PGRP-Iα, PGRP-Iβ (also referred to as PGLYRP-1, PGLYRP-2, PGLYRP-3, PGLYRP-4). Of these, only PGRP-L has amidase activity (Gelius et al. 2003; Kim et al. 2003). Mammalian PGRPs are expressed in a variety of tissues including bone marrow (PGRP-S), skin and intestinal tract (PGRP-Iα, PGRP-Iβ) and liver (PGRP-L; Kang et al. 1998; Lo et al. 2003; Lu et al. 2006; Mathur et al. 2004). Unlike the insect PGRPs, the non-catalytic mammalian PGRPs are bacteriocidal (Cho et al. 2005; Dziarski et al. 2003; Gelius et al. 2003; Liu et al. 2000; Lu et al. 2006; Tydell et al. 2002; Wang et al. 2003). Mice deficient for PGRP-S display increased susceptibility to intraperitoneal infections with non-pathogenic Gram-positive bacteria (Dziarski et al. 2003). PGRP-S is present in neutrophil granules and is involved in the intracellular killing of bacteria. It is also found associated with DNA nets released by activated neutrophils, where it is implicated in direct bacterial killing, acting synergistically with lysozyme (Cho et al. 2005; Liu et al. 2000). PGRP-Iα and -Iβ are secreted as di-sulfide hetero- and homodimers that are bactericidal against both pathogenic and non-pathogenic Gram-positive bacteria but are only bacteriostatic against other normal-flora bacteria. They are also bacteriostatic against some Gram-negative bacteria (Lu et al. 2006). The bacteriocidal and bacteriostatic mechanisms of the mammalian PGRPs are not yet clearly defined. PGRP-L is produced by the liver and secreted into the bloodstream (Zhang et al. 2005). However, the physiologic role of PGRP-L is not clear. It might prevent excessive inflammation following bacterial infection by digesting PGN, a known inflammatory agent. It is also possible that, by hydrolyzing PGNs, PGRP-L generates ligands for particular innate immune receptors, especially NOD1 (Fritz et al. 2007; Uehara et al. 2006).

2.1 Peptidoglycan

Drosophila recognizes bacteria by detecting specific forms of bacterial PGN via the PGRP receptors. PGN is a polymeric glycopeptide that forms the cell wall of most bacteria. PGN contains long glycan chains composed of alternating residues of N-acetylglucosamine and N-acetylmuramic acid (MurNAc), with short stem-peptides of alternating L- and D-amino acids attached to the lactyl group of MurNAc. These stem-peptides can be further cross-linked to each other by short peptide bridges (Mengin-Lecreulx and Lemaitre 2005). PGN from Gram-negative bacteria and certain Gram-positive bacteria (e.g. *Bacillus* spp, *L. monoctyogenes*) contains *meso*-diaminopimelic acid (DAP) at the third position of the stem-peptide chain, while other Gram-positive PGN contains lysine instead at this position. Also, the structure and degree of cross-bridging peptides is highly variable. The difference in the amino acid at position 3 in the stem-peptide along with the amount and type of cross-linking accounts for much of the variability in the structure of PGN produced by different bacteria (Schleifer and Kandler 1972). Another major difference between Gram-positive and Gram-negative bacteria is the localization of the PGN in the cell wall. Gram-negative bacteria include a thin layer of PGN, which is concealed in the periplasmic space between the inner and outer membranes. In contrast, Gram-positive bacteria usually contain a thick, multilayered PGN cell wall at their surface.

The Toll pathway is activated by lysine-type PGN, while the IMD pathway is activate by DAP-type PGN (Kaneko et al. 2004; Leulier et al. 2003). Also, the IMD pathway is activated by both polymeric DAP-type PGN and a monomeric fragment of DAP-type PGN, known as tracheal cytotoxin (TCT; Kaneko et al. 2004; Stenbak et al. 2004). TCT is a disaccharide tetra-peptide fragment of DAP-type PGN with a 1,6-anhydro–arranged muramic acid that is released in large quantities by some Gram-negative bacteria, like *Bordetella pertussis*, *Neisseria gonorrhoeae*, and *Vibrio fischeri* (Cookson et al. 1989; Goldman et al. 1982; Melly et al. 1984; Rosenthal 1979). TCT is linked to the cytopathology caused by *Bordetella* and *Neisseria* infection, and it is implicated in the developmental tissue degeneration caused by successful symbiosis of the squid *Euprymna scolopes* with *V. fischeri* (Koropatnick et al. 2004).

Recent data show that PGRPs employ several strategies to recognize and discriminate between different types of PGN. One strategy discriminates between DAP versus lysine in the third position of the stem-peptide, while another detects the presence or absence of a cross-bridging peptide. The C-terminal PGRP domain of PGRP-Iα binds lys-type PGN, with or without a penta-glycine cross-bridge (as seen in *S. aureus*; Schleifer and Kandler 1972). PGRP-S and PGRP-LCx preferentially recognize uncross-linked DAP-type PGN (Swaminathan et al. 2006; Takehana et al. 2002). PGRP-IαC forms several van der Waals contacts with the lysine through two amino acids, Asn236 and Phe237. By comparison, the corresponding residues that recognize DAP-type PGN are Gly-Thr (in PGRP-LCx, LE, human PGRP-S). Swap experiments demonstrated that these positions are involved in determining lysine versus

DAP-type binding properties. In addition, the structures of PGRP-LC or PGRP-LE bound to TCT indicate that another important residue making specific contact with the DAP residue is a highly conserved arginine found at the bottom of the PGN docking groove. The side-chain carboxylate of DAP forms a bidendate salt bridge with the guanidium group of this Arg (residue 254 in PGRP-LE, 413 in -LC). All PGRPs that have the conserved Gly-Thr also have this Arg residue. Thus, these residues likely function together to stabilize the interaction with DAP-type PGN (Chang et al. 2006; Guan and Mariuzza 2007; Lim et al. 2006).

2.2 NF-κB Proteins

The Rel/NF-κB proteins are a family of highly conserved transcription factors that control expression of genes involved in innate and adaptive immunity, inflammation, cell proliferation and apoptosis in both mammals and insects (Dutta et al. 2006; Hayden et al. 2006; Karin 2006; Silverman and Maniatis 2001). NF-κB proteins share a highly conserved 300-amino-acid N-terminal domain called the Rel-homology domain (RHD) which mediates DNA binding, dimerization and interaction with inhibitory IκB proteins. The RHD may also contain motifs for nuclear localization and transactivation (Perkins et al. 1997; Schmitz et al. 1995). NF-κB proteins are retained in the cytoplasm of unstimulated cells by the inhibitory IκB proteins. In response to immune challenge, IκB proteins are degraded, releasing the NF-κB transcription factors, which then translocate to the nucleus to activate target gene expression. In mammals, NF-κB family members include RelA (p65), RelB, c-Rel, p50/p105 (NF-κB1), and p52/p100 (NF-κB2), while *Drosophila* encodes three family members: DIF (Ip et al. 1993; Manfruelli et al. 1999; Meng et al. 1999), Dorsal (Reichhart et al. 1993; Roth et al. 1989; Steward 1987; Steward et al. 1984), and Relish (Dushay et al. 1996; Hedengren et al. 1999; see Fig. 1). DIF and Dorsal are similar to mammalian RelA (p65). They are retained in the cytoplasm by Cactus, the only member of the IκB protein family in *Drosophila* (Geisler et al. 1992; Lemaitre et al. 1996). In contrast, the compound protein Relish resembles mammalian NF-κB precursors p100 and p105 with an N-terminal RHD and a C-terminal IκB-like domain. Relish is sequestered in the cytoplasm through this C-terminal domain. After the discovery of the transcription factor NF-κB in mammals in 1986, promoters of the antimicrobial peptide genes in insects were also found to have κB-like sequences, suggesting that they are transcriptionally regulated by NF-κB-like transcription factors (Engstrom et al. 1993; Kappler et al. 1993; Reichhart et al. 1992). NF-κB transcription factors can regulate gene expression by binding as dimers to these κB sites (Engstrom et al. 1993; Gross et al. 1996). Although the most common NF-κB complex in mammalian cells is a p50–p65 heterodimer, it is likely that the *Drosophila* NF-κB factors preferentially form homodimers (Chen et al. 1998a, b; Engstrom et al. 1993; Tanji et al. 2007; Wirth and Baltimore 1988).

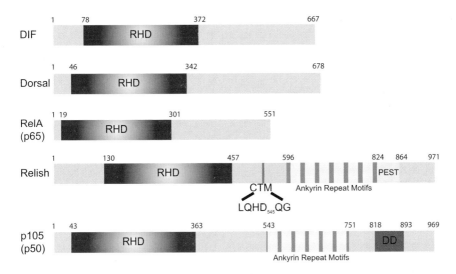

Fig. 1 The NF-κB family in *Drosophila* and mammals. *Drosophila* encodes three NF-κB family members. DIF and Dorsal are similar to mammalian RelA (p65) and Relish is similar to NF-κB precursor p105. All NF-κB family members share an N-terminal Rel-homology domain (RHD) that mediates DNA binding, dimerization, and interaction with IκB proteins. In contrast to mammalian NF-κB precursors, which require proteasome degradation to generate active NF-κB transcription factors, Relish processing depends on caspase-mediated cleavage. Cleavage occurs after residue D545, within a caspase target motif (CTM), generating a stable N-terminal transcription factor that translocates to the nucleus and a stable C-terminal IκB-like product that remains in the cytoplasm. *DD* Death domain, *PEST* domain rich in proline, glutamic acid, serine, and threonine

3 The Toll Pathway

The Toll pathway responds to Gram-positive bacterial and fungal infections (Lemaitre et al. 1996). Unlike human Toll-like receptors (TLRs) *Drosophila* Toll does not directly bind pathogens or microbe-derived compounds. Instead, Toll is a cytokine receptor, activated by the serum protein Spätzle. Spätzle is produced as a pro-protein, with a disulfide-linked dimeric structure. In order to activate the Toll pathway, pathogens activate serine protease cascades that culminate in Spätzle cleavage, liberating the mature Toll ligand (C-terminal 106 amino acids; Hu et al. 2004; Weber et al. 2003).

Recognition of Gram-positive bacteria involves the receptors PGRP-SA and PGRP-SD (Bischoff et al. 2004; Gobert et al. 2003; Michel et al. 2001; Pili-Floury et al. 2004). In addition, PGRP-SA functions in a complex with Gram-negative binding protein 1 (GNBP-1), which is a PGN processing enzyme (See Fig. 2). Both the receptors PGRP-SA and PGRP-SD recognize lysine-type PGN, but probably with slightly different specificities. Recognition of the *Micrococcus luteus* requires the PGRP-SA/GNBP-1 complex. Flies lacking either of these receptors fail to

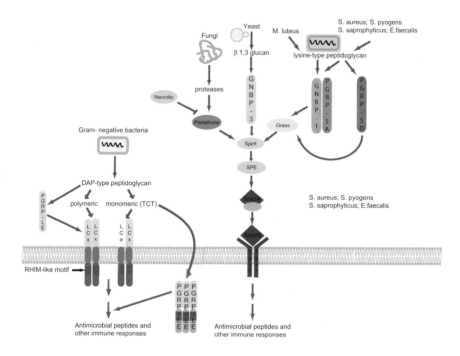

Fig. 2 Immune recognition in *Drosophila*. *Right:* Gram-positive bacteria and fungi activate the Toll pathway. Unlike mammalian TLRs, the *Drosophila* Toll receptor functions as a receptor for the protein Spätzle, which functions like a cytokine. Spätzle is found in the hemolymph in a proform prior to infection. Infection with Gram-positive bacteria or fungi leads to the activation of serine protease cascades and the cleavage of Spätzle. Distinct receptors are involved in recognizing different microbes and microbial products. PGRP-SA together with GNBP-1 is required to recognize certain Gram-positive bacteria, while PGRP-SD is sufficient to recognize other Gram-positive bacteria, and GNBP-3 is needed to recognize yeast. The PGRPs recognize bacterial PGN, GNBP-1 is thought to process PGN for presentation to PGRP-SA, and GNBP-3 recognizes beta-glucans from the fungal cell wall. These receptors, in turn, are believed to activate various serine protease cascades, as indicated. In contrast, proteases released from entomopathogenic fungi may directly cleave and activate other host serine proteases, in particular Persephone. All of these serine protease cascades converge on protease Spirit, which in turn leads to the activation of the Spätzle processing enzyme (SPE) that directly cleaves Spätzle generating the active Toll ligand. *Left:* The IMD pathway is triggered by Gram-negative bacteria and other microbes with DAP-type peptidoglycan. Different receptors are involved in the recognition of DAP-type peptidoglycan, depending on its structure and location. PGPR-LC is a cell surface receptor that recognizes PGN in the extracellular milieu. *PGRP-LC* encodes three distinct receptors via alternative splicing. The recognition of polymeric PGN requires only one PGRP-LC isoform, PGRP-LCx. In contrast, the recognition of a monomeric disaccharide tetrapeptide fragment of DAP-type PGN, known as TCT, requires both PGRP-LCx and PGRP-LCa. These receptors form a TCT-induced heterodimer. In addition, another receptor, PGRP-LE, is present within cells where it can recognize small fragments of PGN that gain access to intracellular compartments. Surprisingly, PGRP-LE, or at least the PGRP-domain of PGRP-LE, is also found the hemolymph, where it is believed to function as circulating receptor that binds PGN and presents it to the cell surface receptor PGRP-LC. The molecular mechanisms of signal transduction from these receptors to the intracellular components of the IMD pathway are not yet clear. However, PGRP-LC and PGRP-LE both have a RHIM-like motif that is critical for signaling

induce antimicrobial peptide gene expression following *M. luteus* infection and are highly susceptible to this microbe (Michel et al. 2001). In contrast, Gram-positive bacteria like *Staphylococcus aureus*, *Streptococcus pyogenes*, *Sta. saprophyticus*, and *Enterococcus faecalis* are recognized by either PGRP-SA or PGRP-SD and only the double *PGRP-SA*, *PGRP-SD* mutant is susceptible to infection with these bacteria (Bischoff et al. 2004). These results suggest that PGRP-SD recognizes a peptidoglycan-derived structure found in *Sta. aureus*, *Str. pyogenes*, *Sta. saprophyticus* and *E. faecalis* but not in *M. luteus*.

The exact function of GNBP-1 in the recognition of lys-type PGN is still under active investigation. GNBP-1 recognizes certain lysine-type PGN (e.g. *M. luteus* but not *Sta. aureus* PGN), and cleaves it into smaller muropeptides (Filipe et al. 2005; Wang et al. 2006). The minimal structure that activates the Toll pathway is a muropeptide dimer, composed of two disaccharide-tetrapeptides cross-linked via a penta-peptide bridge. In fact, GNBP-1 generates these small active PGN fragments from polymeric *M. luteus* PGN. But, GNBP-1 is also capable of digesting PGN fragments even further, into monomers, which are inactive. The interaction between GNBP-1 and PGRP-SA is enhanced by the presence of hydrolyzed PGN fragments. Another recent report demonstrated that in vitro digestion of *Sta. aureus* PGN with lysozyme-generated PGN fragments that activate the Toll and phenoloxidase pathways (in *Drosophila* and *Tenebrio*, respectively; Park et al. 2007), suggesting that other enzymes (in addition to GNBP-1) are capable of processing various lysine-type PGN for presentation to PGRP-SA. *Drosophila* encodes eight lysozyme homologs as well as five GNBP-related genes, perhaps some of these are involved in processing *Sta. aureus* PGN.

Detection of fungal infections relies on two sensor systems that are partially redundant (see Fig. 2). Fungal polysaccharides are recognized by the receptor GNBP-3 (Gottar et al. 2006). In vitro, GNBP-3 binds the yeast *Candida albicans* as well as curdlan, an insoluble polymer of β-(1,3)-glucan, which is present in the yeast cell wall, but not bacterial PGN. A second pathway, which requires a serine protease known as persephone (PSH), defines a second fungal recognition pathway. Live, entomopathogenic molds, such as *B. bassiana* and *M. anisopliae* stimulate the PSH pathway in addition to the GNBP-3-dependent pathway, while yeast or killed molds activate only the GNBP-3-dependent pathway. Instead of relying on a pattern recognition receptor, the PSH-dependent pathway is probably stimulated directly by pathogen-produced proteases, such as PR1A, which are released by pathogenic fungi to breakthrough the host cuticle. These results suggest that the *Drosophila* innate immune system can directly recognize virulence factors, analogous to the plant defense system.

Once activated, PGRP-SA/GNBP-1, PGRP-SD, GNBP-3 or PSH leads to Spätzle cleavage by activating serine protease cascades. During embryonic development a cascade of CLIP domain serine proteases leads to Spätzle activation. Mutants for these proteases (*snake*, *easter*) mount a wild-type immune response, indicating that these proteases are not required for the immune response (Lemaitre et al. 1996). A genetic screen led to the identification of PSH, which is homologous to Snake and is required for the cleavage of Spätzle in response to entomopathogenic fungal infections (Ligoxygakis et al. 2002). The *psh* mutants were first discovered as suppressors of the *necrotic* (*nec*) phenotype. *nec* encodes a serine protease inhibitor of the serpin family and lack of *nec* leads to constitutive activation of the Toll pathway in a *psh*-dependent manner (Levashina et al. 1999; Ligoxygakis et al. 2002). Another serine protease, Grass, is required only for the resistance to Gram-positive bacterial infection (Kambris et al. 2006). Recent studies showed that all these protease pathways appear to converge on two chymotrypsin-like serine proteases: Spirit and the Spätzle-processing enzyme (SPE). Spirit is thought to be the protease that cleaves and activates SPE, although this has not been directly demonstrated, while SPE directly cleaves pro-Spätzle, releasing the active C106 fragment. Both *Spirit* and *SPE* are required to resist both Gram-positive and fungal infections (Jang et al. 2006; Kambris et al. 2006).

Spätzle binding induces dimerization of the Toll receptor. Although the ligand is a symmetric dimer, biophysical studies indicate that the Spätzle-induced Toll dimer is asymmetrical (Weber et al. 2003). It is not yet clear whether the asymmetric aspect of the ligand-induced Toll dimer is critical for the activation of intracellular signaling. Dimerization of the Toll receptor is believed to recruit a pre-existing Myd88/Tube complex (see Fig. 3). Furthermore this complex associates with the kinase Pelle (the homolog of mammalian IRAK; Sun et al. 2004). The assembly of the resulting receptor complex occurs via two distinct functional domains. While the interaction between Toll and Myd88 occurs via their Toll/IL-1R (TIR) domains, Myd88, Tube, and Pelle interact in a trimeric complex via death domains (DD) found in each protein (Sun et al. 2002a, b; Tauszig-Delamasure et al. 2002; Towb et al. 1998). Although the DDs of these proteins are necessary for their interactions, Myd88 and Pelle do not interact directly; Tube acts as the core of the trimeric complex (Sun et al. 2002a). Thus the activated Toll receptor interacts directly with Myd88, which interacts with Tube, which ultimately recruits the kinase Pelle. Similar IRAK-kinase recruitment via an adapter complex is seen in mammalian Myd88-dependent TLR signaling.

Drosophila TNF-receptor-associated factor 2 (dTRAF2), the homolog of mammalian TRAF6, may also play a role in Toll signaling; however its role is unclear. In transiently transfected *Drosophila* cells, Pelle interacts with dTRAF2 and co-expression of Pelle and dTRAF2 synergistically activates the Toll pathway target gene *Drosomycin* (Shen et al. 2001). However RNAi to *dTraf2* shows no suppression of antimicrobial peptide gene expression after stimulation of the Toll or IMD pathways (Sun et al. 2002a; Zhou et al. 2005). In adult flies, overexpression of dTRAF2 is able to induce antimicrobial peptide gene expression and nuclear translocation of DIF as well as Relish. Interestingly, *dTraf2* null larvae exhibited reduced, but not abolished, levels of antimicrobial peptide gene expression following *Escherichia coli* infection (Cha et al. 2003). These data suggested that dTRAF2 may function in both the IMD and Toll pathways, but bypass mechanisms may be present which circumvent dTRAF2 in both cases.

Infections by Gram-positive bacteria and fungi culminate in the nuclear translocation of NF-κB proteins DIF and/or Dorsal. DIF is the main regulator of Toll signaling in both adults and larvae, whereas Dorsal is specifically required for the

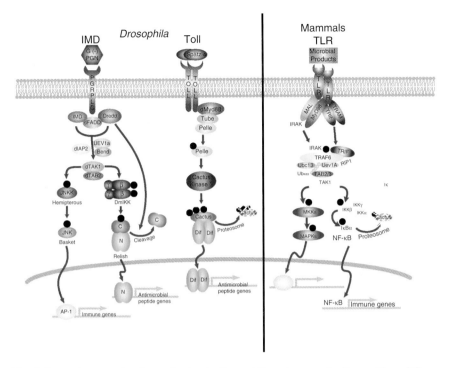

Fig. 3 Innate immune signal transduction in *Drosophila* and mammals. Recognition of Gram-negative PGN by the PGRP-LC receptor activates the IMD pathway (*left*). After activation of the receptor, a complex of FADD, IMD, and Dredd are required for signaling to the MAPKKK TAK1. TAK1 is likely to function with its partner TAB2. TAK1/TAB2 are responsible for the activation of both the JNK and NF-κB/Relish branches of the IMD pathway. NF-κB signaling requires the *Drosophila* IKK complex while JNK signaling involves the JNKK and JNK homologs *hemipterous* and *basket*. The NF-κB/Relish branch is critical for the induction of antimicrobial peptide gene expression, while the role of the JNK pathway remains less clear and controversial. IMD signal transduction is most similar to the mammalian MyD88-independent, Trif-dependent TLR signaling pathway (*right*), especially in the involvement of a RIP1-like gene IMD, the involvement of TAK1 and IKK homologs, and the activation of both NF-κB and MAPK pathways. In contrast, Toll signaling is more similar to the mammalian MyD88-dependent pathway. Activation of the Toll receptor by Spätzle leads to the recruitment of the dMyd88/Tube/Pelle adaptor complex. Pelle, an IRAK-like kinase, then signals to Cactus, probably through an unidentified kinase. Like mammalian IκBs, phosphorylation of Cactus leads to its ubiquitination and proteosome-mediated degradation. Then, the NF-κB protein DIF (or Dorsal) translocates into the nucleus, activating various antimicrobial peptide genes and other immune responsive genes

immune response in larvae. Dorsal was first identified for its role in dorso-ventral patterning in the developing embryo (Santamaria and Nusslein-Volhard 1983). The intracellular signaling components that lead to activation of Dorsal are the same in both early embryo and in the immune response (Drier and Steward 1997). DIF/Dorsal is sequestered in the cytoplasm by its interaction with the IκB protein

Cactus. The six-ankyrin repeats of Cactus are required for this interaction. In the embryo, Cactus and Dorsal are found in a complex of dorsal homodimer interacting with one molecule of Cactus (Isoda and Nusslein-Volhard 1994). Upon signaling, Cactus is degraded and DIF or Dorsal translocates to the nucleus (Belvin et al. 1995; Bergmann et al. 1996; Gillespie and Wasserman 1994; Reach et al. 1996; Wu and Anderson 1998). Cactus degradation, like IκBs, is controlled by phosphorylation and ubiquitin/protesome-mediated degradation. Initially, serines 74, 78, 82, and 83, in a region similar to the IκBα phosphorylation site, were thought to regulate signal-dependent degradation of Cactus (Bergmann et al. 1996; Reach et al. 1996). In contrast, later studies found that the N-terminal 125 amino acids are critical for signal-induced Cactus degradation, but the IκBα-like target motif between residues 74 and 83 is dispensable for degradation (Fernandez et al. 2001). Instead, Fernandez et al. (2001) identified another IκBα-like target motif around serine 116 that is also sufficient for degradation. Serines 74, 78, and 116 must all be changed to alanine to block Cactus degradation in the embryo. In addition, phosphorylation of the PEST domain, found at the C-terminus of Cactus, is implicated in its signal-independent degradation (Liu et al. 1997). However, neither of the two *Drosophila* IKK-related kinases (IKKε, IKKβ) is required for Toll-mediated Cactus phosphorylation and degradation. Although *Drosophila* IKKβ can phosphorylate Cactus in vitro (Kim et al. 2000), it is not required for *Drosomycin* expression in cells or in flies (Lu et al. 2001; Rutschmann et al. 2000; Silverman et al. 2000). Although the sequence motifs that are phosphorylated are very similar to those critical for IκBα phosphorylation in human cells, the kinase that phosphorylates Cactus is yet to be identified. Once phosphorylated, Cactus is likely ubiquitinated via the Slimb-SCF E3-ligase complex. *Drosophila* embryos mutant for *slimb*, the βTrCP homolog, are unable to activate the Dorsal target genes *twist* and *snail* (Spencer et al. 1999). Interestingly, Cactus degradation is required but not sufficient for efficient nuclear translocation of Dorsal during development (Bergmann et al. 1996).

Degradation of Cactus and nuclear translocation of DIF (and Dorsal) leads directly to the transcriptional induction of many immune responsive genes (De Gregorio et al. 2001, 2002a; Irving et al. 2001). For example, the well characterized AMP genes *Defensin*, *Drosomycin*, *Cecropin* and *Metchnikowin* are activated by Toll signaling. The promoter/enhancer regions of all these AMP genes include κB-sites where DIF or Dorsal bind (Senger et al. 2004). In addition, Toll signaling leads to the activation of other less well characterized genes, some of which may be AMPs while others may control different facets of the immune response. In fact, Toll signaling is linked to the activation of the cellular immune response and the proliferation of hemocytes (Qiu et al. 1998; Zettervall et al. 2004). Also, many components of the Toll pathway are regulated by Toll signaling (De Gregorio et al. 2002b; Lemaitre et al. 1996). Most notably, Cactus is up-regulated in response to immune challenge via the Toll pathway. This generates a negative feedback loop to down-modulate the cascade (Nicolas et al. 1998).

Coactivators that function with *Drosophila* NF-κB proteins have not been extensively studied. One study reported that dTRAP80 is required for DIF-induced transcriptional activation of *Drosomycin* in S2 cells (Park et al. 2003). Also, Helicase89B, a SNF2-like ATPase, is involved in activation of antimicrobial peptides in both the Toll and IMD pathways, and is thought to link NF-κB factors to the basal transcription machinery (Yagi and Ip 2005). Another study demonstrated that *Drosophila* CBP is a coactivator for Dorsal, and Dorsal-dependent activation of *twist* requires *nejire*, the CBP encoding gene (Akimaru et al. 1997).

Post-translational modifications are major regulators of transcription factors. Both NF-κB and IκB proteins are subject to various modifications. For example, in embryos it was demonstrated that Dorsal is multiply and dynamically phosphorylated during Toll signaling (Gillespie and Wasserman 1994). Phosphorylation of serine 312 is implicated in Dorsal stability, and phosphorylation of serine 317 is critical for optimal nuclear translocation of Dorsal in the embryo (Drier et al. 1999). The kinases responsible for these modifications are not known yet. One candidate might be the *Drosophila* atypical protein kinase C (ζPKC), which in cell culture is required for the Toll-signaling pathway but does not affect Cactus degradation. ζPKC can phosphorylate DIF in vitro (Avila et al. 2002). The nature and function of this phosphorylation event is yet to be identified.

Toll and IMD pathways are thought to be activated independently and initiate specific responses to different microorganisms. However, some AMPs are activated by both Toll and IMD pathways. Tanji et al. (2007) recently showed that some antimicrobial peptide genes have distinct κB elements in their enhancer region (e.g. *Drosomycin*) that respond to either Relish or DIF, and optimal gene induction occurs only when both the Toll (DIF) and IMD (Relish) pathways are activated, suggesting synergistic regulation of AMPs by two pathways (Tanji et al. 2007).

4 The IMD Pathway

The IMD pathway is potently activated by DAP-type PGN derived from Gram-negative bacteria and certain Gram-positive bacteria, such as *Bacillus* spp. Initially, it was believed that LPS activated the IMD pathway (Silverman et al. 2000; Werner et al. 2003). However, this did not account for the activation of the IMD pathway by certain Gram-positive bacteria (Kaneko et al. 2004; Lemaitre et al. 1997; Leulier et al. 2003). Subsequently, it was demonstrated that the commercial LPS preparations often used to stimulate the IMD pathway, in animals or cell lines, are contaminated with PGN, and it is this PGN that activates the IMD pathway (Kaneko et al. 2004; Leulier et al. 2003; Werner et al. 2003). Highly purified, PGN-free LPS does not stimulate IMD signaling in flies or fly cells.

Recognition of DAP-type PGN involves the receptors PGRP-LC and PGRP-LE (Choe et al. 2002; Gottar et al. 2002; Leulier et al. 2003; Ramet et al. 2002; Takehana et al. 2002; see Fig. 2). *PGRP-LC* encodes three alternatively spliced transcripts *PGRP-LCa, -LCx, -LCy*. All three isoforms encode single-pass transmembrane cell surface receptors. They each have distinct extracellular domains, which include a PGRP motif, anchored to the identical transmembrane and cytoplasmic domains (Werner et al. 2000). PGRP-LE encodes only one protein, which lacks both a signal sequence and a transmembrane domain. Although *PGRP-LC* null flies, which lack all three isoforms, induce dramatically reduced levels of antimicrobial peptide gene expression following infection with Gram-negative bacteria, such as *Escherichia coli* and *Agrobacterium tumefaciens*, they are not particularly susceptible to infection with all Gram-negative bacteria. For example, *PGRP-LC* mutants are sensitive to *A. tumefaciens*, *Erwinia carotovora carotovora*, and *Enterobacter cloacae*, but not *E. coli* and *B. megaterium*. (Choe et al. 2002; Gottar et al. 2002; Takehana et al. 2004). In contrast, mutants that abolish signaling through the IMD pathway, such as null alleles in IKK genes (see below), are highly susceptible to all Gram-negative bacteria. Therefore, it was hypothesized that another receptor must also recognize and respond to Gram-negative bacteria. Moreover, it was suggested that relatively low levels of antimicrobial peptide gene induction, as observed in *PGRP-LC* mutants, are sufficient to protect against infection with many Gram-negative bacteria. Genetic experiments suggested that PGRP-LE is the alternate receptor for the IMD pathway. Double *PGRP-LC*, *PGRP-LE* mutants are hypersusceptible to most Gram-negative bacteria, similar to other null mutants in the IMD pathway, and these double mutants do not induce detectable levels of antimicrobial peptide genes following infection. Overexpression of either PGRP-LC or PGRP-LE, in flies or in cell culture, is sufficient to drive AMP expression through the IMD pathway. PGRP-LE overexpression also activates the phenoloxidase cascade (Park et al. 2007).

Why might flies have two receptor systems (or perhaps four, if one considers the three *PGRP-LC* splice isoforms) for the recognition of bacteria and the activation of the IMD pathway? It appears that these receptors serve to recognize different forms of DAP-type PGN and to protect distinct niches. Monomeric and polymeric forms of DAP-type PGN are recognized by different receptors. In cell culture and in flies, only PGRP-LCx is required for recognizing polymeric PGN (isolated from *E. coli*). In contrast, both PGRP-LCx and -LCa are required in cultured cells for recognition of the monomeric fragment of DAP-type PGN known as TCT (Kaneko et al. 2004; Stenbak et al. 2004) The role of PGRP-LCy in microbial recognition is still unknown.

TCT binds PGRP-LCx directly, and then this ligand/receptor complex interacts with PGRP-LCa (Chang et al. 2005; Mellroth et al. 2003). The crystal structure of TCT bound to the ectodomains of PGRP-LCx and -LCa has been solved. TCT binds in the deep PGN binding cleft of PGRP-LCx, typical of PGRP–muropeptide interactions. The disaccharide unit of TCT makes important contributions to the interactions between PGRP-LCx (bound to TCT) and PGRP-LCa (Chang et al. 2006).

In adult flies, the recognition of monomeric TCT is even more complex. *PGRP-LC* null flies induce antimicrobial peptide gene expression following injection of TCT, but not after injection of polymeric *E. coli* PGN. *PGRP-LE* mutants respond normally to both monomeric and polymeric PGN. However, double *PGRP-LC*, *PGRP-LE* mutants fail to respond to TCT. Thus, in adult flies TCT can be recognized by either PGRP-LC or PGRP-LE. As mentioned previously, PGRP-LE lacks a transmembrane domain and a signal peptide, and is likely an intracellular receptor that recognizes small fragments of PGN-like TCT. These small PGN fragments may be able to gain access to PGRP-LE within the cell, while larger polymeric PGNs can only stimulate the cell surface receptor PGRP-LC. In support of this model, overexpression of PGRP-LE in the malphigian tubules (an immune-responsive kidney-like organ) triggers IMD signaling in a cell autonomous manner. Malphigian tubules ex vivo responded to TCT primarily through PGRP-LE and independently of the cell surface receptor PGRP-LC, and PGRP-LE was detected within these cells (Kaneko et al. 2006). The PGRP domain of PGRP-LE binds TCT with a K_d of about 27 nM, and TCT induces the formation of PGRP-LE multimers (Kaneko et al. 2006; Lim et al. 2006). The interactions responsible for TCT-induced PGRP-LE multimerization are very similar, in molecular detail, to those responsible for the TCT-mediated PGRP-LCx/LCa dimer. Because PGRP-LCa cannot bind TCT in a typical PGN binding cleft (the LCa cleft is occluded; Chang et al. 2005, 2006), the LC complex is limited to a dimeric form, while PGRP-LE forms a head-to-tail multimer, with each subunit binding to TCT and interacting with another subunit (Lim et al. 2006).

In addition to its role as intracellular receptor, several findings argue that PGRP-LE also functions outside the cell. When overexpressed in the fat body, PGRP-LE stimulates the IMD pathway in a cell non-autonomous manner. And, the PGRP domain of PGRP-LE (PRGP-LEpg) is found in the hemolymph (the insect sera). It is hypothesized that PRGP-LEpg binds PGN in the hemolymph and presents it to the cell surface receptor PGRP-LC, analogous to CD14/LPS interactions in mammals. Supporting this model, overexpression of PGRP-LE in the fat body induces IMD signaling in a manner that depends in part on PGRP-LC, and expression of PGRP-LEpg in cell culture leads to an enhancement of the PGRP-LC-mediated response to TCT. Although several lines of evidence strongly suggest that PGRP-LEpg is found in the hemolymph, it is not clear how PGRP-LE is released from cells.

The molecular mechanism by which PGN binding to either PGRP-LC or PGRP-LE leads to activation of the IMD pathway is still unclear. The cytoplasmic domain of PGRP-LC is responsible for initiating this signal transduction cascade (Choe et al. 2005). Epistatic experiments suggest that the *imd* protein functions immediately downstream of PGRP-LC and upstream of all other known members of the pathway. IMD is a death domain protein similar to mammalian receptor interacting protein 1 (RIP1; Georgel et al. 2001) and immunoprecipitation experiments showed that PGRP-LC and IMD interact (Choe et al. 2005; Kaneko et al. 2006). Kaneko et al. (2006) identified a RIP homotypic interaction motif (RHIM)-like domain that is crucial for signaling by PGRP-LC (when over-expressed or following infection; Kaneko et al. 2006). The RHIM domain, a motif of approximately 35 amino acids, was first identified in mammalian RIP1, RIP3, and in the adaptor protein Trif (Meylan et al. 2004; Sun et al. 2002b). The RHIM of Trif interacts with RIP1 and RIP3; and RIP1 and RIP3 also interact with each other through the RHIM domain. The Trif–RIP1 interaction is implicated in TLR3-induced NF-κB activation (Meylan et al. 2004). Likewise, the RHIM-like domain of PGRP-LC is critical for signaling. However, the PGRP-LC RHIM-like domain is not necessary for the interaction between PGRP-LC and IMD. Instead, PGRP-LC interacts with IMD via a region that is not required for signaling (Kaneko et al. 2006). Thus, the PGRP-LC/IMD interaction appears be superfluous for the activation of the pathway. Although the N-terminal signaling domains of PGRP-LC and PGRP-LE are not homologous, a RHIM-like motif was also identified in PGRP-LE. Mutation of the PGRP-LE RHIM-like motif blocks the signaling induced by forced expression of this intracellular receptor. The mechanism by which the RHIM-like domains of PGRP-LC and -LE function to transduce IMD signaling remains unclear. Perhaps the RHIM-like domain interacts with some unidentified component of the pathway.

Downstream of PGRP-LC and the *imd* protein, signal transduction through the IMD pathway leads to the *Drosophila* TAK1 homolog and then activation of the *Drosophila* IKK complex (Silverman and Maniatis 2001; Silverman et al. 2003; Vidal et al. 2001; see Fig. 3). The molecular mechanisms involved in signaling to TAK1 are still unclear, although RNAi-based experiments in cultured cells suggest that ubiquitination may play a key role. Work by Zhou and colleagues indicated that the E2 ubiquitin conjugating enzyme complex of dUEV1A and Bendless (the *Drosophila* Ubc13 homolog) functions downstream of IMD yet upstream of TAK1 in the IMD pathway (Zhou et al. 2005). The mammalian homologs of this E2 complex, Uev1A and Ubc13, are responsible for K63-polyubiquitination. Unlike K48-polyubiquitination, which leads to proteasomal degradation, K63-polyubiquitin chains are often regulatory and used to recruit and activate other signaling components. Thus, it is highly probable that K63-polyubiquitination plays an important role in the IMD signaling pathway between IMD and dTAK1. To date however, no K63-polyubiquinated protein has been identified in the IMD pathway.

Also, the E3 ligase involved in the IMD pathway remains elusive. Recently the *Drosophila* inhibitor of apoptosis protein 2 (dIAP2) was identified as a member of the IMD signaling pathway (Gesellchen et al. 2005; Kleino et al. 2005; Leulier et al. 2006; Valanne et al. 2007). Similar to other E3 proteins, dIAP2 contains a RING domain which is required for IMD signaling (Huh et al. 2007). Although dIAP2 has not yet been epistatically placed in the IMD signaling cascade, it is a good candidate to act as the E3 ligase, along with the dUEV1A/Bendless E2 complex. In addition, dTRAF2 could function as an E3 in the IMD pathway in some circumstances, as described above in the Toll signaling discussion (Cha et al. 2003).

It was also shown that the apical caspase DREDD plays a role between IMD and TAK1, perhaps functioning as an E3-ligase accessory factor (Zhou et al. 2005). Together, the E2/E3 complex of dUEV1A, Bendless, dIAP2 (and/or dTRAF2), and perhaps DREDD may mediate the K63-polyubiquitination of some unidentified member of the IMD pathway. This ubiquitinated protein is likely critical for signaling to TAK1, the next component in the pathway.

TAK1 may function in a complex with the *Drosophila* TAB2 homolog (Zhuang et al. 2006). Similar to mammalian TAB2, which was originally identified as a TAK1 binding protein, *Drosophila* TAB2 contains a conserved K63 polyubiquitin binding domain (Wang et al. 2001; Zhou et al. 2005), lending more credibility to the notion that ubiquitination plays a crucial role in IMD signaling. Signaling by the TAK1/TAB2 complex leads to the simultaneous induction of two downstream branches of the IMD pathway, which culminate in JNK or NF-κB/Relish activation (Silverman et al. 2003).

The JNK arm of the IMD pathway is activated by the TAK1-mediated signaling to Hemipterous, the *Drosophila* MKK7/JNKK homolog (Chen et al. 2002; Holland et al. 1997; Sluss et al. 1996). Hemipterous then goes on to phosphorylate the *basket* protein (JNK), which activates *Drosophila* AP-1. Signaling through the IMD/JNK pathway has been linked to the up-regulation of wound repair and stress response genes (Boutros et al. 2002; Silverman et al. 2003). Yet, the precise role that JNK signaling plays in the IMD pathway is controversial. Several reports have concluded that JNK signaling is not involved in AMP gene induction. Instead, AMP gene expression relies entirely on the NF-κB/Relish branch of the IMD pathway (Boutros et al. 2002; Silverman et al. 2003). In fact, an unidentified product of the Relish branch of the IMD pathway was proposed to inhibit JNK signaling (Park et al. 2004), while the JNK pathway was proposed to directly inhibit AMP gene expression by recruiting histone deacetylases (Kim et al. 2005). However, Delaney and colleagues (2006) concluded that the TAK1/JNK branch of the IMD pathway is critical for AMP gene induction, at least in clones of JNK-deficient cells within the larval fat body (Delaney et al. 2006). The role of the JNK pathway in antimicrobial gene expression remains controversial and further work will be necessary to clarify whether JNK has a positive and/or, negative role in the process.

In parallel to JNK activation TAK1 is also required for induction of the NF-κB/Relish branch of the IMD pathway, through activation of the *Drosophila* IKK complex (Silverman et al. 2003; Vidal et al. 2001). The *Drosophila* IKK complex contains two subunits: a catalytic kinase subunit encoded by *ird5* (IKKβ) and a regulatory subunit encoded by *kenny* (IKKγ; Rutschmann et al. 2000; Silverman et al. 2000). In S2 cells, it was clearly demonstrated that the IKK complex is activated rapidly following immune stimulation and this activation requires TAK1 (Silverman et al. 2001, 2003). Activated IKK complex can directly phosphorylate Relish.

Relish is a bipartite protein similar to mammalian NF-κB precursors p100 and p105. It contains an N-terminal Rel homology domain (RHD) and an inhibitory IκB domain with six ankyrin repeats that holds the protein in the cytoplasm. Upon infection with Gram-negative bacteria, *Relish* expression is strongly induced in adult flies (Dushay et al. 1996). *Relish* mutant flies show extreme sensitivity to infections and fail to induce antimicrobial genes after bacterial infection (Hedengren et al. 1999). Although the *Relish* locus encodes an embryo specific isoform, *Relish* does not seem to have a role in development because homozygous Relish mutants are viable and fertile.

In mammals, the NF-κB precursors p100 and p105 are processed by the proteasome and their C-terminus is degraded to produce p50 and p52, respectively. This processing is regulated by phosphorylation of C-terminal serine residues, which leads to ubiquitination and partial proteasome degradation of the C-terminus (Perkins 2007). In contrast, Relish processing does not depend on proteasomal degradation. Relish is endoproteolytically cleaved by a caspase, producing an N-terminal RHD transcription factor module that translocates to the nucleus to activate immune genes, and a stable C-terminal domain that remains in the cytoplasm (Stöven et al. 2000). Relish cleavage occurs after residue D545, within a typical caspase target motif. In addition to its role upstream in the IMD pathway (mentioned above), DREDD also appears to function downstream in the pathway and is likely the caspase that cleaves Relish. DREDD and Relish physically interact in cell culture and *Dredd* RNAi prevents antimicrobial peptide gene expression induced by an activated allele of TAK1 (Zhou et al. 2005). *Dredd* mutants fail to cleave Relish, induce AMP gene expression, and are highly sensitive to Gram-negative bacterial infections (Leulier et al. 2000; Stöven et al. 2003).

The mechanisms involved in the signal-dependent cleavage and activation of Relish are uncertain. Relish is phosphorylated in a signal-dependent manner by the *Drosophila* IKK complex. The C-terminus of Relish is required for both its phosphorylation and cleavage (Stöven et al. 2003), suggesting phosphorylation and cleavage are linked. However, more studies are required to define how phosphorylation might control cleavage. Another possibility is raised by the report from Delaney et al. (2006), who showed that TAK1 is not required for Relish cleavage (in vivo or in cultured cells), while it was previously reported that the IKK complex, which is activated by TAK1, is necessary for cleavage. This suggested that the IKK complex may control Relish cleavage independently of phosphorylation. In this case, the mechanism(s) by which TAK1-dependent, IKK-mediated phosphorylation regulate Relish remain(s) mysterious.

While Relish is activated by caspase-dependent cleavage, the ubiquitin-proteasome pathway may target Relish for destruction. It was found that inhibiting the SCF E3–ubiquitin–ligase complex caused constitutive expression of antimicrobial peptide genes and increased levels of Relish (Khush et al. 2002). Thus, the ubiquitin proteosome pathway may be important for down-regulating the IMD pathway.

5 Down-Regulation of the IMD Pathway by PGRP Amidases

Recent studies demonstrated that level of IMD signaling is also down-modulated by the catalytic PGRPs (Bischoff et al. 2006; Zaidman-Remy et al. 2006). PGRP-LB and PGRP-SC are amidases that remove the stem-peptide from the glycan backbone of PGN and make it significantly less immunostimulatory (Kaneko et al. 2004; Mellroth and Steiner 2006; Mellroth et al. 2003; Zaidman-Remy et al. 2006). PGRP-LB digests only DAP-type PGN, whereas PGRP-SC digests both DAP-type and lys-type PGN (Mellroth et al. 2003; Zaidman-Remy et al. 2006). Depletion of PGRP-SC1/2 or PGRP-LB by RNAi leads to higher induction of *Diptericin* after infection. PGRP-LB is a secreted protein and its expression is up-regulated by the IMD pathway. The presence of PGRP-LB in the hemolymph provides a negative feed-back loop that regulates the degree of immune activation in response to bacterial infection. PGRP-LB may also regulate the immune reactivity of *Drosophila* to environmental bacteria. These results show that local IMD immune activation in the gut and trachea is held in check by PGRP-LB, and the IMD pathway is activated only when the scavenger activity of PGRP-LB is saturated or inactivated. Amidase PGRPs may also prevent potentially severe consequences to host tissues because of prolonged immune activity. In support of this idea, it has been shown that

PGRP-SC1/2 RNAi larvae infected with bacteria show increased developmental defects and lethality (Bischoff et al. 2006).

6 JAK/STAT Pathway

While the two NF-κB signaling pathways, IMD and Toll, are largely responsible for antimicrobial peptide gene induction, another evolutionarily conserved pathway known as the Janus kinase/signal transducers and activators of transcription (JAK/STAT) pathway also plays a significant role in the *Drosophila* immune response. The JAK/STAT pathway plays important roles in hematopoiesis (Zeidler et al. 2000) and in the induction of Tep genes and stress-inducible Turandot (Tot) genes (Ekengren and Hultmark 2001; Lagueux et al. 2000). Similar to the Toll pathway, the JAK/STAT signaling pathway is also necessary for proper development of *Drosophila*. It was first identified in flies due to its role in embryonic segmentation (Binari and Perrimon 1994).

The main components of this pathway are the ligand, Unpaired (UPD), the receptor Domeless (DOME), the JAK (Hopscotch/HOP), and the STAT (STAT92E/Marelle). The ligand UPD is a secreted cytokine and has a signal sequence and several potential N-linked glycosylation sites. Recently, analysis of the *Drosophila* genome revealed the presence of two other *upd*-like cytokine-encoding genes near the *upd* locus (known as *upd2, upd3*; Agaisse et al. 2003; Hombria and Brown 2002).

Signal transduction in the *Drosophila* JAK/STAT pathway is thought to occur in a fashion similar to that seen in canonical mammalian JAK/STAT signaling. Using this pathway as a model, it is believed that HOP, the only *Drosophila* JAK homolog, associates intracellularly with the JAK/STAT receptor DOME (Binari and Perrimon 1994; Zeidler et al. 2000). Upon binding its ligand Unpaired, DOME is then believed to dimerize. This dimerization allows the transphosphorylation and activation of HOP, which are now in close proximity. Activated HOP kinases then proceed to phosphorylate the DOME receptor, allowing recruitment of the STAT92E via its SH2 domain. STAT92E, a homolog of human STAT5, is the only STAT in flies. It contains a SH2 domain as well as a DNA-binding domain and a single highly conserved C-terminal tyrosine residue (Brownell et al. 1996; Hou et al. 1996; Yan et al. 1996). After binding the phosphorylated Domeless receptor via an SH2 domain, STAT92E is also phosphorylated by HOP on this C-terminal tyrosine. Phosphorylation of this tyrosine residue allows for dimerization and nuclear translocation of the STAT92E (Agaisse et al. 2003). Recently, two groups identified new components of the JAK/STAT pathway using genome-wide RNAi screens in *Drosophila* cells stimulated with UPD (Baeg et al. 2005; Muller et al. 2005). Baeg et al. (2005) identified RanBP3 and RanBP10 as negative regulators of JAK/STAT pathway possibly controlling nucleocytoplasmic shuttling of STAT92E. In addition, Muller et al. (2005) identified a bromo-domain containing protein similar to mammalian BRWD3, which is implicated in B-cell leukemia. Both groups also discovered that Ptp61F, a protein tyrosine phosphatase, is induced by JAK/STAT signaling and is a negative regulator of the pathway. The exact functions of these proteins and their role in *Drosophila* immunity remain to be characterized.

Among the known JAK/STAT targets in the *Drosophila* immune response are the Tot genes. Studies in both adult and larvae showed that bacterial infection as well as various stresses result in the induction of different subsets of Tot genes (Ekengren and Hultmark 2001). In adults, injection of *E. cloacae* leads to expression of all Tot genes (*TotA, -B, -E, -F, -M, -X, -Z*) to varying degrees. Larvae appear much less responsive and only show expression of a subset of Tot genes under comparable conditions (*TotA, -F, -X, -Z*). Similar to infection, when adults are heat-stressed all Tot genes are induced; however oxidative stress, via paraquat, induces only *TotA, TotX*, and *TotZ*. In larvae, UV irradiation leads to the expression of all Tot genes except *TotM* and *TotX*.

The regulation of *TotA* by JAK/STAT signaling has been best characterized. Infection induced *TotA* expression does not require the ligand *upd1* (Agaisse et al. 2003). Instead, genetic experiments suggest that *upd3* expression is induced upon infection in the hemocytes and acts as a cytokine to stimulate *TotA* expression in the fat body, indicating that UPD3 is one of the ligands required for activation of the JAK/STAT pathway in immunity (Agaisse et al. 2003). However, induction of *TotA* may not simply involve the JAK/STAT pathway because other data indicate that both Relish signaling, via the IMD pathway, and the MAPKK-kinase dMEKK1 are required for *TotA* induction (Agaisse et al. 2003; Brun et al. 2006).

In addition to *TotA*, another target of the JAK/STAT pathway is *Tep1* (Lagueux et al. 2000). *Tep1* is one of six TEP genes in *Drosophila* which are related to complement thioester proteins and are thought to function as microbial opsonins (Cherry and Silverman 2006). Lagueux et al. (2000) found that *Tep1* is up-regulated in a *hopscotch* (JAK)-dependent manner in infected larvae. Also, gain-of-function mutants of *hopscotch* constitutively express *Tep1*.

JAK/STAT signaling also plays a significant role in the *Drosophila* antiviral immune response. Dostert and colleagues identified a group of genes that were induced upon infection with *Drosophila* C virus (DCV, a picornavirus), but not via the Toll or IMD pathways. Several genes induced by DCV infection, such as *vir-1*, contain putative STAT binding sites, and DCV infection induces STAT DNA-binding activity in flies (Dostert et al. 2005). Using mutant flies they demonstrated that expression of a number of these virally induced genes required *hop*. Work in mice also indicates that STAT1 is necessary for innate anti-viral immunity (Dupuis et al. 2003; Karst et al. 2003). Thus, the JAK/STAT pathway likely represents an evolutionary conserved signaling pathway that responds to viral pathogens.

7 Concluding Remarks

Drosophila melanogaster is a powerful model for the study of innate immunity, as the molecular, genetic, and genomic tools available for research in this system are unparalleled. The *Drosophila* model has provided insights into many insect host-defense mechanisms including phagocytosis and NF-κB signaling pathways. Furthermore, these studies have highlighted similarities between insect and mammalian innate immunity. For example, the discovery of the role of Toll in the *Drosophila* immune response (Lemaitre et al. 1996) led directly to the identification of Toll-like receptors (TLRs) in mammals (Lemaitre 2004). Many components of the insect and mammalian innate immune signaling pathways are also highly conserved. For example, the *Drosophila* Toll pathway has a great deal of similarity to the MyD88-dependent TLR pathway, whereas the IMD pathway is more similar to the MyD88-independent, Trif-dependent pathway (Fig. 3).

Besides these similarities, *Drosophila* has obvious differences from mammalian innate immune pathways. In particular, most Toll-like receptors are implicated in directly detecting microbial products, whereas *Drosophila* Toll recognizes a cytokine, the mature cleaved Spätzle. Although many of the receptor proximal signaling events are similar, downstream signaling events in the Toll and IMD pathways show significant differences compared with the TLR pathways. For example, the IKK-related kinases in *Drosophila* (IKKβ, IKKε) seem to have no function in the Toll pathway. Also, activation of the NF-κB factor Relish by caspase cleavage is, so far, unique to insects.

For a long time, *Drosophila* immunity was believed to solely depend on non-specific innate responses. Surprisingly, two recent reports suggest that insects may have some form of immunologic memory and adaptation. Pham et al. (2007) found that priming *Drosophila* with sub-lethal doses of *Streptococcus pneumoniae* or *Beauveria bassiana* protects flies when they encounter a second lethal dose of the same microbe, but does not protect against lethal infection with other microbes. This suggests that *Drosophila* immunity includes a mechanism for specific immunological memory (Pham et al. 2007). Another important discovery illuminating the molecular complexity of *Drosophila* immunity is the *Dscam* gene, which alternatively spliced into a predicted ~ 18 000 isoforms in the fat body and hemocytes. This highly diversified protein repertoire appears to function as opsonins and/or phagocytic receptors with variable specificity for different microbes (Dong et al. 2006; Watson et al. 2005). It is not yet clear whether the diversified receptors generated by Dscam relate to the specific memory observed by Pham and colleagues. However, these findings show that *Drosophila* has more complex defense mechanisms then originally believed, some of which are comparable with mammalian adaptive immunity, raising interesting questions to be addressed in the future.

References

Agaisse H, Petersen UM, Boutros M, Mathey-Prevot B, Perrimon N (2003) Signaling role of hemocytes in *Drosophila* JAK/STAT-dependent response to septic injury. Dev Cell 5:441–450

Akimaru H, Hou DX, Ishii S (1997) *Drosophila* CBP is required for dorsal-dependent twist gene expression. Nat Genet 17:211–214

Asling B, Dushay MS, Hultmark D (1995) Identification of early genes in the *Drosophila* immune response by PCR-based differential display: the Attacin A gene and the evolution of attacin-like proteins. Insect Biochem Mol Biol 25:511–518

Avila A, Silverman N, Diaz-Meco MT, Moscat J (2002) The *Drosophila* atypical protein kinase C-ref(2)p complex constitutes a conserved module for signaling in the toll pathway. Mol Cell Biol 22:8787–8795

Baeg GH, Zhou R, Perrimon N (2005) Genome-wide RNAi analysis of JAK/STAT signaling components in *Drosophila*. Genes Dev 19:1861–1870

Belvin MP, Jin Y, Anderson KV (1995) Cactus protein degradation mediates *Drosophila* dorsal-ventral signaling. Genes Dev 9:783–793

Bergmann A, Stein D, Geisler R, Hagenmaier S, Schmid B, Fernandez N, Schnell B, Nusslein-Volhard C (1996) A gradient of cytoplasmic Cactus degradation establishes the nuclear localization gradient of the dorsal morphogen in *Drosophila*. Mech Dev 60:109–123

Binari R, Perrimon N (1994) Stripe-specific regulation of pair-rule genes by hopscotch, a putative Jak family tyrosine kinase in *Drosophila*. Genes Dev 8:300–312

Bischoff V, Vignal C, Boneca IG, Michel T, Hoffmann JA, Royet J (2004) Function of the *Drosophila* pattern-recognition receptor PGRP-SD in the detection of Gram-positive bacteria. Nat Immunol 5:1175–1180

Bischoff V, Vignal C, Duvic B, Boneca IG, Hoffmann JA, Royet J (2006) Downregulation of the *Drosophila* immune response by peptidoglycan-recognition proteins SC1 and SC2. PLoS Pathog 2:e14

Boman HG, Nilsson I, Rasmuson B (1972) Inducible antibacterial defence system in *Drosophila*. Nature 237:232–235

Boutros M, Agaisse H, Perrimon N (2002) Sequential activation of signaling pathways during innate immune responses in *Drosophila*. Dev Cell 3:711–722

Brennan CA, Anderson KV (2004) *Drosophila*: the genetics of innate immune recognition and response. Annu Rev Immunol 22:457–483

Brey PT (1998) The contributions of the Pasteur school of insect immunity in molecular mechanisms of immune responses in insects. Chapman & Hall, London

Brownell JE, Zhou J, Ranalli T, Kobayashi R, Edmondson DG, Roth SY, Allis CD (1996) Tetrahymena histone acetyltransferase A: a homolog to yeast Gcn5p linking histone acetylation to gene activation. Cell 84:843–851

Brun S, Vidal S, Spellman P, Takahashi K, Tricoire H, Lemaitre B (2006) The MAPKKK Mekk1 regulates the expression of Turandot stress genes in response to septic injury in *Drosophila*. Genes Cells 11:397–407

Cha GH, Cho KS, Lee JH, Kim M, Kim E, Park J, Lee SB, Chung J (2003) Discrete functions of TRAF1 and TRAF2 in *Drosophila melanogaster* mediated by c-Jun N-terminal kinase and NF-kappaB-dependent signaling pathways. Mol Cell Biol 23:7982–7991

Chang CI, Ihara K, Chelliah Y, Mengin-Lecreulx D, Wakatsuki S, Deisenhofer J (2005) Structure of the ectodomain of *Drosophila* peptidoglycan-recognition protein LCa suggests a molecular mechanism for pattern recognition. Proc Natl Acad Sci USA 102:10279–10284

Chang CI, Chelliah Y, Borek D, Mengin-Lecreulx D, Deisenhofer J (2006) Structure of trachael cytotoxin in complex with a heterodimeric pattern-recognition receptor. Science 311:1761–1764

Chaves-Carballo E (2005) Carlos Finlay and yellow fever: triumph over adversity. Mil Med 170:881–885

Chen FE, Huang DB, Chen YQ, Ghosh G (1998a) Crystal structure of p50/p65 heterodimer of transcription factor NF-kappaB bound to DNA. Nature 391:410–413

Chen W, White MA, Cobb MH (2002) Stimulus-specific requirements for MAP3 kinases in activating the JNK pathway. J Biol Chem 277:49105–49110

Chen YQ, Ghosh S, Ghosh G (1998b) A novel DNA recognition mode by the NF-kappa B p65 homodimer. Nat Struct Biol 5:67–73

Cherry S, Silverman N (2006) Host–pathogen interactions in *Drosophila*: new tricks from an old friend. Nat Immunol 7:911–917

Cho JH, Fraser IP, Fukase K, Kusumoto S, Fujimoto Y, Stahl GL, Ezekowitz RAB (2005) Human peptidoglycan recognition protein S is an effector of neutrophil-mediated innate immunity. Blood 106:2551–2558

Choe KM, Werner T, Stöven S, Hultmark D, Anderson KV (2002) Requirement for a peptidoglycan recognition protein (PGRP) in Relish activation and antibacterial immune responses in *Drosophila*. Science 296:359–362

Choe KM, Lee H, Anderson KV (2005) *Drosophila* peptidoglycan recognition protein LC (PGRP-LC) acts as a signal-transducing innate immune receptor. Proc Natl Acad Sci USA 102:1122–1126

Cookson BT, Cho HL, Herwaldt LA, Goldman WE (1989) Biological activities and chemical composition of purified tracheal cytotoxin of *Bordetella pertussis*. Infect Immun 57:2223–2229

De Gregorio E, Spellman PT, Rubin GM, Lemaitre B (2001) Genome-wide analysis of the *Drosophila* immune response by using oligonucleotide microarrays. Proc Natl Acad Sci USA 98:12590–12595

De Gregorio E, Han SJ, Lee WJ, Baek MJ, Osaki T, Kawabata S, Lee BL, Iwanaga S, Lemaitre B, Brey PT (2002a) An immune-responsive Serpin regulates the melanization cascade in *Drosophila*. Dev Cell 3:581–592

De Gregorio E, Spellman PT, Tzou P, Rubin GM, Lemaitre B (2002b) The Toll and IMD pathways are the major regulators of the immune response in *Drosophila*. EMBO J 21:2568–2579

Delaney JR, Stoven S, Uvell H, Anderson KV, Engstrom Y, Mlodzik M (2006) Cooperative control of *Drosophila* immune responses by the JNK and NF-kappaB signaling pathways. EMBO J 25:3068–3077

Dimarcq JL, Hoffmann D, Meister M, Bulet P, Lanot R, Reichhart JM, Hoffmann JA (1994) Characterization and transcriptional profiles of a *Drosophila* gene encoding an insect defensin. A study in insect immunity. Eur J Biochem 221:201–209

Dong Y, Taylor HE, Dimopoulos G (2006) AgDscam, a hypervariable immunoglobulin domain-containing receptor of the *Anopheles gambiae* innate immune system. PLoS Biol 4:e229

Dostert C, Jouanguy E, Irving P, Troxler L, Galiana-Arnoux D, Hetru C, Hoffmann JA, Imler JL (2005) The Jak-STAT signaling pathway is required but not sufficient for the antiviral response of *Drosophila*. Nat Immunol 6:946–953

Drier EA, Steward R (1997) The dorsoventral signal transduction pathway and the Rel-like transcription factors in *Drosophila*. Semin Cancer Biol 8:83–92

Drier EA, Huang LH, Steward R (1999) Nuclear import of the *Drosophila* Rel protein Dorsal is regulated by phosphorylation. Genes Dev 13:556–568

Dupuis S, Jouanguy E, Al-Hajjar S, Fieschi C, Al-Mohsen IZ, Al-Jumaah S, Yang K, Chapgier A, Eidenschenk C, Eid P, et al (2003) Impaired response to interferon-alpha/beta and lethal viral disease in human STAT1 deficiency. Nat Genet 33:388–391

Dushay MS, Åsling B, Hultmark D (1996) Origins of immunity: Relish, a compound Rel-like gene in the antibacterial defense of *Drosophila*. Proc Natl Acad Sci USA 93:10343–10347

Dutta J, Fan Y, Gupta N, Fan G, Gelinas C (2006) Current insights into the regulation of programmed cell death by NF-kappaB. Oncogene 25:6800–6816

Dziarski R, Platt KA, Gelius E, Steiner H, Gupta D (2003) Defect in neutrophil killing and increased susceptibility to infection with non-pathogenic Gram-positive bacteria in peptidoglycan recognition protein-S (PGRP-S)-deficient mice. Blood 102:689–697

Ekengren S, Hultmark D (1999) *Drosophila* cecropin as an antifungal agent. Insect Biochem Mol Biol 29:965–972

Ekengren S, Hultmark D (2001) A family of Turandot-related genes in the humoral stress response of *Drosophila*. Biochem Biophys Res Commun 284:998–1003

Engstrom Y, Kadalayil L, Sun SC, Samakovlis C, Hultmark D, Faye I (1993) kappa B-like motifs regulate the induction of immune genes in *Drosophila*. J Mol Biol 232:327–333

Fernandez NQ, Grosshans J, Goltz JS, Stein D (2001) Separable and redundant regulatory determinants in Cactus mediate its dorsal group dependent degradation. Development 128:2963–2974

Ferrandon D, Jung AC, Criqui M, Lemaitre B, Uttenweiler-Joseph S, Michaut L, Reichhart J, Hoffmann JA (1998) A drosomycin-GFP reporter transgene reveals a local immune response in *Drosophila* that is not dependent on the Toll pathway. EMBO J 17:1217–1227

Filipe SR, Tomasz A, Ligoxygakis P (2005) Requirements of peptidoglycan structure that allow detection by the *Drosophila* Toll pathway. EMBO Rep 6:327–333

Fritz JH, Le Bourhis L, Sellge G, Magalhaes JG, Fsihi H, Kufer TA, Collins C, Viala J, Ferrero RL, Girardin SE, Philpott DJ (2007) Nod1-mediated innate immune recognition of peptidoglycan contributes to the onset of adaptive immunity. Immunity 26:445–459

Geisler R, Bergmann A, Hiromi Y, Nusslein-Volhard C (1992) *cactus*, a gene involved in dorsoventral pattern formation of *Drosophila*, is related to the I kappa B gene family of vertebrates. Cell 71:613–621

Gelius E, Persson C, Karlsson J, Steiner H (2003) A mammalian peptidoglycan recognition protein with N-acetylmuramoyl-L-alanine amidase activity. Biochem Biophys Res Commun 306:988–994

Georgel P, Naitza S, Kappler C, Ferrandon D, Zachary D, Swimmer C, Kopczynski C, Duyk G, Reichhart JM, Hoffmann JA (2001) Drosophila immune deficiency (IMD) is a death domain protein that activates antibacterial defense and can promote apoptosis. Dev Cell 1:503–514

Gesellchen V, Kuttenkeuler D, Steckel M, Pelte N, Boutros M (2005) An RNA interference screen identifies Inhibitor of Apoptosis Protein 2 as a regulator of innate immune signalling in *Drosophila*. EMBO Rep 6:979–984

Gillespie SK, Wasserman SA (1994) Dorsal, a *Drosophila* Rel-like protein, is phosphorylated upon activation of the transmembrane protein Toll. Mol Cell Biol 14:3559–3568

Gobert V, Gottar M, Matskevich AA, Rutschmann S, Royet J, Belvin M, Hoffmann JA, Ferrandon D (2003) Dual activation of the *Drosophila* Toll pathway by two pattern recognition receptors. Science 302:2126–2130

Goldman WE, Klapper DG, Baseman JB (1982) Detection, isolation, and analysis of a released *Bordetella pertussis* product toxic to cultured tracheal cells. Infect Immun 36:782–794

Gottar M, Gobert V, Michel T, Belvin M, Duyk G, Hoffmann JA, Ferrandon D, Royet J (2002) The *Drosophila* immune response against Gram-negative bacteria is mediated by a peptidoglycan recognition protein. Nature 416:640–644

Gottar M, Gobert V, Matskevich AA, Reichhart JM, Wang C, Butt TM, Belvin M, Hoffmann JA, Ferrandon D (2006) Dual detection of fungal infections in *Drosophila* via recognition of glucans and sensing of virulence factors. Cell 127:1425–1437

Gross I, Georgel P, Kappler C, Reichhart JM, Hoffmann JA (1996) *Drosophila* immunity: a comparative analysis of the Rel proteins dorsal and Dif in the induction of the genes encoding diptericin and cecropin. Nucleic Acids Res 24:1238–1245

Guan R, Mariuzza RA (2007) Peptidoglycan recognition proteins of the innate immune system. Trends Microbiol 15:127–134

Hayden MS, West AP, Ghosh S (2006) NF-kappaB and the immune response. Oncogene 25:6758–6780

Hedengren M, Åsling B, Dushay MS, Ando I, Ekengren S, Wihlborg M, Hultmark D (1999) Relish, a central factor in the control of humoral but not cellular immunity in *Drosophila*. Mol Cell 4:827–837

Heimpel AM, Harshbarger JC (1965) Symposium on microbial insecticides. V. Immunity in insects. Bacteriol Rev 29:397–405

Holland PM, Suzanne M, Campbell JS, Noselli S, Cooper JA (1997) MKK7 is a stress-activated mitogen-activated protein kinase kinase functionally related to hemipterous. J Biol Chem 272:24994–24998

Hombria JC, Brown S (2002) The fertile field of *Drosophila* Jak/STAT signalling. Curr Biol 12: R569–R575

Hou XS, Melnick MB, Perrimon N (1996) Marelle acts downstream of the *Drosophila* HOP/JAK kinase and encodes a protein similar to the mammalian STATs. Cell 84:411–419

Hu X, Yagi Y, Tanji T, Zhou S, Ip YT (2004) Multimerization and interaction of Toll and Spatzle in *Drosophila*. Proc Natl Acad Sci USA 101:9369–9374

Huh JR, Foe I, Muro I, Chen CH, Seol JH, Yoo SJ, Guo M, Park JM, Hay BA (2007) The Drosophila inhibitor of apoptosis (IAP) DIAP2 is dispensable for cell survival, required for the innate immune response to gram-negative bacterial infection, and can be negatively regulated by the reaper/hid/grim family of IAP-binding apoptosis inducers. J Biol Chem 282:2056–2068

Hultmark D (2003) *Drosophila* immunity: paths and patterns. Curr Opin Immunol 15:12–19

Hultmark D, Engstrom A, Andersson K, Steiner H, Bennich H, Boman HG (1983) Insect immunity. Attacins, a family of antibacterial proteins from *Hyalophora cecropia*. EMBO J 2:571–576

Ip YT, Reach M, Engstrom Y, Kadalayil L, Cai H, Gonzalez-Crespo S, Tatei K, Levine M (1993) Dif, a dorsal-related gene that mediates an immune response in *Drosophila*. Cell 75:753–763

Irving P, Troxler L, Heuer TS, Belvin M, Kopczynski C, Reichhart JM, Hoffmann JA, Hetru C (2001) A genome-wide analysis of immune responses in *Drosophila*. Proc Natl Acad Sci USA 98:15119–15124

Isoda K, Nusslein-Volhard C (1994) Disulfide cross-linking in crude embryonic lysates reveals three complexes of the *Drosophila* morphogen dorsal and its inhibitor cactus. Proc Natl Acad Sci USA 91:5350–5354

Jang IH, Chosa N, Kim SH, Nam HJ, Lemaitre B, Ochiai M, Kambris Z, Brun S, Hashimoto C, Ashida M, et al (2006) A Spatzle-processing enzyme required for toll signaling activation in *Drosophila* innate immunity. Dev Cell 10:45–55

Kambris Z, Brun S, Jang IH, Nam HJ, Romeo Y, Takahashi K, Lee WJ, Ueda R, Lemaitre B (2006) *Drosophila* immunity: a large-scale in vivo RNAi screen identifies five serine proteases required for Toll activation. Curr Biol 16:808–813

Kaneko T, Goldman WE, Mellroth P, Steiner H, Fukase K, Kusumoto S, Harley W, Fox A, Golenbock D, Silverman N (2004) Monomeric and polymeric gram-negative peptidoglycan but not purified LPS stimulate the *Drosophila* IMD pathway. Immunity 20:637–649

Kaneko T, Yano T, Aggarwal K, Lim JH, Ueda K, Oshima Y, Peach C, Erturk-Hasdemir D, Goldman WE, Oh BH, et al (2006) PGRP-LC and PGRP-LE have essential yet distinct functions in the *Drosophila* immune response to monomeric DAP-type peptidoglycan. Nat Immunol 7:715–723

Kang D, Liu G, Lundstrom A, Gelius E, Steiner H (1998) A peptidoglycan recognition protein in innate immunity conserved from insects to humans. Proc Natl Acad Sci USA 95:10078–10082

Kappler C, Meister M, Lagueux M, Gateff E, Hoffmann JA, Reichhart JM (1993) Insect immunity. Two 17 bp repeats nesting a kappa B-related sequence confer inducibility to the diptericin gene and bind a polypeptide in bacteria-challenged *Drosophila*. EMBO J 12:1561–1568

Karin M (2006) Nuclear factor-kappaB in cancer development and progression. Nature 441:431–436

Karst SM, Wobus CE, Lay M, Davidson J, Virgin HW (2003) STAT1-dependent innate immunity to a Norwalk-like virus. Science 299:1575–1578

Khush RS, Cornwell WD, Uram JN, Lemaitre B (2002) A ubiquitin-proteasome pathway represses the *Drosophila* immune deficiency signaling cascade. Curr Biol 12:1728–1737

Kim MS, Byun M, Oh BH (2003) Crystal structure of peptidoglycan recognition protein LB from *Drosophila melanogaster*. Nat Immunol 4:787–793

Kim T, Yoon J, Cho H, Lee WB, Kim J, Song YH, Kim SN, Yoon JH, Kim-Ha J, Kim YJ (2005) Downregulation of lipopolysaccharide response in *Drosophila* by negative crosstalk between the AP1 and NF-kappaB signaling modules. Nat Immunol 6:211–218

Kim Y-S, Han S-J, Ryu J-H, Choi K-H, Hong Y-S, Chung Y-H, Perrot S, Raibaud A, Brey PT, Lee WJ (2000) Lipopolysaccharide-activated kinase, an essential component for the induction of the antimicrobial peptide genes in *Drosophila melanogaster* cells. J Biol Chem 275:2071–2079

Kleino A, Valanne S, Ulvila J, Kallio J, Myllymaki H, Enwald H, Stoven S, Poidevin M, Ueda R, Hultmark D, et al (2005) Inhibitor of apoptosis 2 and TAK1-binding protein are components of the *Drosophila* Imd pathway. EMBO J 24:3423–3434

Koropatnick TA, Engle JT, Apicella MA, Stabb EV, Goldman WE, McFall-Ngai MJ (2004) Microbial factor-mediated development in a host–bacterial mutualism. Science 306:1186–1188

Kylsten P, Samakovlis C, Hultmark D (1990) The cecropin locus in *Drosophila*; a compact gene cluster involved in the response to infection. EMBO J 9:217–224

Lagueux M, Perrodou E, Levashina EA, Capovilla M, Hoffmann JA (2000) Constitutive expression of a complement-like protein in toll and JAK gain-of-function mutants of *Drosophila*. Proc Natl Acad Sci USA 97:11427–11432

Lemaitre B (2004) The road to Toll. Nat Rev Immunol 4:521–527

Lemaitre B, Hoffmann J (2007) The host defense of *Drosophila melanogaster*. Annu Rev Immunol 25:697–743

Lemaitre B, Nicolas E, Michaut L, Reichhart JM, Hoffmann JA (1996) The dorsoventral regulatory gene cassette spätzle/Toll/cactus controls the potent antifungal response in *Drosophila* adults. Cell 86:973–983

Lemaitre B, Reichhart JM, Hoffmann JA (1997) *Drosophila* host defense: differential induction of antimicrobial peptide genes after infection by various classes of microorganisms. Proc Natl Acad Sci USA 94:14614–14619

Leulier F, Rodriguez A, Khush RS, Abrams JM, Lemaitre B (2000) The *Drosophila* caspase Dredd is required to resist Gram-negative bacterial infection. EMBO Rep 1:353–358

Leulier F, Parquet C, Pili-Floury S, Ryu JH, Caroff M, Lee WJ, Mengin-Lecreulx D, Lemaitre B (2003) The *Drosophila* immune system detects bacteria through specific peptidoglycan recognition. Nat Immunol 4:478–484

Leulier F, Lhocine N, Lemaitre B, Meier P (2006) The *Drosophila* inhibitor of apoptosis protein DIAP2 functions in innate immunity and is essential to resist gram-negative bacterial infection. Mol Cell Biol 26:7821–7831

Levashina EA, Ohresser S, Bulet P, Reichhart JM, Hetru C, Hoffmann JA (1995) Metchnikowin, a novel immune-inducible proline-rich peptide from *Drosophila* with antibacterial and antifungal properties. Eur J Biochem 233:694–700

Levashina EA, Langley E, Green C, Gubb D, Ashburner M, Hoffmann JA, Reichhart JM (1999) Constitutive activation of Toll-mediated antifungal defense in serpin-deficient *Drosophila*. Science 285:1917–1919

Liehl P, Blight M, Vodovar N, Boccard F, Lemaitre B (2006) Prevalence of local immune response against oral infection in a *Drosophila/Pseudomonas* infection model. PLoS Pathog 2:e56

Ligoxygakis P, Pelte N, Hoffmann JA, Reichhart JM (2002) Activation of *Drosophila* Toll during fungal infection by a blood serine protease. Science 297:114–116

Lim JH, Kim MS, Kim HE, Yano T, Oshima Y, Aggarwal K, Goldman WE, Silverman N, Kurata S, Oh BH (2006) Structural basis for preferential recognition of diaminopimelic acid-type peptidoglycan by a subset of peptidoglycan recognition proteins. J Biol Chem 281:8286–8295

Liu C, Gelius E, Liu G, Steiner H, Dziarski R (2000) Mammalian peptidoglycan recognition protein binds peptidoglycan with high affinity, is expressed in neutrophils, and inhibits bacterial growth. J Biol Chem 275:24490–24499

Liu ZP, Galindo RL, Wasserman SA (1997) A role for CKII phosphorylation of the cactus PEST domain in dorsoventral patterning of the *Drosophila* embryo. Genes Dev 11:3413–3422

Lo D, Tynan W, Dickerson J, Mendy J, Chang HW, Scharf M, Byrne D, Brayden D, Higgins L, Evans C, O'Mahony DJ (2003) Peptidoglycan recognition protein expression in mouse Peyer's Patch follicle associated epithelium suggests functional specialization. Cell Immunol 224:8–16

Lu X, Wang M, Qi J, Wang H, Li X, Gupta D, Dziarski R (2006) Peptidoglycan recognition proteins are a new class of human bactericidal proteins. J Biol Chem 281:5895–5907

Lu Y, Wu LP, Anderson KV (2001) The antibacterial arm of the *Drosophila* innate immune response requires an IkappaB kinase. Genes Dev 15:104–110

Manfruelli P, Reichhart JM, Steward R, Hoffmann JA, Lemaitre B (1999) A mosaic analysis in *Drosophila* fat body cells of the control of antimicrobial peptide genes by the Rel proteins Dorsal and DIF. EMBO J 18:3380–3391

Mathur P, Murray B, Crowell T, Gardner H, Allaire N, Hsu YM, Thill G, Carulli JP (2004) Murine peptidoglycan recognition proteins PglyrpIalpha and PglyrpIbeta are encoded in the epidermal differentiation complex and are expressed in epidermal and hematopoietic tissues. Genomics 83:1151–1163

Mellroth P, Steiner H (2006) PGRP-SB1: an N-acetylmuramoyl L-alanine amidase with antibacterial activity. Biochem Biophys Res Commun 350:994–999

Mellroth P, Karlsson J, Steiner H (2003) A scavenger function for a *Drosophila* peptidoglycan recognition protein. J Biol Chem 278:7059–7064

Melly MA, McGee ZA, Rosenthal RS (1984) Ability of monomeric peptidoglycan fragments from *Neisseria gonorrhoeae* to damage human fallopian-tube mucosa. J Infect Dis 149:378–386

Meng X, Khanuja BS, Ip YT (1999) Toll receptor-mediated *Drosophila* immune response requires Dif, an NF-kB factor. Genes Dev 13:792–797

Mengin-Lecreulx D, Lemaitre B (2005) Structure and metabolism of peptidoglycan and molecular requirements allowing its detection by the *Drosophila* innate immune system. J Endotoxin Res 11:105–111

Meylan E, Burns K, Hofmann K, Blancheteau V, Martinon F, Kelliher M, Tschopp J (2004) RIP1 is an essential mediator of Toll-like receptor 3-induced NF-kappa B activation. Nat Immunol 5:503–507

Michaut L, Fehlbaum P, Moniatte M, Van Dorsselaer A, Reichhart JM, Bulet P (1996) Determination of the disulfide array of the first inducible antifungal peptide from insects: drosomycin from *Drosophila melanogaster*. FEBS Lett 395:6–10

Michel T, Reichhart JM, Hoffmann JA, Royet J (2001) *Drosophila* Toll is activated by Gram-positive bacteria through a circulating peptidoglycan recognition protein. Nature 414:756–759

Muller P, Kuttenkeuler D, Gesellchen V, Zeidler MP, Boutros M (2005) Identification of JAK/STAT signalling components by genome-wide RNA interference. Nature 436:871–875

Nicolas E, Reichhart JM, Hoffmann JA, Lemaitre B (1998) In vivo regulation of the IkappaB homologue cactus during the immune response of *Drosophila*. J Biol Chem 273:10463–10469

Ochiai M, Ashida M (1999) A pattern recognition protein for peptidoglycan. Cloning the cDNA and the gene of the silkworm, *Bombyx mori*. J Biol Chem 274:11854–11858

Park JM, Kim JM, Kim LK, Kim SN, Kim-Ha J, Kim JH, Kim YJ (2003) Signal-induced transcriptional activation by Dif requires the dTRAP80 mediator module. Mol Cell Biol 23:1358–1367

Park JM, Brady H, Ruocco MG, Sun H, Williams D, Lee SJ, Kato T Jr, Richards N, Chan K, Mercurio F, et al (2004) Targeting of TAK1 by the NF-kappa B protein Relish regulates the JNK-mediated immune response in *Drosophila*. Genes Dev 18:584–594

Park JW, Kim CH, Kim JH, Je BR, Roh KB, Kim SJ, Lee HH, Ryu JH, Lim JH, Oh BH, et al (2007) Clustering of peptidoglycan recognition protein-SA is required for sensing lysine-type peptidoglycan in insects. Proc Natl Acad Sci USA 104:6602–6607

Perkins ND (2007) Integrating cell-signalling pathways with NF-kappaB and IKK function. Nat Rev Mol Cell Biol 8:49–62

Perkins ND, Felzien LK, Betts JC, Leung K, Beach DH, Nabel GJ (1997) Regulation of NF-kappaB by cyclin-dependent kinases associated with the p300 coactivator. Science 275:523–527

Pham LN, Dionne MS, Shirasu-Hiza M, Schneider DS (2007) A specific primed immune response in *Drosophila* is dependent on phagocytes. PLoS Pathog 3:e26

Pili-Floury S, Leulier F, Takahashi K, Saigo K, Samain E, Ueda R, Lemaitre B (2004) In vivo RNA interference analysis reveals an unexpected role for GNBP1 in the defense against Gram-positive bacterial infection in *Drosophila* adults. J Biol Chem 279:12848–12853

Qiu P, Pan PC, Govind S (1998) A role for the *Drosophila* Toll/Cactus pathway in larval hematopoiesis. Development 125:1909–1920

Ramet M, Manfruelli P, Pearson A, Mathey-Prevot B, Ezekowitz RA (2002) Functional genomic analysis of phagocytosis and identification of a *Drosophila* receptor for *E. coli*. Nature 416:644–648

Reach M, Galindo RL, Towb P, Allen JL, Karin M, Wasserman SA (1996) A gradient of cactus protein degradation establishes dorsoventral polarity in the *Drosophila* embryo. Dev Biol 180:353–364

Reichhart JM, Meister M, Dimarcq JL, Zachary D, Hoffmann D, Ruiz C, Richards G, Hoffmann JA (1992) Insect immunity: developmental and inducible activity of the *Drosophila* diptericin promoter. EMBO J 11:1469–1477

Reichhart JM, Georgel P, Meister M, Lemaitre B, Kappler C, Hoffmann JA (1993) Expression and nuclear translocation of the rel/NF-kappa B-related morphogen dorsal during the immune response of *Drosophila*. C R Acad Sci III 316:1218–1224

Rosenthal RS (1979) Release of soluble peptidoglycan from growing gonococci: hexaminidase and amidase activities. Infect Immun 24:869–878

Roth S, Stein D, Nusslein-Volhard C (1989) A gradient of nuclear localization of the dorsal protein determines dorsoventral pattern in the *Drosophila* embryo. Cell 59:1189–1202

Rutschmann S, Jung AC, Zhou R, Silverman N, Hoffmann JA, Ferrandon D (2000) Role of *Drosophila* IKK gamma in a toll-independent antibacterial immune response. Nat Immunol 1:342–347

Samakovlis C, Kimbrell DA, Kylsten P, Engstrom A, Hultmark D (1990) The immune response in *Drosophila*: pattern of cecropin expression and biological activity. Embo J 9:2969–2976

Samakovlis C, Åsling B, Boman HG, Gateff E, Hultmark D (1992) In vitro induction of cecropin genes – an immune response in a *Drosophila* blood cell line. Biochem Biophys Res Commun 188:1169–1175

Santamaria P, Nusslein-Volhard C (1983) Partial rescue of dorsal, a maternal effect mutation affecting the dorso-ventral pattern of the *Drosophila* embryo, by the injection of wild-type cytoplasm. EMBO J 2:1695–1699

Schleifer KH, Kandler O (1972) Peptidoglycan types of bacterial cell walls and their taxonomic implications. Bacteriol Rev 36:407–477

Schmitz ML, Santos Silva MA dos, Baeuerle PA (1995) Transactivation domain 2 (TA2) of p65 NF-kappa B. Similarity to TA1 and phorbol ester-stimulated activity and phosphorylation in intact cells. J Biol Chem 270:15576–15584

Senger K, Armstrong GW, Rowell WJ, Kwan JM, Markstein M, Levine M (2004) Immunity regulatory DNAs share common organizational features in *Drosophila*. Mol Cell 13:19–32

Shen B, Liu H, Skolnik EY, Manley JL (2001) Physical and functional interactions between *Drosophila* TRAF2 and Pelle kinase contribute to Dorsal activation. Proc Natl Acad Sci USA 98:8596–8601

Silverman N, Zhou R, Stöven S, Pandey N, Hultmark D, Maniatis T (2000) A *Drosophila* IkappaB kinase complex required for Relish cleavage and antibacterial immunity. Genes Dev 14:2461–2471

Silverman N, Maniatis T (2001) NF-kappaB signaling pathways in mammalian and insect innate immunity. Genes Dev 15:2321–2342

Silverman N, Zhou R, Erlich RL, Hunter M, Bernstein E, Schneider D, Maniatis T (2003) Immune activation of NF-kappaB and JNK requires *Drosophila* TAK1. J Biol Chem 278:48928–48934

Sluss HK, Han Z, Barrett T, Davis RJ, Ip YT (1996) A JNK signal transduction pathway that mediates morphogenesis and an immune response in *Drosophila*. Genes Dev 10:2745–2758

Spencer E, Jiang J, Chen ZJ (1999) Signal-induced ubiquitination of IkBa by the F-box protein Slimb/bTrCP. Genes Dev 13:284–294

Steiner H, Hultmark D, Engstrom A, Bennich H, Boman HG (1981) Sequence and specificity of two antibacterial proteins involved in insect immunity. Nature 292:246–248

Steinhaus EA (1940) The microbiology of insects: with special reference to the biologic relationships between bacteria and insects. Bacteriol Rev 4:17–57

Stenbak CR, Ryu JH, Leulier F, Pili-Floury S, Parquet C, Herve M, Chaput C, Boneca IG, Lee WJ, Lemaitre B, Mengin-Lecreulx D (2004) Peptidoglycan molecular requirements allowing detection by the *Drosophila* immune deficiency pathway. J Immunol 173:7339–7348

Steward R (1987) Dorsal, an embryonic polarity gene in *Drosophila*, is homologous to the vertebrate proto-oncogene, c-rel. Science 238:692–694

Steward R, McNally FJ, Schedl P (1984) Isolation of the dorsal locus of *Drosophila*. Nature 311:262–265

Stöven S, Ando I, Kadalayil L, Engström Y, Hultmark D (2000) Activation of the *Drosophila* NF-kB factor Relish by rapid endoproteolytic cleavage. EMBO Rep 1:347–352

Stöven S, Silverman N, Junell A, Hedengren-Olcott M, Erturk D, Engstrom Y, Maniatis T, Hultmark D (2003) Caspase-mediated processing of the *Drosophila* NF-kappaB factor Relish. Proc Natl Acad Sci USA 100:5991–5996

Sun H, Bristow BN, Qu G, Wasserman SA (2002a) A heterotrimeric death domain complex in Toll signaling. Proc Natl Acad Sci USA 99:12871–12876

Sun H, Towb P, Chiem DN, Foster BA, Wasserman SA (2004) Regulated assembly of the Toll signaling complex drives *Drosophila* dorsoventral patterning. EMBO J 23:100–110

Sun SC, Asling B, Faye I (1991) Organization and expression of the immunoresponsive lysozyme gene in the giant silk moth, *Hyalophora cecropia*. J Biol Chem 266:6644–6649

Sun X, Yin J, Starovasnik MA, Fairbrother WJ, Dixit VM (2002b) Identification of a novel homotypic interaction motif required for the phosphorylation of receptor-interacting protein (RIP) by RIP3. J Biol Chem 277:9505–9511

Swaminathan CP, Brown PH, Roychowdhury A, Wang Q, Guan R, Silverman N, Goldman WE, Boons GJ, Mariuzza RA (2006) Dual strategies for peptidoglycan discrimination by peptidoglycan recognition proteins (PGRPs). Proc Natl Acad Sci USA 103:684–689

Takehana A, Katsuyama T, Yano T, Oshima Y, Takada H, Aigaki T, Kurata S (2002) Overexpression of a pattern-recognition receptor, peptidoglycan-recognition protein-LE, activates imd/relish-mediated antibacterial defense and the prophenoloxidase cascade in *Drosophila* larvae. Proc Natl Acad Sci USA 99:13705–13710

Takehana A, Yano T, Mita S, Kotani A, Oshima Y, Kurata S (2004) Peptidoglycan recognition protein (PGRP)-LE and PGRP-LC act synergistically in *Drosophila* immunity. EMBO J 23:4690–4700

Tanji T, Hu X, Weber AN, Ip YT (2007) Toll and IMD pathways synergistically activate innate immune response in *Drosophila*. Mol Cell Biol 27:4578–4588

Tauszig-Delamasure S, Bilak H, Capovilla M, Hoffmann JA, Imler JL (2002) *Drosophila* MyD88 is required for the response to fungal and Gram-positive bacterial infections. Nat Immunol 3:91–97

Towb P, Galindo RL, Wasserman SA (1998) Recruitment of Tube and Pelle to signaling sites at the surface of the *Drosophila* embryo. Development 125:2443–2450

Tydell CC, Yount N, Tran D, Yuan J, Selsted ME (2002) Isolation, characterization, and antimicrobial properties of bovine oligosaccharide-binding protein. A microbicidal granule protein of eosinophils and neutrophils. J Biol Chem 277:19658–19664

Tzou P, Ohresser S, Ferrandon D, Capovilla M, Reichhart JM, Lemaitre B, Hoffmann JA, Imler JL (2000) Tissue-specific inducible expression of antimicrobial peptide genes in *Drosophila* surface epithelia. Immunity 13:737–748

Tzou P, Reichhart JM, Lemaitre B (2002) Constitutive expression of a single antimicrobial peptide can restore wild-type resistance to infection in immunodeficient *Drosophila* mutants. Proc Natl Acad Sci USA 99:2152–2157

Uehara A, Fujimoto Y, Kawasaki A, Kusumoto S, Fukase K, Takada H (2006) Meso-diaminopimelic acid and meso-lanthionine, amino acids specific to bacterial peptidoglycans, activate human epithelial cells through NOD1. J Immunol 177:1796–1804

Valanne S, Kleino A, Myllymaki H, Vuoristo J, Ramet M (2007) Iap2 is required for a sustained response in the *Drosophila* Imd pathway. Dev Comp Immunol (in press)

Vidal S, Khush RS, Leulier F, Tzou P, Nakamura M, Lemaitre B (2001) Mutations in the *Drosophila* dTAK1 gene reveal a conserved function for MAPKKKs in the control of rel/NF-kappaB-dependent innate immune responses. Genes Dev 15:1900–1912

Wang C, Deng L, Hong M, Akkaraju GR, Inoue J, Chen ZJ (2001) TAK1 is a ubiquitin-dependent kinase of MKK and IKK. Nature 412:346–351

Wang L, Weber AN, Atilano ML, Filipe SR, Gay NJ, Ligoxygakis P (2006) Sensing of Gram-positive bacteria in *Drosophila*: GNBP1 is needed to process and present peptidoglycan to PGRP-SA. EMBO J 25:5005–5014

Wang ZM, Li X, Cocklin RR, Wang M, Fukase K, Inamura S, Kusumoto S, Gupta D, Dziarski R (2003) Human peptidoglycan recognition protein-L is an N-acetylmuramoyl-L-alanine amidase. J Biol Chem 278:49044–49052

Watson FL, Puttmann-Holgado R, Thomas F, Lamar DL, Hughes M, Kondo M, Rebel VI, Schmucker D (2005) Extensive diversity of Ig-superfamily proteins in the immune system of insects. Science 309:1874–1878

Weber AN, Tauszig-Delamasure S, Hoffmann JA, Lelievre E, Gascan H, Ray KP, Morse MA, Imler JL, Gay NJ (2003) Binding of the *Drosophila* cytokine Spatzle to Toll is direct and establishes signaling. Nat Immunol 4:794–800

Werner T, Liu G, Kang D, Ekengren S, Steiner H, Hultmark D (2000) A family of peptidoglycan recognition proteins in the fruit fly *Drosophila melanogaster*. Proc Natl Acad Sci USA 97:13772–13777

Werner T, Borge-Renberg K, Mellroth P, Steiner H, Hultmark D (2003) Functional diversity of the *Drosophila* PGRP-LC gene cluster in the response to lipopolysaccharide and peptidoglycan. J Biol Chem 278:26319–26322

Wicker C, Reichhart JM, Hoffmann D, Hultmark D, Samakovlis C, Hoffmann JA (1990) Insect immunity. Characterization of a *Drosophila* cDNA encoding a novel member of the diptericin family of immune peptides. J Biol Chem 265:22493–22498

Wirth T, Baltimore D (1988) Nuclear factor NF-kappa B can interact functionally with its cognate binding site to provide lymphoid-specific promoter function. EMBO J 7:3109–3113

Wu LP, Anderson KV (1998) Regulated nuclear import of Rel proteins in the *Drosophila* immune response. Nature 392:93–97

Yagi Y, Ip YT (2005) Helicase89B is a Mot1p/BTAF1 homologue that mediates an antimicrobial response in *Drosophila*. EMBO Rep 6:1088–1094

Yan R, Small S, Desplan C, Dearolf CR, Darnell JE Jr (1996) Identification of a Stat gene that functions in *Drosophila* development. Cell 84:421–430

Yoshida H, Kinoshita K, Ashida M (1996) Purification of a peptidoglycan recognition protein from hemolymph of the silkworm, *Bombyx mori*. J Biol Chem 271:13854–13860

Zaidman-Remy A, Herve M, Poidevin M, Pili-Floury S, Kim MS, Blanot D, Oh BH, Ueda R, Mengin-Lecreulx D, Lemaitre B (2006) The *Drosophila* amidase PGRP-LB modulates the immune response to bacterial infection. Immunity 24:463–473

Zeidler MP, Bach EA, Perrimon N (2000) The roles of the *Drosophila* JAK/STAT pathway. Oncogene 19:2598–2606

Zettervall CJ, Anderl I, Williams MJ, Palmer R, Kurucz E, Ando I, Hultmark D (2004) A directed screen for genes involved in *Drosophila* blood cell activation. Proc Natl Acad Sci USA 101:14192–14197

Zhang Y, Fits L van der, Voerman JS, Melief MJ, Laman JD, Wang M, Wang H, Wang M, Li X, Walls CD, et al (2005) Identification of serum N-acetylmuramoyl-l-alanine amidase as liver peptidoglycan recognition protein 2. Biochim Biophys Acta 1752:34–46

Zhou R, Silverman N, Hong M, Liao DS, Chung Y, Chen ZJ, Maniatis T (2005) The role of ubiquitnation in *Drosophila* innate immunity. J Biol Chem 280:34048–34055

Zhuang ZH, Sun L, Kong L, Hu JH, Yu MC, Reinach P, Zang JW, Ge BX (2006) *Drosophila* TAB2 is required for the immune activation of JNK and NF-kappaB. Cell Signal 18:964–970

Chapter 4
Cellular Immune Responses in *Drosophila melanogaster*

Adrienne Ivory, Katherine Randle, and Louisa Wu(✉)

1 Introduction 74
2 Encapsulation 75
   2.1 Recognition Centers on Membrane Differences 76
   2.2 Lamellocyte Proliferation: Necessary for Successful Encapsulation Response ... 78
   2.3 Adhesion Requires Integrins, Rac, and Rho 78
   2.4 Encapsulation Terminates with the Formation of Basement Membrane 82
3 Phagocytosis 83
   3.1 Proteins Opsonize Invading Bacteria and Fungi to Promote Phagocytosis 84
   3.2 Transmembrane and Circulating Peptidoglycan Recognition Proteins are Involved in the Recognition of Bacteria 85
   3.3 Receptors with Scavenger-Like Activity Recognize a Variety of Microbes 86
   3.4 Phagocytosis Requires Reorganization of the Actin Cytoskeleton 88
   3.5 Engulfed Pathogens are Degraded in Phagolysosomes 89
   3.6 Interactions Between Cellular and Humoral Immune Responses 90
References 91

Abstract The ability of blood cells, known as hemocytes, to detect and eliminate pathogens is vital to the *Drosophila* immune response. Various pathogens that can subvert the cellular immune response are often lethal to the fly. For example, parasitoid wasps deposit their eggs with chemicals targeting *Drosophila* hemocytes. These chemicals increase parasitoid success. Similarly, when hemocyte counts are drastically lowered through mutations like *domino*, mutant larvae are vulnerable to large-scale colonization by live bacteria. Further, the inhibition of phagocytic ability in hemocytes leads to a dramatic increase in susceptibility to *Escherichia coli* infection in flies lacking a humoral response. This chapter discusses our current understanding of encapsulation and phagocytosis, two cellular immune responses important for defense against parasites and bacteria. Both responses initiate with recognition, followed by activation of the blood cells, and finish with either encapsulation or uptake of the microbe. Recent works from many laboratories have used

---

Center for Biosystems Research, University of Maryland Biotechnology Institute, 5115 Plant Sciences Bldg, College Park, MD 20742, USA, *wul@umbi.umd.edu*

H. Heine (ed.), *Innate Immunity of Plants, Animals, and Humans.*
Nucleic Acids and Molecular Biology 21.
© Springer-Verlag Berlin Heidelberg 2008

whole-genome RNAi screens, forward genetic screens, and fluorescent visualization of cellular processes to identify old and new players in these cellular immune responses.

1 Introduction

Three types of blood cells have been identified in *Drosophila*: lamellocytes, plasmatocytes, and crystal cells. Lamellocytes are usually observed in *Drosophila* only if the organism has certain mutations or is encapsulating foreign material within the organism (Rizki and Rizki 1992). During the encapsulation response, lamellocytes differentiate in the lymph glands and then migrate to the site of encapsulation, where they flatten and spread to create a barrier around the encapsulated material. This immune response is dependent on the lamellocytes. Researchers have identified a lineage of *Drosophila* that naturally lacks lamellocytes, and the larvae are unable to encapsulate parasitic wasp eggs (Eslin and Doury 2006). Plasmatocytes comprise the vast majority (~95%) of hemocytes in both larvae and adults. These hemocytes recognize and phagocytose invading microorganisms. They appear early in embryonic development and multiply in the larval stages. Plasmatocytes circulate throughout the animal in order to detect pathogens and activate the immune response when necessary. At the end of the third larval instar, the larval lymph glands release additional blood cells into the hemolymph. Many embryonic and larval hemocytes persist through metamorphosis to eventually become a population of sessile cells clustered around the dorsal blood vessel of the adult fly (Holz et al. 2003). Despite the change in localization, adult plasmatocytes carry out the same macrophage-like functions as larval hemocytes; it has been shown that both larval and adult blood cells are capable of effectively phagocytosing a variety of pathogens in less than thirty minutes (Fig. 1; Elrod-Erickson et al. 2000). Crystal cells are a small proportion of hemocytes, and most (~95%) do not circulate (Carton et al. 1986). These cells release

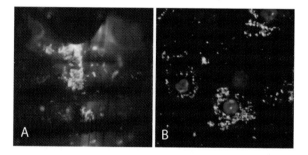

Fig. 1 *Drosophila* hemocytes act as macrophages. FITC-conjugated *E. coli* particles are phagocytosed by sessile hemocytes along the adult dorsal blood vessel (**A**) and by circulating hemocytes in larvae (**B**). Nuclei in (**B**) are stained with Hoechst 33258

reactive oxygen molecules and are involved in melanization (Rizki and Rizki 1959), which is not covered in this chapter. This chapter discusses encapsulation and phagocytosis.

2 Encapsulation

Encapsulation and nodulation refer to the end result from the action of plasmatocytes and lamellocytes surrounding a foreign object and an aggregate of bacteria respectively. Some researchers distinguish between the two, stating that blood cells adhere to large parasites in encapsulation, but adhere to each other in the presence of microorganisms in nodulation (Johansson 1999). As the two look the same in ultrastructure (Ratcliffe and Gagen 1976, 1977), they are treated as one phenomenon here and referred to as encapsulation. This cellular immune response appears to use some genes also associated with the humoral response. For example, mutations in both the Toll and JAK/STAT pathways are associated with the encapsulation of tissue in the absence of parasitoids or bacteria (Fig. 2; Zettervall et al. 2004).

The encapsulation response is vital for *Drosophila* to survive parasitization by wasps. The importance of this response is shown by the multiple methods parasites have evolved to countermand encapsulation: some parasitic wasps inject factors that attack encapsulation components, most notably lamellocytes (Labrosse et al. 2003, 2005a, b). *Drosophila* resistance to these parasitic actions against the encapsulation response reflects the ongoing evolutionary race between parasite and host. Two specific genes, *Resistance to Leptopilina boulardi* (*Rlb*) and

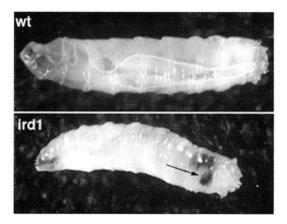

Fig. 2 Encapsulation of self tissue. Melanotic masses arise in *Drosophila* larvae with the mutation *ird1*, which affects the Imd and Toll signaling pathways. A wild-type (*wt*) larva without melanotic tumors is shown for comparison

Resistance to Asobara tabida (*Rat*) have been identified that are involved in variation in the ability to resist, or successfully encapsulate, oviposited parasites (Carton and Nappi 1997; Orr and Irving 1997). *Rlb* confers resistance to *L. boulardi* and is dominant. *Rat* confers resistance to *A. tabida* and is also dominant. As the two genes map to different regions of the second chromosome, the two systems of parasite resistance are likely distinct (Carton and Nappi 1997; Benassi et al. 1998; Hita et al. 1999; Poirie et al. 2000; Carton et al. 2005).

Encapsulation occurs in *Drosophila* larvae in approximately 24 h following parasite egg deposition (Russo et al. 1996). The process is often described in four stages: recognition of foreign matter, binding, adhesion, and termination. This discussion describes four different stages: recognition and binding, lamellocyte proliferation, adhesion and cell spreading, and termination.

2.1 Recognition Centers on Membrane Differences

Not much is known about pathogen recognition in encapsulation. As in phagocytosis, pattern recognition receptors (PRRs) appear to recognize differences from normal host cell membranes. Research on pathogen recognition has concentrated on host recognition of cell wall components specific to microbes, such as peptidoglycan and lipopolysaccharide (LPS). Research on altered-self cell recognition has centered on host recognition of changes in the basal membrane. Few PRRs that stimulate encapsulation have been identified in *Drosophila*, but research findings in other insects may be applicable. These studies have found specific receptors important for encapsulation that may have homologs in *Drosophila*.

The integrin ligand Peroxidasin has been identified as a PRR in encapsulation in arthropods. Peroxidasin is an exracellular matrix (ECM) protein, theorized to mediate cell adhesion to ECM (Nelson et al. 1994). It is synthesized in blood cells and released during degranulation (Johansson and Soderhall 1988). In the crayfish *Pacifastacus leniusculus*, peroxidasin is secreted in an inactive form, then activated by LPS or β-1,3-glucans. The active form binds to an integrin to initiate adhesion (Holmblad et al. 1997). *Drosophila* peroxidasin is expressed exclusively in hemocytes and exists as a homotrimer with a peroxidase domain and motifs common to ECM-associated proteins (Nelson et al. 1994). Peroxidasin homologs can be found in several different species, including mollusks (Weis et al. 1996), echinoderms (Gesualdo et al. 1997), and *Caenorhabditis elegans* (Wilson et al. 1994), suggesting that peroxidasin function may be conserved.

Research in the tobacco hornworm *Manduca sexta* has identified a set of recognition receptors called Immulectins. Immulectins are members of the calcium-dependent (C-type) lectin family, and contain two carbohydrate-recognition domains (CRDs; Yu et al. 1999). Thirty-two C-type lectin-like genes have been identified in the *Drosophila* genome (Dodd and Drickamer 2001), but none have been positively identified as immulectin homologs thus far. In *M. sexta*, three immulectins have been characterized in vivo and in vitro. Immulectin-1 (IML-1) enhances encapsulation (Ling and Yu 2006), and Immulectin-2 (IML-2) is involved in encapsulation and melanization (Yu and Kanost 2004). Experiments with Immulectin-4 (IML-4) suggest that these receptors may recognize LPS (Yu et al. 2006). Other studies have shown that commercially available LPS is often contaminated with peptidoglycan, another cell wall component (Kaneko et al. 2004). Thus, IML-4 may be binding to either LPS or peptidoglycan.

Recognition of altered or damaged self appears centered on changes in the basement membrane. In *Drosophila*, Rizki and Rizki (1979) characterized this phenomenon in their *tu-w* mutants. The mutants were named for the presence of melanotic masses, or melanized encapsulates, in the absence of pathogens. Investigation of the phenotype showed that it resulted from encapsulation of fat body cells with cell membrane abnormalities (Rizki and Rizki 1979).

Other researchers hypothesized that the induction of lamellocyte proliferation during sterile wounding may arise following recognition of damage to cell membranes at the wound site (Markus et al. 2005). Recent research in self-recognition has focused on the shielding of self membranes from encapsulation by membrane-bound carbohydrates and sialic acid. In other insects, host cells are more likely to be encapsulated when they lose these groups (Lackie 1980; Karacali et al. 2000).

Binding requires recognition, making it more difficult to differentiate between mechanisms for each. General studies of encapsulation of beads with different attached functional groups have been performed in the moth *Pseudoplusia includens*, and have identified chemical properties and functional groups that make foreign matter more readily encapsulated. Specifically, foreign matter is readily encapsulated when it is cationic, or linked to sulfonic groups or some amino groups. Matter with attached carbohydrate moieties is not readily encapsulated, supporting hypotheses that carbohydrates shield membranes from recognition or binding by host receptors (Lackie 1980). Additional experiments have supported a dual role for humoral and cellular receptors in recognition. Specifically, *P. includens* plasmatocytes can encapsulate matter with sulfonic groups in vitro, but require pre-incubation of the matter in plasma in order to encapsulate matter with various amino groups (Lavine and Strand 2001).

In *Drosophila*, 6–16h after egg deposition, an electron-dense discontinuous layer appears to stick to the parasite egg and to spread (Russo et al. 1996). Studies in other insects have suggested that this layer is secreted by granulocytes, the rough equivalent of *Drosophila* crystal cells, and other studies in insects have suggested that granulocyte binding is necessary for encapsulation (Pech and Strand 1996). Unfortunately, plasmatocytes may cause granulocytes to apoptose, making their exact role in early encapsulation more difficult to determine (Pech and Strand 2000). This layer appears necessary for encapsulation, but its exact function is unknown.

Approximately 16–24h after egg deposition, plasmatocytes aggregate around the egg and form the next layer, which forms all over the egg chorion, including regions not covered by the electron-dense layer (Russo et al. 1996). These observations, combined with the studies in *P. includens*, suggest that proteins in *Drosophila* hemolymph may recognize the parasite egg, surround it, and facilitate plasmatocyte binding. These proteins could include peroxidasin and immulectins, but their identity has not been conclusively established.

2.2 Lamellocyte Proliferation: Necessary for Successful Encapsulation Response

The primary hemocytes involved in encapsulation are plasmatocytes and lamellocytes. Plasmatocytes comprise much of the early layers of immune cells surrounding parasite eggs, and phagocytose proteins from the wasp long gland that are deposited in the *Drosophila* hemocoel with the parasitic egg. Lamellocytes are present in the earlier cell layers, and these are the cells that flatten and form the later layers during encapsulation (Russo et al. 1996). Lamellocytes usually make up less than 5% of circulating hemocytes (Rizki and Rizki 1992), but increase in number following wounding and parasitism. This proliferation results from mitosis of lamellocyte precursors in the hematopoietic organ (Sorrentino et al. 2002). Recent studies have shown that lamellocyte proliferation is targeted by parasites. Virulent wasps parasitoids induced atrophy of hematopoietic organs (Chiu et al. 2001; Moreau et al. 2003), indicating that the proliferation response is likely necessary for a successful immune response. Further evidence of the necessity of lamellocyte proliferation was found in a comparison of virulent and avirulent parasites in *Drosophila*. Lamellocyte proliferation was greater after egg deposition by avirulent parasites than after sterile wounding or egg deposition by virulent parasites. Interestingly, the lamellocyte proliferation following sterile wounding is greater than that following egg deposition by virulent parasites (Russo et al. 1996). This result suggests that virulent parasites deposit their eggs with factors that block proliferation.

2.3 Adhesion Requires Integrins, Rac, and Rho

At 16–24 h after egg deposition, plasmatocytes surround the parasite egg and spread. At 24 h after egg deposition, large numbers of lamellocytes are observed around the partially encapsulated egg. Extreme flattening is observed as the cells spread to form the outer layers in encapsulation (Russo et al. 1996). The mechanisms of this process involve integrins and the downstream small GTPases Rac and Rho, but may also involve other adhesion receptors and their downstream components.

In mammals, four families of receptors are implicated in adhesion regulation, namely: selectins, cadherins, the Ig (immunoglobulin) superfamily, and integrins (Baggiolini et al. 1997). Examples of each receptor have been found in *Drosophila*. Selectins are involved in vertebrate white blood cell adhesion, and contain both C-type lectin and epidermal growth factor (EGF) domains. Thirty-two C-type lectin-like genes have been identified in the *Drosophila* genome (Dodd and Drickamer 2001). A specific study of three C-type lectins, *Drosophila* lectin 1

4 Cellular Immune Responses in *Drosophila melanogaster*

(DL1), *Drosophila* lectin 2 (DL2), and *Drosophila* lectin 3 (DL3) showed them to be secreted into the hemolymph by the fat body, but it is not known whether they play any role in encapsulation (Tanji et al. 2006). Cadherins and Ig superfamily proteins are involved in cell–cell adhesion. A recent genome analysis identified 12 new *Drosophila* cadherin homologs while confirming five others (Hynes and Zhao 2000). The same study identified approximately 150 genes with Ig domains. Additionally, an Ig-related opsin has been shown to inhibit aggregation in encapsulation experiments in *P. includens* (Ladendorff and Kanost 1991; Bettencourt et al. 1997). Of these four families, integrins have received the most attention in *Drosophila* and insect encapsulation research (Lavine and Strand 2001), possibly due to the availability of mutations and their known role in the activation of Rho and Rac for regulation of key cytoskeletal processes (Price et al. 1998).

Like cadherins and Ig-superfamily proteins, integrins are theorized to have evolved in the metazoic era (Hynes and Zhao 2000) and are found across various phyla, including mammals, crustaceans (Wiegand et al. 2000), sponges (Brower et al. 1997), and insects (Pech and Strand 1995). They are heterodimers of independent α and β subunits. Each heterodimer usually contains large extracellular domains, two type-I transmembrane domains, and two C-terminal cytoplasmic tails (Hynes 1992). Two β subunits and five α subunits have been identified in *Drosophila* (Gotwals et al. 1994; Stark et al. 1997; Grotewiel et al. 1998; Hynes and Zhao 2000). Integrins are involved in focal adhesions, most notably to the ECM, through both outside-in and inside-out signaling (Hynes and Zhao 2000). Of the integrins, βPS integrin has been shown to be necessary for encapsulation in *Drosophila* (Irving et al. 2005).

Integrin ligands include collagens (Vuorio and de Crombrugghe 1990), laminins (Timpl and Brown 1994), fibronectin (Gonzalez-Amaro and Sanchez-Madrid 1999), and plasma proteins (Altieri 1995). All of these ligands contain an Arg-Gly-Asp (RGD) motif (Ruoslahti 1996), and RGD peptides inhibit cell spreading (Davids and Yoshino 1998), hemocyte aggregation, nodule formation, and the encapsulation of microbe aggregates (Pech and Strand 1996). Peroxidasin, a PRR mentioned earlier, appears to stimulate encapsulation by binding to integrin. Another integrin ligand, peroxinectin, stimulates encapsulation and phagocytosis in other arthropod cells (Kobayashi et al. 1990), but has only been identified as a cell adhesion protein in *Drosophila*. Found originally in crayfish, peroxinectin appears similar to animal peroxidases (Johansson et al. 1995). A homologous gene has been found in *Drosophila* (Ng et al. 1992), and it will be interesting to determine whether it plays a similar role in encapsulation.

2.3.1 Cell Spreading and Adhesion are Mediated by Rho and Rac

Cell spreading and later adhesion appear to be primarily mediated by Rho GTPases, a family of proteins within the Ras-related small GTP-binding protein (G protein) superfamily. Rho GTPases are bound to guanine triphosphate (GTP) in their active form and guanine diphosphate when inactivated. They appear to be downstream effectors of integrins and are regulated through their dependence on GTP by two families of enzymes: guanine nucleotide exchange factors (GEFs) and GTPase activating proteins (GAPs). GEFs activate Rho GTPases by phophorylating them, thus turning Rho-GDP or Rac-GDP into Rho-GTP or Rac-GTP. GAPs inactivate Rho GTPases by the reverse process: dephosphorylating Rho-GTP or Rac-GTP to turn it into Rho-GDP or Rac-GDP (for a review, see Hall 1998). Recent studies of parasitic virulence underscore the importance of Rho GTPases and their regulators in encapsulation: a virulence factor used by some wasps appears to be a Rho-GAP (Labrosse et al. 2005a).

Downstream of integrins, two specific Rho GTPases, Rac and Rho, cause cell spreading and then adhesion by working in opposition. Rac promotes cell spreading through lamellipodia extension, Rho inhibits lamellipodia extension and promotes cell contractility (Xu et al. 2003). Because of the importance of cell spreading and adhesion to cardiovascular disease and cancer, much of the elucidation of the roles of Rac and Rho has been done in mammalian cells. As Rho GTPases and their downstream components have been found not only in *Drosophila*, but also in organisms as diverse as protozoans (Arias-Romero et al. 2006; De Melo et al. 2006), yeast (Barale et al. 2006), and *Arabidopsis* (Gu et al. 2006), it seems likely that these cellular functions are conserved.

2.3.2 Cell Spreading Requires Rac

In mammalian models, Rac stimulates cell spreading in the formation of stable lamellipodia through several different cascades that affect actin polymerization (Fig. 3). Phosphatidylinositol-3-kinase (PI3K) promotes the production of PIP3, a membrane component, that binds to GEFs that activate Rac and integrate it into

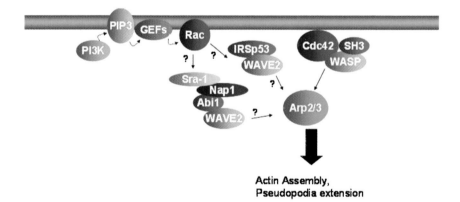

Fig. 3 Rac and Cdc42 stimulate pseudopodia extension through cascades affecting actin polymerization. Phosphatidylinositol-3-kinase (PI3K) promotes the production of PIP3, a membrane component, that binds to GEFs that activate Rac, and integrate it into the cell membrane. Rac activates WAVE2, possibly through IRSp53 or perhaps through Sra-1, which is bound to Nap. Nap activates Abl interactor (Abi), which binds to WAVE2 (also called suppressor of cAMP receptor (SCAR). WAVE2 activates the actin-related protein (Arp)2/3 complex that regulates lamellipodia extension as in phagocytosis. The pathway can also be activated by Cdc42, in a trimolecular complex with Cdc42 and WASP

the cell membrane (Hall 1998; Rickert et al. 2000; Chung et al. 2001). The signaling events downstream of Rac are less clear. Some research suggests that Rac is bound to the amino terminus of insulin receptor tyrosine kinase substrate p53 (IRSp53), which is then bound via a carboxy-terminal Src-homology-3 domain to WAVE2 (also called suppressor of cAMP receptor, SCAR; Bear et al. 1998), to form a trimolecular complex (Miki et al. 2000). WAVE2 is a ubiquitously expressed member of the Wiskott–Aldrich syndrome protein (WASP) family (Suetsugu et al. 1999), which activates the actin-related protein (Arp)2/3 complex that regulates lamellipodia extension, as also occurs in phagocytosis (Snapper et al. 2001). Other studies, however, have shown that IRSp53 is not needed for Rac activation of WAVE in vivo, and that Rac activates WAVE2 via a different signal transduction cascade. Specifically, Rac appears to bind to Sra-1, which is bound to Nap. Nap then activates Abi (Abl interactor), which binds WAVE2 (Innocenti et al. 2004).

Another cascade downstream of Rac involves PI(4)P5-K and PtdIns(4,5)P2. This cascade leads to the uncapping of the capping protein from the end of actin filaments. Rac also activates PAKs (p21-activated kinase), which appear to induce Lim kinase to inhibit depolymerization of actin filaments by cofilin (Fig. 4). Rho also appears to activate Lim kinase, but this activation drives only a small amount of actin polymerization (Machesky and Hall 1997).

Rac has three homologs in *Drosophila*: Rac1 (Luo et al. 1994; Kaufmann et al. 1998), Rac2 (Hariharan et al. 1995), and Mig-2-like (Mtl; Hakeda-Suzuki et al. 2002). Rac1 has been shown to signal through Lim kinase to inhibit cofilin (Raymond et al. 2004; Chen et al. 2005) and activate the Arp2/3 complex as described earlier. While Rac1 has not been shown to activate the PI(4)P5-K cascade, it does activate an additional MAP kinase cascade that leads to the activation of the Jun kinase, Basket. Activation of both Basket and Arp2/3 is required for stable lamellipodia formation (Williams et al. 2006). Rac2 appears to redundantly

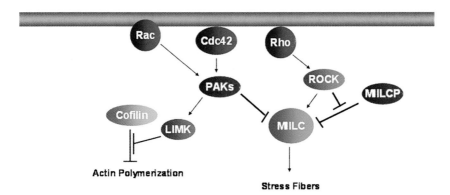

Fig. 4 Rho has the opposite effect of Rac and Cdc42. Rho appears to modulate bundle assembly through Rho kinase (ROCK), which promotes the activity of myosin light chain (MLC) and prevents its inhibition by blocking myosin light chain phosphorylase (MLCP) from phosphorylating MLC. Rac and Cdc42 inhibit MLC through p21-activated kinases (PAKs). PAKs activated by Rac and Cdc42 also induce Lim kinase to inhibit depolymerization of actin filaments by cofilin

control lamellipodia formation with Rac1 (Paladi and Tepass 2004). The PAK and WASP pathways associated with Rac can also be activated by Cdc42, which has a homolog in *Drosophila*. Mtl resembles Cdc42 in sequence, but it appears to stimulate the same pathways as Rac (Newsome et al. 2000).

2.3.3 Later Adhesion Requires Rho

While Rac induces the formation of focal complexes (Kiosses et al. 2001), it promotes adhesion mostly through cell spreading and integrin clustering (D'Souza-Schorey et al. 1998). Rho appears to be more involved than Rac in later stages of adhesion. Rho appears to have the opposite effects of Rac: where Rac promotes spreading and lamellipodia extension, Rho promotes contractility and inhibits lamellipodia; where Rac promotes actin polymerization, Rho promotes actin depolymerization (Sander et al. 1999). Both Rac and Rho are necessary for this stage of encapsulation. The cell is able to effect cell spreading and adhesion by regulating the balance of the Rho GTPases Rac, Cdc42, and Rho (Hirose et al. 1998; Moorman et al. 1999; Rottner et al. 1999).

Rho and its *Drosophila* homolog RhoA are involved in later phase adhesion in the clustering of focal complexes and in the transmission of tension to integrin ligation sites by promoting cell contractility (Chrzanowska-Wodnicka and Burridge 1996; Kimura et al. 1996). The tension induces aggregation of cadherins and integrins and causes them to align with the ends of stress fibers (Chrzanowska-Wodnicka and Burridge 1996; Rottner et al. 1999; Fox et al. 2005). Through this process, the focal complexes become more permanent focal adhesions.

Finally, Rho causes cell contractility. Some of the contractility may come from the phosphorylation of myosin light chain, which Rho promotes through PAK (Sanders et al. 1999). Despite evidence that Rho activates LIMK to inhibit cofilin, the amount of actin polymerization it induces is relatively small. Overall, Rho is associated with actin chain disassembly. Rho appears to modulate bundle assembly through Rho kinase and homologs of Diaphanous (Watanabe et al. 1999).

2.4 Encapsulation Terminates with the Formation of Basement Membrane

The formation of a basement membrane around the encapsulated mass is observed across several species of insects including *Drosophila* (Ball et al. 1987; Chain et al. 1992; Russo et al. 1996; Watanabe et al. 1999). The mechanism of membrane formation has not been elucidated. Research in the moth *Pseudoplusia includens* has unearthed evidence that granulocytes may be necessary in termination, but similar work has not been done with crystal cells. Termination is often followed by melanization, a response that encompasses the production several reactive oxygen molecules that are toxic to many pathogens. Melanization occurs in many but not all cases of encapsulation, and also has been observed as a response to wounding and infection by bacteria.

Future studies to identify host components important for the different steps in encapsulation may include visualization of localized processes using fluorescently tagged proteins and the testing of candidate genes identified from genomic studies using RNA interference. These approaches have been successful in identifying genes important for phagocytosis in *Drosophila*, as the next section describes.

3 Phagocytosis

Phagocytosis, the uptake of microbes into a phagocyte, is believed to be a highly conserved process with ancient origins. *Drosophila* hemocytes phagocytose microbes in a manner similar to other macrophages. Given the likely conservation of the genes involved in phagocytosis, *Drosophila* has been used as a model system. The advent of RNA interference (RNAi) has allowed the dissection of this phenomenon in vitro using *Drosophila* cell lines, such as S2 cells, which may be derived from larval hemocytes. This approach has been used to study host–pathogen interactions with a number of microorganisms (Ramet et al. 2002; Agaisse et al. 2005; Cheng et al. 2005; Philips et al. 2005; Stroschein-Stevenson et al. 2006). Gene expression can be knocked down by incubating the cells in the presence of double-stranded RNA (dsRNA). The advantage of these RNAi screens is that a whole-genome approach can be taken and it is possible to study partial phenotypes that may be difficult to assess in vivo (Clemens et al. 2000). The disadvantage of this technique is that it may overlook extracellular components and signaling events occurring in other tissues that are important for the physiological response in the animal. RNAi in vivo in *Drosophila* is another reverse genetic approach (Kocks et al. 2005). Transgenic flies have been created which contain an inverted repeat of the gene to be silenced; when transcribed, this sequence gives rise to a hairpin-loop RNA that induces RNAi against the gene of interest. This approach allows for a stable, heritable form of RNAi that can be temporally and spatially controlled through the use of specific GAL4 drivers (Brand and Perrimon 1993; Kennerdell and Carthew 2000).

As RNAi can have variable penetrance in organisms, studies of phagocytosis with ex vivo phagocytes and in vivo studies with mutants are important for a full understanding of the role of these genes in phagocytosis. In ex vivo studies, hemocytes are collected directly from *Drosophila* for study (Pearson et al. 2003; Bettencourt et al. 2004). Studies in vivo use a classic genetic approach with the characterization of mutants affected in phagocytosis (Elrod-Erickson et al. 2000; Garver et al. 2006).

3.1 Proteins Opsonize Invading Bacteria and Fungi to Promote Phagocytosis

Several proteins have been discovered in *Drosophila* which share sequence similarities with the mammalian complement C3/α2-macroglobulin superfamily of proteins, and are predicted to be secreted into hemolymph (Lagueux et al. 2000). These proteins, named thiolester proteins (TEPs), share a predicted signal peptide and a unique C-terminal region featuring six cysteine residues in conserved positions. These predicted proteins all contain a highly variable central region, indicating the potential for great diversity and specificity of activity. TEP II, for example, has five different splice forms. A microarray study of genes up-regulated upon infection in adult flies detected CG4823, a fly homolog of the complement-binding protein α2-macroglobulin (De Gregorio et al. 2001). This gene may encode a protein that binds to opsonized microbes to trigger phagocytosis. In the mosquito *Anopheles gambiae*, aTEP-I has been shown to opsonize and promote the phagocytosis of *Escherichia coli* and other Gram-negative bacteria (Levashina et al. 2001). A recent study demonstrated that *Drosophila* TEPs promote phagocytosis of a number of microbes and demonstrate some binding specificity. RNAi studies with S2 cells have shown that TEP II is specifically involved in the phagocytosis of *E. coli,* and cells without TEP III are partially impaired in the uptake of *Staphylococcus aureus*. This same study characterized macroglobulin complement-related (Mcr), another member of the α2-macroglobulin/complement protein family, which binds directly to and promotes the phagocytosis of *Candida albicans*, but not *E. coli, S. aureus*, or latex beads. In further support of an opsonin-like role, S2 cells secrete Mcr into the media before exposure to *C. albicans*, indicating that the protein may circulate in the fly hemolymph to survey for the presence of yeast cells (Stroschein-Stevenson et al. 2006).

A member of the Ig superfamily of proteins also has a potential role in opsonization (Watson et al. 2005). Down syndrome cell adhesion molecule (Dscam) contains a transmembrane domain, a signal peptide, and ten Ig domains (Schmucker et al. 2000). A microarray study of Dscam transcripts produced by differential mRNA splicing suggested that Dscam has 18 000 alternate splice forms, with 15 specific to hemocytes. Multiple splice forms of the Ig2, Ig3, and Ig7 domains were detected. While this result indicated that much of the diversity of Dscam splicing may be important for its role in the development of the *Drosophila* central nervous system, the 15 splice forms of Dscam may be important for recognizing diverse pathogens in the *Drosophila* immune response. A secreted form of Dscam was detected both in S2 media and larval hemolymph. Additionally, Dscam was shown to bind directly to bacteria, and RNAi in vivo and in vitro against Dscam impaired phagocytosis of *E. coli* by 30%. As the study only looked at Dscam activity in relation to *E. coli* phagocytosis, the potential role of Dscam diversity for differential recognition of specific pathogens remains to be explored (Watson et al. 2005). Further research may elucidate the role Dscam plays in immune response specificity, perhaps leading to the identification of a novel adaptive immune response in animals.

3.2 Transmembrane and Circulating Peptidoglycan Recognition Proteins are Involved in the Recognition of Bacteria

Peptidoglycan recognition proteins (PGRPs) are a highly conserved family of proteins. Proteins in the PGRP family have been classified into long (PGRP-L) or short (PGRP-S) forms. Both forms include membrane-bound and secreted proteins. (For a review of PGRP structure and activity, see Werner et al. 2000; Steiner 2004.) Both types recognize bacterial peptidoglycan (PGN) and contain a C-terminal domain that bears similarity to N-acetylmuramoyl-L-alanine amidases. It is believed that this amidase activity acts as a scavenger to degrade free PGN and limit an immune response (Mellroth et al. 2003). Several PGRPs have been identified as receptors in immune-responsive tissues. PGRP recognition of PGN leads to a cellular response of phagocytosis by hemocytes, or a humoral response, with the synthesis of antimicrobial peptides (AMPs) by the fat body.

One family member, PGRP-LC, was originally identified for its role in the humoral immune response. This receptor acts upstream of the Imd signaling pathway and triggers the production of AMPs such as diptericin in response to infection by Gram-negative bacteria (Choe et al. 2002; Gottar et al. 2002; Ramet et al. 2002; Kaneko et al. 2004). RNAi of PGRP-LC in S2 cells partially inhibits phagocytosis of *E. coli*, decreasing uptake by up to 40%. PGRP-LC mutants were also reported in this study to have an increased susceptibility to infection by *E. coli*, but not by *S. aureus* (Ramet et al. 2002). It is unclear if this is due to the phagocytic defect, as the PGRP-LC mutation also affects the humoral response. The PGRP-LC mutant, *ird7*, has been found to phagocytose both *E. coli* and *S. aureus* in vivo (Garver et al. 2006). One possible reason for the different findings in vitro and in vivo may be because of PGRP-LE. PGRP-LE appears to play a role both redundant and complementary to PGRP-LC in the recognition of Gram-negative peptidoglycan, and it is not expressed in hemocytes or S2 cells (Kaneko et al. 2006). Hence this may be the reason why a phagocytosis phenotype can be seen in S2 cells, but not in vivo.

Two short PGRP family members have also been implicated in the cellular immune response to Gram-positive bacteria. PGRP-SA (*semmelweis*) mutants exhibited a marked and specific decrease (~75% of animals showing no phagocytosis) in phagocytosis of *S. aureus* (Garver et al. 2006). Similarly, the PGRP-SC1a mutant *picky* is also specifically impaired in its ability to phagocytose *S. aureus*. Flies carrying the *picky* mutation failed to take up fluorescently-labeled *S. aureus* into their hemocytes and exhibited increased susceptibility to *S. aureus* infection. Furthermore, expression level of Drosomycin, the AMP target gene of the Toll pathway was virtually undetectable. PGRP-SA mutants showed similar humoral response impairments (Michel et al. 2001). This suggested that PGRP-SA and PGRP-SC1a are both important for Toll signaling and phagocytosis. The phagocytosis of *S. aureus* and survival to infection depended upon the amidase activity of PGRP-SC1a, suggesting that, in addition to a scavenger function, the amidase activity may be important for clearance of *S. aureus* and the ability to survive infection. The amidase activity may allow PGRP-SC1a to opsonize Gram-positive bacteria, or perhaps cleavage of PGN polymers into smaller units is necessary to trigger recognition of the bacteria (Garver et al. 2006). PGN monomers are effective activators of the humoral immune response (Kaneko et al. 2004), so PGN cleavage may be a prerequisite or play a role in activating the Toll and Imd signaling pathways (Filipe et al. 2005).

It is particularly interesting that these PGRPs are important for activity in both humoral and cellular immune responses and that these responses are specific to either Gram-positive or Gram-negative bacteria. These proteins may act in recognition events upstream of both phagocytosis and the AMP signaling pathways; alternatively, there may be an interdependence between these two branches of the innate immune system, with an optimal AMP response depending on a cellular immune response.

3.3 Receptors with Scavenger-Like Activity Recognize a Variety of Microbes

PGRPs allow for specific recognition of Gram-positive versus Gram-negative bacteria and are important for activating both cellular and humoral immune resoponses. The first *Drosophila* PRR shown to trigger phagocytosis was the scavenger-receptor protein, dSR-CI. Chinese hamster ovary cells transfected with dSR-CI gained the ability to bind both *E. coli* and *S. aureus*, and the RNAi of dSR-CI in S2 cells decreased cell association with bacteria by 25% (Ramet et al. 2001). The predicted dSR-CI protein is a type-I transmembrane protein with extracellular CCP (complement control protein) and MAM (meprin, A5, PTPmu) family domains (Pearson et al. 1995) which allow direct binding to bacteria (Ramet et al. 2001). A later study also indicated that dSR-CI was involved in initiating the uptake of bacteria by S2 cells (Kocks et al. 2005).

A novel receptor, Eater, has been identified and is required for the recognition and uptake of both *E. coli* and *S. aureus* in vivo and in S2 cells. RNAi of Eater led to a partial (up to 75%) decrease in the ability of S2 cells to phagocytose both bacteria. Similar, but less dramatic results were reported in *eater* null flies. The humoral response was unaffected by the *eater* mutation, indicating that *eater*'s effects may be specific to the cellular immune response. The predicted protein is a transmembrane receptor with an extracellular domain containing EGF-like repeats followed by a variable N-terminal region. The N-terminus of the Eater protein directly recognizes and binds *S. aureus*, the yeast *Candida silvativa*, and the Gram-negative bacterium *Serratia marcescens* (Kocks et al. 2005). It is likely that Eater acts redundantly or in conjunction with dSR-CI and PGRP-LC to recognize *E. coli* and *S. aureus* and initiate phagocytosis. Indeed, the earlier reports characterizing dSR-CI and PGRP-LC activity noted this possibility and suggested that other receptors might be necessary for full activity (Ramet et al. 2001, 2002). RNAi of all three components or characterization of a fly mutant for all three genes may reveal whether these receptors are acting in a redundant fashion.

A CD36-like receptor, Croquemort, was originally characterized as recognizing apoptotic cells destined for uptake by plasmatocytes during embryogenesis. The putative protein is an integral membrane protein with putative protein kinase C (PKC) phosphorylation sites and several cysteine clusters (Franc et al. 1996). A study of phagocytosis of bacteria by S2 cells revealed that Croquemort also binds to and signals for the uptake of *S. aureus*, but not *E. coli* (Fig. 5); RNAi of Croquemort decreased uptake of *S. aureus* by 35% (Stuart et al. 2005).

Another CD36-related receptor, Peste, was discovered in a whole-genome RNAi study of S2 cells exposed to *Mycobacterium fortuitum* and *M. smegmatis*. Knockdown of Peste in S2 cells drastically reduced uptake of *M. fortuitum* and *M. smegmatis*, but did not affect the response to *E. coli* or *S. aureus* (Philips et al. 2005). Peste is involved in the recognition of another intracellular pathogen,

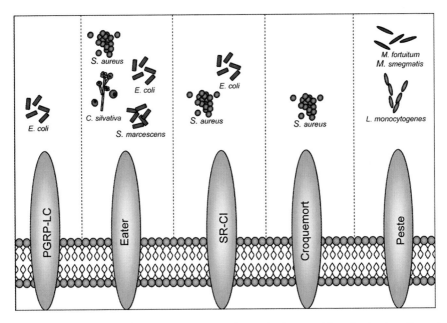

Fig. 5 Various transmembrane receptors recognize and trigger uptake of microbes. Several *Drosophila* PRRs which initiate phagocytosis have been identified. PGRP-LC specifically binds *E. coli*. The scavenger-like receptor Eater recognizes *E. coli*, *S. aureus*, *S. marcescens*, and *C. silvativa*; and another scavenger-like receptor, dSR-CI, has been shown to bind both *E. coli* and *S. aureus*. These three receptors may act cooperatively to initiate the phagocytosis of *E. coli*. Two members of the CD36 family have also been described as having roles in phagocytosis. Croquemort specifically binds to *S. aureus*, and Peste binds to the intracellular pathogens *L. monocytogenes*, *M. fortuitum*, and *M. smegmatis*

Listeria monocytogenes (Agaisse et al. 2005). Future studies may show what common feature allows for the recognition of both *Mycobacteria* and *Listeria* by Peste. One surprising result from the RNAi screens with *Listeria* was that no unique receptor for *Listeria* was identified (Agaisse et al. 2005; Cheng et al. 2005). This could be due to the inherent limitations of RNAi tissue culture screens. In the future, it will be exciting to determine whether Peste and other receptors identified in RNAi screens play similar roles in vivo.

3.4 Phagocytosis Requires Reorganization of the Actin Cytoskeleton

As in other animal cells, phagocytosis in plasmatocytes begins by the extension of actin-rich pseudopodia that surround and engulf the targeted microbe. This phenomenon has been observed in vitro in the S2 and mbn-2 cell lines (Pearson et al. 2003; Johansson et al. 2005; Stroschein-Stevenson et al. 2006). The formation of pseudopodia is dependent on remodeling the cytoskeleton through extensive actin polymerization and depolymerization, and this process is extremely well characterized. The process has three main phases: the nucleation of actin filaments, elongation, and eventually the bundling and stabilization of the filaments (Vignjevic et al. 2003).

Nucleation of actin filaments in animal cells occurs under the control of the Arp2/3 complex, which is believed to allow pseudopodia formation by stabilizing actin filaments at the leading edge of the cell. This complex has been shown by fluorescent and electron microscopy to localize to the pointed ends of microfilaments in *Xenopus* keratocytes (Svitkina and Borisy 1999). Furthermore, a study in the microorganism *Acanthamoeba* demonstrated that the Arp2/3 complex caps the pointed ends of microfilaments with high affinity (Mullins et al. 1998). Components of the Arp2/3 complex have been identified in RNAi studies for genes required for phagocytosis of both *M. fortuitum* (Philips et al. 2005) and *L. monocytogenes* (Agaisse et al. 2005) in *Drosophila* S2 cells. Hence, it is likely this process is conserved in *Drosophila* hemocytes.

In *Drosophila*, the Arp2/3 complex is primarily activated by D-SCAR, a member of the WASp/SCAR protein family. RNAi of SCAR in *Drosophila* cell lines drastically reduces actin staining at the leading edge of pseudopodia (Biyasheva et al. 2004) and SCAR is believed to be stabilized by another complex that includes the Abelson interacting protein (Abi; Kunda et al. 2003; Rogers et al. 2003). D-WASp may also play a role in this process, though its functional relationship to SCAR is unclear. The two proteins may be redundant, as RNAi of each gene in *Drosophila* cell lines produces only a partial phenotype of impaired phagocytosis and pseudopod formation (Pearson et al. 2003; Biyasheva et al. 2004). WASp is activated by a signaling pathway involving Rho proteins, including Cdc42, Rac 1, and Rac 2. Cdc42 has been detected in several screens for genes required for the uptake of bacteria, *Mycobacteria*, and yeast (Rogers et al. 2003; Agaisse et al. 2005; Philips et al. 2005; Stroschein-Stevenson et al. 2006).

Once actin filament nucleation has been achieved under the control of the Arp2/3 complex, the extension of pseudopodia is accomplished through the elongation and stabilization of the filaments. Radical cytoskeletal remodeling via the polymerization and depolymerization of actin occurs and depends upon the actin monomer-associated proteins profilin/chickadee and cofilin/twinstar, respectively. Both genetic and proteomic studies of phagocytosis in *Drosophila* cell lines have detected these proteins ex vivo and in vitro (Pearson et al. 2003; Loseva and Engstrom 2004; Philips et al. 2005). They have also been shown to be required for pseudopodia extension during S2 cell migration (Rogers et al. 2003), so it is likely that they play a similar role in *Drosophila* phagocytosis.

In order to provide cellular extensions with strength and stability, the elongated microfilaments are tightly associated through the interactions of cross-linking proteins (Vignjevic et al. 2003). In an RNAi screen, *short stop*, which is related to plectin was found to be important for the phagocytosis of Gram-positive and Gram-negative bacteria by S2 cells (Ramet et al. 2002). Plectin cross-links intermediate filaments with the cytoskeleton in mammalian cells (Svitkina et al. 1998). Genetic and proteomic screens of *Drosophila* cell lines have found that effective phagocytosis requires several actin-binding and actin-capping proteins, including Spire and Annexin IX (Loseva and Engstrom 2004; Johansson et al. 2005; Philips et al. 2005).

3.5 Engulfed Pathogens are Degraded in Phagolysosomes

In animal cells, pathogens and other extracellular material contained in phagosomes are eventually degraded in a phagolysosome. The process of phagosome maturation is not completely understood, but it is believed that phagosomes acquire proteolytic enzymes and a lower pH through a series of fleeting interactions with endocytic compartments (Desjardins et al. 1994). This process has not been studied in *Drosophila*, but it seems likely that the fundamental process will resemble that of phagocytosis and endocytosis in mammalian cells. Various studies of host/pathogen interactions in *Drosophila* S2 cells have shown that genes involved in vesicle trafficking and fusion are required for an effective cellular immune response. Transcripts for proteins involved in vesicle formation, including clathrin and members of the COPI complex, have been detected in RNAi studies of cells phagocytosing various types of bacteria (Ramet et al. 2002; Philips et al. 2005; Stroschein-Stevenson et al. 2006). These studies and others have also identified genes involved in vesicle docking and fusion, including the t-SNAREs Snap and Syntaxin as well as the small GTPases Rab 2 and Rab 5, as being required for processing phagocytosed pathogens (Agaisse et al. 2005; Philips et al. 2005; Stroschein-Stevenson et al. 2006). An RNAi screen of phagocytosis of *L. monocytogenes* by S2 cells showed that vesicular trafficking complexes that interact with late endosomes are necessary for the cellular immune response (Cheng et al. 2005). Furthermore, it has been reported that CTS proteases, which are most active at pH 4.5, localize to phagosomes in S2 cells

(Kocks et al. 2003) and lysosomal proteases such as cathepsins are modified and activated in mbn-2 cells upon exposure to lipopolysaccharide (Loseva and Engstrom 2004).

3.6 Interactions Between Cellular and Humoral Immune Responses

Although the *Drosophila* cellular and humoral immune responses were originally believed to be largely independent (Braun et al. 1998), it appears that there may be a complex and as yet poorly defined relationship between the two processes. Studies of the *Drosophila* immune response in which flies are naturally infected by allowing microbes to enter the host via the digestive tract led to a number of insights into how hemocytes contribute to the initiation of the humoral response. It has been suggested that hemocytes use cytokine-like molecules to mediate between the gut, where ingested pathogens are first detected, and the fat body, which regulates the systemic humoral response. Upon natural infection by the Gram-negative bacteria, *Erwinia carotovora*, larvae affected by mutations which result in the absence of blood cells have much lower levels of AMP expression as compared with wild-type flies (Basset et al. 2000).

One possible signaling molecule in this process is nitric oxide (NO). Pharmacological inhibition of nitric oxide synthase (NOS) results in reduced expression of diptericin upon infection by Gram-negative bacteria. Furthermore, introduction of NO via S-nitroso-N-acetylpenicillamine (SNAP) is enough to induce diptericin expression in the absence of infection (Foley and O'Farrell 2003).

Another possible signaling molecule is unpaired 3 (upd3), a ligand that activates the JAK/STAT pathway. upd3 binds to the putative transmembrane protein Domeless (Dome), which shares structural similarities with vertebrate cytokine class I receptors and the IL-3 family of receptors (Brown et al. 2001). This receptor has been shown to activate the JAK/STAT response in the *Drosophila* fat body upon infection with *E. coli* and *Micrococcus luteus* in S2 cells, possibly in conjunction with signaling via the Imd pathway (Agaisse et al. 2003). The ligand for Dome, upd3, is produced in hemocytes and is up-regulated upon septic injury (Agaisse et al. 2003; Johansson et al. 2006). Taken together, these results indicate that upd3 may be acting as the cytokine produced by hemocytes upon infection to induce the humoral response via the Domeless receptor in the fat body.

The advantages of the *Drosophila* system make it an excellent system for the study of cellular immune responses. Advances in recent years include the identification of many new receptors important for phagocytosis of different microbes through both genomic and genetic approaches. One challenge for the future will be to understand the role of these genes in phagocytosis in vivo, through the characterization of mutants. It will also be exciting to understand how components of the cellular response might activate or interact with the humoral responses. The encapsulation studies hint at the complexity required for a coordinated cellular response against parasites. As information from microarray and proteomic studies comes in, we may see similar functional approaches (transgenic RNAi expression or generation of mutations) used to uncover the role that newly identified genes may play in encapsulation.

References

Agaisse H, Burrack LS, et al (2005) Genome-wide RNAi screen for host factors required for intracellular bacterial infection. Science 309:1248–1251

Agaisse H, Petersen UM, et al (2003) Signaling role of hemocytes in *Drosophila* JAK/STAT-dependent response to septic injury. Dev Cell 5:441–450

Altieri DC (1995) Leukocyte interaction with protein cascades in blood coagulation. Curr Opin Hematol 2:41–46

Arias-Romero LE, Jesus Almaraz-Barrera M de, et al (2006) EhPAK2, a novel p21-activated kinase, is required for collagen invasion and capping in *Entamoeba histolytica*. Mol Biochem Parasitol 149:17–26

Baggiolini M, Dewald B, et al (1997) Human chemokines: an update. Annu Rev Immunol 15:675–705

Ball EE, Couet HG de, et al (1987) Haemocytes secrete basement membrane components in embryonic locusts. Development 99:255–259

Barale S, McCusker D, et al (2006) Cdc42p GDP/GTP cycling is necessary for efficient cell fusion during yeast mating. Mol Biol Cell 17:2824–2838

Basset A, Khush RS, et al (2000) The phytopathogenic bacteria *Erwinia carotovora* infects *Drosophila* and activates an immune response. Proc Natl Acad Sci USA 97:3376–3381

Bear JE, Rawls JF, et al (1998) SCAR, a WASP-related protein, isolated as a suppressor of receptor defects in late Dictyostelium development. J Cell Biol 142:1325–1335

Benassi V, Frey F, et al (1998) A new specific gene for wasp cellular immune resistance in *Drosophila*. Heredity 80:347–352

Bettencourt R, Lanz-Mendoza H, et al (1997) Cell adhesion properties of hemolin, an insect immune protein in the Ig superfamily. Eur J Biochem 250:630–637

Bettencourt R, Asha H, et al (2004) Hemolymph-dependent and -independent responses in *Drosophila* immune tissue. J Cell Biochem 92:849–863

Biyasheva A, Svitkina T, et al (2004) Cascade pathway of filopodia formation downstream of SCAR. J Cell Sci 117:837–848

Brand AH, Perrimon N (1993) Targeted gene expression as a means of altering cell fates and generating dominant phenotypes. Development 118:401–415

Braun A, Hoffmann JA, et al (1998) Analysis of the *Drosophila* host defense in domino mutant larvae, which are devoid of hemocytes. Proc Natl Acad Sci USA 95:14337–14342

Brower DL, Brower SM, et al (1997) Molecular evolution of integrins: genes encoding integrin beta subunits from a coral and a sponge. Proc Natl Acad Sci USA 94:9182–9187

Brown S, Hu N, et al (2001) Identification of the first invertebrate interleukin JAK/STAT receptor, the *Drosophila* gene domeless. Curr Biol 11:1700–1705

Carton Y, Nappi AJ (1997) *Drosophila* cellular immunity against parasitoids. Parasitol Today 13:218–227

Carton Y, Bouletreau M, et al (1986) The *Drosophila* parasitoid wasps. In: Ashburner M, Carson HL, Thompson JN (eds) The genetics and biology of *Drosophila*. Academic, New York, pp 347–394

Carton Y, Nappi AJ, et al (2005) Genetics of anti-parasite resistance in invertebrates. Dev Comp Immunol 29:9–32

Chain BM, Leyshon-Sorland K, et al (1992) Haemocyte heterogeneity in the cockroach *Periplaneta americana* analysed using monoclonal antibodies. J Cell Sci 103:1261–1267

Chen GC, Turano B, et al (2005) Regulation of Rho and Rac signaling to the actin cytoskeleton by paxillin during *Drosophila* development. Mol Cell Biol 25:979–987

Cheng LW, Viala JP, et al (2005) Use of RNA interference in *Drosophila* S2 cells to identify host pathways controlling compartmentalization of an intracellular pathogen. Proc Natl Acad Sci USA 102:13646–13651

Chiu H, Sorrentino RP, et al (2001) Suppression of the *Drosophila* cellular immune response by *Ganaspis xanthopoda*. Adv Exp Med Biol 484:161–167

Choe KM, Werner T, et al (2002) Requirement for a peptidoglycan recognition protein (PGRP) in Relish activation and antibacterial immune responses in *Drosophila*. Science 296:359–362

Chrzanowska-Wodnicka M, Burridge K (1996) Rho-stimulated contractility drives the formation of stress fibers and focal adhesions. J Cell Biol 133:1403–1415

Chung CY, Funamoto S, et al (2001) Signaling pathways controlling cell polarity and chemotaxis. Trends Biochem Sci 26:557–566

Clemens JC, Worby CA, et al (2000) Use of double-stranded RNA interference in *Drosophila* cell lines to dissect signal transduction pathways. Proc Natl Acad Sci USA 97:6499–6503

D'Souza-Schorey C, Boettner B, et al (1998) Rac regulates integrin-mediated spreading and increased adhesion of T lymphocytes. Mol Cell Biol 18:3936–3946

Davids BJ, Yoshino TP (1998) Integrin-like RGD-dependent binding mechanism involved in the spreading response of circulating molluscan phagocytes. Dev Comp Immunol 22:39–53

De Gregorio E, Spellman PT, et al (2001) Genome-wide analysis of the *Drosophila* immune response by using oligonucleotide microarrays. Proc Natl Acad Sci USA 98:12590–12595

De Melo LD, Eisele N, et al (2006) TcRho1, the *Trypanosoma cruzi* Rho homologue, regulates cell-adhesion properties: evidence for a conserved function. Biochem Biophys Res Commun 345:617–622

Desjardins M, Huber LA, et al (1994) Biogenesis of phagolysosomes proceeds through a sequential series of interactions with the endocytic apparatus. J Cell Biol 124:677–688

Dodd RB, Drickamer K (2001) Lectin-like proteins in model organisms: implications for evolution of carbohydrate-binding activity. Glycobiology 11:71R–79R

Elrod-Erickson M, Mishra S, et al (2000) Interactions between the cellular and humoral immune responses in *Drosophila*. Curr Biol 10:781–784

Eslin P, Doury G (2006) The fly *Drosophila* subobscura: a natural case of innate immunity deficiency. Dev Comp Immunol 30:977–983

Filipe SR, Tomasz A, et al (2005) Requirements of peptidoglycan structure that allow detection by the *Drosophila* Toll pathway. EMBO Rep 6:327–333

Foley E, O'Farrell PH (2003) Nitric oxide contributes to induction of innate immune responses to gram-negative bacteria in *Drosophila*. Genes Dev 17:115–125

Fox DT, Homem CC, et al (2005) Rho1 regulates *Drosophila* adherens junctions independently of p120ctn. Development 132:4819–4831

Franc NC, Dimarcq JL, et al (1996) Croquemort, a novel *Drosophila* hemocyte/macrophage receptor that recognizes apoptotic cells. Immunity 4:431–443

Garver LS, Wu J, et al (2006) The peptidoglycan recognition protein PGRP-SC1a is essential for Toll signaling and phagocytosis of *Staphylococcus aureus* in *Drosophila*. Proc Natl Acad Sci USA 103:660–665

Gesualdo I, Aniello F, et al (1997) Molecular cloning of a peroxidase mRNA specifically expressed in the ink gland of *Sepia officinalis*. Biochim Biophys Acta 1353:111–117

Gonzalez-Amaro R, Sanchez-Madrid F (1999) Cell adhesion molecules: selectins and integrins. Crit Rev Immunol 19:389–429

Gottar M, Gobert V, et al (2002) The *Drosophila* immune response against Gram-negative bacteria is mediated by a peptidoglycan recognition protein. Nature 416:640–644

Gotwals PJ, Paine-Saunders SE, et al (1994) *Drosophila* integrins and their ligands. Curr Opin Cell Biol 6:734–739

Grotewiel MS, Beck CD, et al (1998) Integrin-mediated short-term memory in *Drosophila*. Nature 391:455–460

Gu Y, Li S, et al (2006) Members of a novel class of *Arabidopsis* Rho guanine nucleotide exchange factors control Rho GTPase-dependent polar growth. Plant Cell 18:366–381

Hakeda-Suzuki S, Ng J, et al (2002) Rac function and regulation during *Drosophila* development. Nature 416:438–442

Hall A (1998) Rho GTPases and the actin cytoskeleton. Science 279:509–514

Hariharan IK, Hu KQ, et al (1995) Characterization of rho GTPase family homologues in *Drosophila* melanogaster: overexpressing Rho1 in retinal cells causes a late developmental defect. EMBO J 14:292–302

Hirose M, Ishizaki T, et al (1998) Molecular dissection of the Rho-associated protein kinase (p160ROCK)-regulated neurite remodeling in neuroblastoma N1E-115 cells. J Cell Biol 141:1625–1636

Hita MT, Poirie M, et al (1999) Genetic localization of a *Drosophila melanogaster* resistance gene to a parasitoid wasp and physical mapping of the region. Genome Res 9:471–481

Holmblad T, Thornqvist PO, et al (1997) Identification and cloning of an integrin beta subunit from hemocytes of the freshwater crayfish *Pacifastacus leniusculus*. J Exp Zool 277:255–61

Holz A, Bossinger B, et al (2003) The two origins of hemocytes in *Drosophila*. Development 130:4955–4962

Hynes RO (1992) Integrins: versatility, modulation, and signaling in cell adhesion. Cell 69:11–25

Hynes RO, Zhao Q (2000) The evolution of cell adhesion. J Cell Biol 150:F89–F96

Innocenti M, Zucconi A, et al (2004) Abi1 is essential for the formation and activation of a WAVE2 signalling complex. Nat Cell Biol 6:319–327

Irving P, Ubeda JM, et al (2005) New insights into *Drosophila* larval haemocyte functions through genome-wide analysis. Cell Microbiol 7:335–350

Johansson KC, Metzendorf C, et al (2005) Microarray analysis of immune challenged *Drosophila* hemocytes. Exp Cell Res 305:145–155

Johansson KC, Soderhall K, et al (2006) Diptericin expression in bacteria infected *Drosophila* mbn-2 cells – effect of infection dose and phagocytosis. Insect Mol Biol 15:57–62

Johansson MW (1999) Cell adhesion molecules in invertebrate immunity. Dev Comp Immunol 23:303–315

Johansson MW, Soderhall K (1988) Isolation and purification of a cell adhesion factor from crayfish blood cells. J Cell Biol 106:1795–1803

Johansson MW, Lind MI, et al (1995) Peroxinectin, a novel cell adhesion protein from crayfish blood. Biochem Biophys Res Commun 216:1079–1087

Kaneko T, Goldman WE, et al (2004) Monomeric and polymeric gram-negative peptidoglycan but not purified LPS stimulate the *Drosophila* IMD pathway. Immunity 20:637–649

Kaneko T, Yano T, et al (2006) PGRP-LC and PGRP-LE have essential yet distinct functions in the *Drosophila* immune response to monomeric DAP-type peptidoglycan. Nat Immunol 7:715–723

Karacali S, Deveci R, et al (2000) Adhesion of hemoctyes to desialylated prothoracic glands of *Galleria mellonella* (Lepidoptera) in the larval stage. Invert Reprod Dev 37:167–170

Kaufmann N, Wills ZP, et al (1998) *Drosophila* Rac1 controls motor axon guidance. Development 125:453–461

Kennerdell JR, Carthew RW (2000) Heritable gene silencing in *Drosophila* using double-stranded RNA. Nat Biotechnol 18:896–898

Kimura K, Ito M, et al (1996) Regulation of myosin phosphatase by Rho and Rho-associated kinase (Rho-kinase). Science 273:245–248

Kiosses WB, Shattil SJ, et al (2001) Rac recruits high-affinity integrin alphavbeta3 to lamellipodia in endothelial cell migration. Nat Cell Biol 3:316–20

Kobayashi M, Johansson MW, et al (1990) The 76kDa cell-adhesion factor from crayfish haemocytes promotes encapsulation in vitro. Cell Tissue Res 260:113–118

Kocks C, Maehr R, et al (2003) Functional proteomics of the active cysteine protease content in *Drosophila* S2 cells. Mol Cell Proteomics 2:1188–1197

Kocks C, Cho JH, et al (2005) Eater, a transmembrane protein mediating phagocytosis of bacterial pathogens in *Drosophila*. Cell 123:335–346

Kunda P, Craig G, et al (2003) Abi, Sra1, and Kette control the stability and localization of SCAR/WAVE to regulate the formation of actin-based protrusions. Curr Biol 13:1867–1875

Labrosse C, Carton Y, et al (2003) Active suppression of *D. melanogaster* immune response by long gland products of the parasitic wasp *Leptopilina boulardi*. J Insect Physiol 49:513–522

Labrosse C, Eslin P, et al (2005a) Haemocyte changes in *D. melanogaster* in response to long gland components of the parasitoid wasp *Leptopilina boulardi*: a Rho-GAP protein as an important factor. J Insect Physiol 51:161–170

Labrosse C, Stasiak K, et al (2005b) A RhoGAP protein as a main immune suppressive factor in the *Leptopilina boulardi* (Hymenoptera, Figitidae) – *Drosophila melanogaster* interaction. Insect Biochem Mol Biol 35:93–103

Lackie AM (1980) Invertebrate immunity. Parasitology 80:393–412

Ladendorff NE, Kanost MR (1991) Bacteria-induced protein P4 (hemolin) from *Manduca sexta*: a member of the immunoglobulin superfamily which can inhibit hemocyte aggregation. Arch Insect Biochem Physiol 18:285–300

Lagueux M, Perrodou E, et al (2000) Constitutive expression of a complement-like protein in toll and JAK gain-of-function mutants of *Drosophila*. Proc Natl Acad Sci USA 97:11427–11432

Lavine MD, Strand MR (2001) Surface characteristics of foreign targets that elicit an encapsulation response by the moth *Pseudoplusia includens*. J Insect Physiol 47:965–974

Levashina EA, Moita LF, et al (2001) Conserved role of a complement-like protein in phagocytosis revealed by dsRNA knockout in cultured cells of the mosquito, *Anopheles gambiae*. Cell 104:709–718

Ling E, Yu XQ (2006) Cellular encapsulation and melanization are enhanced by immulectins, pattern recognition receptors from the tobacco hornworm *Manduca sexta*. Dev Comp Immunol 30:289–299

Loseva O, Engstrom Y (2004) Analysis of signal-dependent changes in the proteome of *Drosophila* blood cells during an immune response. Mol Cell Proteomics 3:796–808

Luo L, Liao YJ, et al (1994) Distinct morphogenetic functions of similar small GTPases: *Drosophila* Drac1 is involved in axonal outgrowth and myoblast fusion. Genes Dev 8:1787–1802

Machesky LM, Hall A (1997) Role of actin polymerization and adhesion to extracellular matrix in Rac- and Rho-induced cytoskeletal reorganization. J Cell Biol 138:913–926

Markus R, Kurucz E, et al (2005) Sterile wounding is a minimal and sufficient trigger for a cellular immune response in *Drosophila* melanogaster. Immunol Lett 101:108–111

Mellroth P, Karlsson J, et al (2003) A scavenger function for a *Drosophila* peptidoglycan recognition protein. J Biol Chem 278:7059–7064

Michel T, Reichhart JM, et al (2001) *Drosophila* Toll is activated by Gram-positive bacteria through a circulating peptidoglycan recognition protein. Nature 414:756–759

Miki H, Yamaguchi H, et al (2000) IRSp53 is an essential intermediate between Rac and WAVE in the regulation of membrane ruffling. Nature 408:732–735

Moorman JP, Luu D, et al (1999) A balance of signaling by Rho family small GTPases RhoA, Rac1 and Cdc42 coordinates cytoskeletal morphology but not cell survival. Oncogene 18:47–57

Moreau SJ, Eslin P, et al (2003) Comparative study of the strategies evolved by two parasitoids of the genus *Asobara* to avoid the immune response of the host, *Drosophila melanogaster*. Dev Comp Immunol 27:273–282

Mullins RD, Heuser JA, et al (1998) The interaction of Arp2/3 complex with actin: nucleation, high affinity pointed end capping, and formation of branching networks of filaments. Proc Natl Acad Sci USA 95:6181–6186

Nelson RE, Fessler LI, et al (1994) Peroxidasin: a novel enzyme-matrix protein of *Drosophila* development. EMBO J 13:3438–3447

Newsome TP, Schmidt S, et al (2000) Trio combines with dock to regulate Pak activity during photoreceptor axon pathfinding in *Drosophila*. Cell 101:283–294

Ng SW, Wiedemann M, et al (1992) Molecular characterization of a putative peroxidase gene of *Drosophila melanogaster*. Biochim Biophys Acta 1171:224–228

Orr HA, Irving S (1997) The genetics of adaptation: the genetic basis of resistance to wasp parasitism in *Drosophila melanogaster*. Evolution 51:1877–1885

Paladi M, Tepass U (2004) Function of Rho GTPases in embryonic blood cell migration in *Drosophila*. J Cell Sci 117:6313–6326

Pearson A, Lux A, et al (1995) Expression cloning of dSR-CI, a class C macrophage-specific scavenger receptor from *Drosophila melanogaster*. Proc Natl Acad Sci USA 92:4056–4060

Pearson AM, Baksa K, et al (2003) Identification of cytoskeletal regulatory proteins required for efficient phagocytosis in *Drosophila*. Microbes Infect 5:815–824

Pech LL, Strand MR (1995) Effects of basement membranes on the behavior of hemocytes from *Pseudoplusia includens*: development of an in vitro encapsulation assay. J Insect Physiol 41:801–807

Pech LL, Strand MR (1996) Granular cells are required for encapsulation of foreign targets by insect haemocytes. J Cell Sci 109:2053–2060

Pech LL, Strand MR (2000) Plasmatocytes from the moth *Pseudoplusia includens* induce apoptosis of granular cells. J Insect Physiol 46:1565–1573

Philips JA, Rubin EJ, et al (2005) *Drosophila* RNAi screen reveals CD36 family member required for mycobacterial infection. Science 309:1251–1253

Poirie M, Frey F, et al (2000) *Drosophila* resistance genes to parasitoids: chromosomal location and linkage analysis. Proc Biol Sci 267:1417–1421

Price LS, Leng J, et al (1998) Activation of Rac and Cdc42 by integrins mediates cell spreading. Mol Biol Cell 9:1863–1871

Ramet M, Pearson A, et al (2001) *Drosophila* scavenger receptor CI is a pattern recognition receptor for bacteria. Immunity 15:1027–1038

Ramet M, Manfruelli P, et al (2002) Functional genomic analysis of phagocytosis and identification of a *Drosophila* receptor for *E. coli*. Nature 416:644–648

Ratcliffe NA, Gagen SJ (1976) Cellular defense reactions of insect hemocytes in vivo: nodule formation and development in *Galleria mellonella* and *Pieris brassicae* larvae. J Invert Pathol 28:373–382

Ratcliffe NA, Gagen SJ (1977) Studies on the in vivo cellular reactions of insects: an ultrastructural analysis of nodule formation in *Galleria mellonella*. Tissue Cell 9:73–85

Raymond K, Bergeret E, et al (2004) A screen for modifiers of RacGAP(84C) gain-of-function in the *Drosophila* eye revealed the LIM kinase Cdi/TESK1 as a downstream effector of Rac1 during spermatogenesis. J Cell Sci 117:2777–2789

Rickert P, Weiner OD, et al (2000) Leukocytes navigate by compass: roles of PI3Kgamma and its lipid products. Trends Cell Biol 10:466–473

Rizki MT, Rizki RM (1959) Functional significance of the crystal cells in the larva of *Drosophila melanogaster*. J Biophys Biochem Cytol 5:235–240

Rizki RM, Rizki TM (1979) Cell interactions in the differentiation of a melanotic tumor in *Drosophila*. Differentiation 12:167–178

Rizki TM, Rizki RM (1992) Lamellocyte differentiation in *Drosophila* larvae parasitized by *Leptopilina*. Dev Comp Immunol 16:103–110

Rogers SL, Wiedemann U, et al (2003) Molecular requirements for actin-based lamella formation in *Drosophila* S2 cells. J Cell Biol 162:1079–1088

Rottner K, Hall A, et al (1999) Interplay between Rac and Rho in the control of substrate contact dynamics. Curr Biol 9:640–648

Ruoslahti E (1996) RGD and other recognition sequences for integrins. Annu Rev Cell Dev Biol 12:697–715

Russo J, Dupas S, et al (1996) Insect immunity: early events in the encapsulation process of parasitoid (*Leptopilina boulardi*) eggs in resistant and susceptible strains of *Drosophila*. Parasitology 112:135–142

Sander EE, Klooster JP ten, et al (1999) Rac downregulates Rho activity: reciprocal balance between both GTPases determines cellular morphology and migratory behavior. J Cell Biol 147:1009–1022

Sanders LC, Matsumura F, et al (1999) Inhibition of myosin light chain kinase by p21-activated kinase. Science 283:2083–2085

Schmucker D, Clemens JC, et al (2000) *Drosophila* Dscam is an axon guidance receptor exhibiting extraordinary molecular diversity. Cell 101:671–684

Snapper SB, Takeshima F, et al (2001) N-WASP deficiency reveals distinct pathways for cell surface projections and microbial actin-based motility. Nat Cell Biol 3:897–904

Sorrentino RP, Carton Y, et al (2002) Cellular immune response to parasite infection in the *Drosophila* lymph gland is developmentally regulated. Dev Biol 243:65–80

Stark KA, Yee GH, et al (1997) A novel alpha integrin subunit associates with betaPS and functions in tissue morphogenesis and movement during *Drosophila* development. Development 124:4583–4594

Steiner H (2004) Peptidoglycan recognition proteins: on and off switches for innate immunity. Immunol Rev 198:83–96

Stroschein-Stevenson SL, Foley E, et al (2006) Identification of *Drosophila* gene products required for phagocytosis of *Candida albicans*. PLoS Biol 4:e4

Stuart LM, Deng J, et al (2005) Response to *Staphylococcus aureus* requires CD36-mediated phagocytosis triggered by the COOH-terminal cytoplasmic domain. J Cell Biol 170:477–485

Suetsugu S, Miki H, et al (1999) Identification of two human WAVE/SCAR homologues as general actin regulatory molecules which associate with the Arp2/3 complex. Biochem Biophys Res Commun 260:296–302

Svitkina TM, Borisy GG (1999) Arp2/3 complex and actin depolymerizing factor/cofilin in dendritic organization and treadmilling of actin filament array in lamellipodia. J Cell Biol 145:1009–1026

Svitkina TM, Verkhovsky AB, et al (1998) Plectin sidearms mediate interactions of intermediate filaments with microtubules and other components of the cytoskeleton. Biol Bull 194:409–410

Tanji T, Ohashi-Kobayashi A, et al (2006) Participation of a galactose-specific C-type lectin in *Drosophila* immunity. Biochem J 396:127–138

Timpl R, Brown JC (1994) The laminins. Matrix Biol 14:275–281

Vignjevic D, Yarar D, et al (2003) Formation of filopodia-like bundles in vitro from a dendritic network. J Cell Biol 160:951–962

Vuorio E, Crombrugghe B de (1990) The family of collagen genes. Annu Rev Biochem 59:837–872

Watanabe N, Kato T, et al (1999) Cooperation between mDia1 and ROCK in Rho-induced actin reorganization. Nat Cell Biol 1:136–143

Watson FL, Puttmann-Holgado R, et al (2005) Extensive diversity of Ig-superfamily proteins in the immune system of insects. Science 309:1874–1878

Weis VM, Small AL, et al (1996) A peroxidase related to the mammalian antimicrobial protein myeloperoxidase in the *Euprymna–Vibrio* mutualism. Proc Natl Acad Sci USA 93:13683–13688

Werner T, Liu G, et al (2000) A family of peptidoglycan recognition proteins in the fruit fly *Drosophila melanogaster*. Proc Natl Acad Sci USA 97:13772–13777

Wiegand C, Levin D, et al (2000) Monoclonal antibody MS13 identifies a plasmatocyte membrane protein and inhibits encapsulation and spreading reactions of *Manduca sexta* hemocytes. Arch Insect Biochem Physiol 45:95–108

Williams MJ, Wiklund ML, et al (2006) Rac1 signalling in the *Drosophila* larval cellular immune response. J Cell Sci 119:2015–2024

Wilson R, Ainscough R, et al (1994) 2.2 Mb of contiguous nucleotide sequence from chromosome III of *C. elegans*. Nature 368:32–38

Xu J, Wang F, et al (2003) Divergent signals and cytoskeletal assemblies regulate self-organizing polarity in neutrophils. Cell 114:201–214

Yu XQ, Kanost MR (2004) Immulectin-2, a pattern recognition receptor that stimulates hemocyte encapsulation and melanization in the tobacco hornworm, *Manduca sexta*. Dev Comp Immunol 28:891–900

Yu XQ, Gan H, et al (1999) Immulectin, an inducible C-type lectin from an insect, *Manduca sexta*, stimulates activation of plasma prophenol oxidase. Insect Biochem Mol Biol 29:585–597

Yu XQ, Ling E, et al (2006) Immulectin-4 from the tobacco hornworm *Manduca sexta* binds to lipopolysaccharide and lipoteichoic acid. Insect Mol Biol 15:119–128

Zettervall CJ, Anderl I, et al (2004) A directed screen for genes involved in *Drosophila* blood cell activation. Proc Natl Acad Sci USA 101:14192–14197

Chapter 5
Immune Reactions in the Vertebrates' Closest Relatives, the Urochordates

Konstantin Khalturin(✉), Ulrich Kürn, and Thomas C.G. Bosch

1 Introduction	100
2 Urochordates are at the Root of Vertebrate Evolution	100
3 Natural History and Ecology of Urochordates	101
4 Immunity in Urochordates	102
4.1 Antimicrobial Peptides from Urochordates	103
4.2 Allorecognition in Urochordates	104
4.3 Complement in Urochordates	106
4.4 Despite the Absence of MHC, Urochordate Blood Contains NK-Like Cells	107
5 Conclusion	108
References	108

Abstract Currently existing urochordates (sea squirts) and vertebrates diverged in evolution around 570 million years (myr) ago. Phylogenetic analyses based on molecular data provide compelling evidence that these animals are the closest living relatives of vertebrates. Urochordares, therefore, are of critical importance for understanding the origin of vertebrate immune system. For a number of species a large body of molecular data is now available. An extensive EST project and the draft genome sequences of *Ciona intestinalis* and *C. savignyi* allow rapid "in silico" searches for immunorelevant molecules. Recent data convincingly demonstrate that urochordates possess nearly full repertoire of vertebrate innate immune system, but totally lack most of the components of an adaptive immunity such as MHC, TCRs and antibodies. In this review we show that knowledge of immunity in lower vertebrate and invertebrate species is now rapidly increasing. Elucidating the details of the origin of the immune systems from a comparative point of view in vertebrate's closest relatives, may finally lead to a better understanding of our own immune system.

Zoological Institute, Christian-Albrechts-University, Olshausenstrasse 40,
24098 Kiel, Germany, *kkhalturin@zoologie.uni-kiel.de*

1 Introduction

Immune systems are generally characterized by their ability to distinguish between self and non-self cells, tissues, or molecules, and to eliminate the non-self (Janeway et al., 2005). While successful strategies for the detection and elimination of pathogens are present at all levels of animal evolution, the possibility of fusion or contamination with cells from one's own species still exists in many marine metazoans (e.g., Porifera, Cnidaria, tunicates) but has practically disappeared in vertebrates. The immune system in vertebrates is a complex and intricate system that can recognize non-self and provide protection from a wide variety of pathogens. While there is a high degree of interconnectivity between its components, the immune system can be loosely divided into two subsystems, the innate and the adaptive immune systems. Innate immunity is a non-specific, inducible response to pathogens. It is immediate in action, yet short-lived. The adaptive immune system, consisting of antibodies, B cells, T cells, and the major histocompatibility antigens, is much more specific, but takes longer to activate. It also features immunological memory and can augment itself to respond more quickly and with greater specificity to future infections of similar pathogens. Innate immunity is present in all phyla, whereas adaptive immunity is present only in jawed vertebrates.

There is a conspicuous paucity of evidence for a gradual transition from the innate immune system of invertebrates to the recombinatorial immune system of vertebrates. Astonishingly, the adaptive immune system appeared quite suddenly, around 450 million years (myr) ago with the emergence of the gnathostomes. Where the genes for the components of the adaptive immune system came from is a mystery. These genes seem to have somehow "jumped" into the genome of a jawed vertebrate about 450 myr ago (Janeway et al. 2005).

In this review we show that knowledge of immunity in lower vertebrate and invertebrate species is now increasing rapidly. Thus, elucidating the details of the origin of the immune systems from a comparative point of view in the vertebrates' closest relatives may finally lead to a rational explanation of immunology's "big bang" – and add to the understanding of our own immune system.

2 Urochordates are at the Root of Vertebrate Evolution

Urochordates (tunicates), cephalochordates (lancelets), and vertebrates (including lamprey and hagfish) constitute the three extant groups of chordate animals (Fig. 1). Phylogenetic analyses based on molecular data provide compelling evidence that urochordates represent the closest living relatives of vertebrates (Delsuc et al. 2006). Currently existing urochordates and vertebrates diverged in evolution around 570 myr ago. Urochordates, therefore, are of critical importance for understanding the origins of vertebrates. They comprise approximately 2800–3000 species and, beside their phylogenetic position at the root of vertebrate evolution, are valuable objects because of the availability of a large body of molecular data.

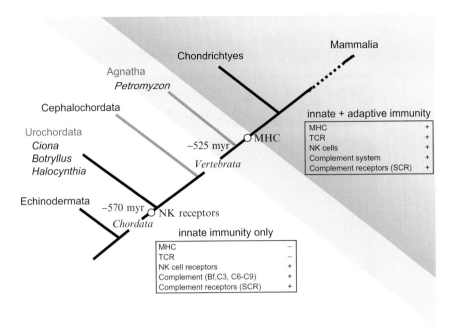

Fig. 1 Urochordates separated from the predecessors of a vertebrate lineage around 570 million years (*myr*) ago and share many features of the innate immune system with vertebrates. However, they do not possess the molecular machinery of the adaptive immune system found in jawed vertebrates (Gnathostomata). The emergence of a recombinatorial adaptive immune system featuring immunoglobulins, TCR and MHC took place sometime during the transition from Agnatha to Gnatostomata. A search for these cardinal elements of a recombinatorial immune system in lymphocyte-like cells of the sea lamprey, *Petromyzon marinus*, has not been fruitful, hence the agnathans remain a *gray zone* in the evolution of adaptive immunity in vertebrates. *Blue background*: phylogenetic groups with both adaptive and innate immune systems

A large expressed sequence tag (EST) project (Satou et al. 2002) and the draft genome sequences of *Ciona intestinalis* (Dehal et al. 2002) and *Ciona savignyi* allow extensive "in silico" searches for immunorelevant molecules.

3 Natural History and Ecology of Urochordates

Urochordates are exclusively marine animals, occurring worldwide in all oceans. The environmental variables most important in determining their distribution are temperature, salinity, light, and hydrodynamics (Lambert 2005). In their dimorphic life cycle, a non-feeding, short-living pelagic larva is followed by the sessile filter-feeding adult ascidian (see Fig. 2). The efficient filtration abilities may be the key to recent population explosions of various urochordate species in regions close to expanding human populations and coastal development, and may be tied to increasing

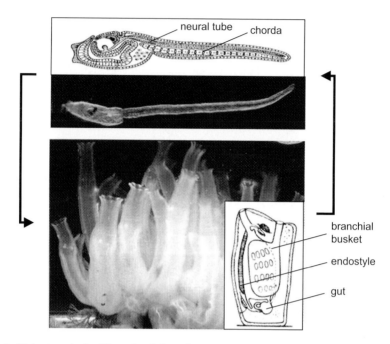

Fig. 2 Main steps in the life cycle of the solitary urochordate *Ciona intestinalis*. The larva of *Ciona* possesses chorda and other features of higher animals, while the morphology of the adult animal is very peculiar

bacterial densities (Lambert 2005). Most native species are very sensitive to pollution and eutrophication from anthropogenic activities (Naranjo et al. 1996). Similar to other marine invertebrates (see for example Bosch, this volume: *The path less explored: innate immune reactions in cnidarians*), the life-history strategies of many urochordates are affected by competition for space and food. Strategies for occupying space include the removal of neighbors by overgrowth or inhibition of settlement, but also by creating dense mono-specific aggregations.

4 Immunity in Urochordates

To prevent invasion or destruction by pathogens and to compete for resources with members of their own species, urochordates rely solely on innate immune mechanisms that include both humoral and cellular responses. Humoral immunity is characterized by antimicrobial agents present in the blood cells and plasma (Taylor et al. 1997; Tincu and Taylor 2004). Cellular immunity is based on cell defense reactions and is mediated by hemocytes, motile cells that phagocytize microbes and secrete soluble antimicrobial and cytotoxic substances into the hemolymph. All of these factors contribute to a self-defense system against invading microorganisms, which can number up to 10^6 bacteria/ml and 10^9 viruses/ml of seawater (Ammerman et al. 1984). The survival of urochordates and all other marine invertebrates in this environment suggests that their innate immune system is effective and robust.

4.1 Antimicrobial Peptides from Urochordates

In all organisms, antimicrobial peptides are a major component of the innate immune defense system providing an immediate and rapid response to invading microorganisms (Zasloff 2002). These peptides generally act by forming pores in microbial membranes or otherwise disrupting membrane integrity, which is facilitated by their amphiphilic structure (see Schröder, this volume: *Antimicrobial peptides as first-line effector molecules of the human innate immune system*).

Much of the work on antimicrobial peptides from the Urochordata has been performed on hemocytes of ascidians of the family Styelidae (for a review, see Tincu and Taylor 2004). Briefly, from the hemocytes of *Styela clava*, the clavanins (a family of four-helical, amphipathic, histidine-rich antimicrobial peptides that contain 23 amino acids and exhibit C-terminal amidation) were purified (Lee et al., 1997a, b; Menzel et al. 2002). Clavanins A to D resemble the magainins, well characterized antimicrobial peptides from the skin of *Xenopus laevis*. Synthetic clavanin A displays antimicrobial activity comparable with that of magainins and cecropins (Lee et al. 1997a, b). In addition to *Escherichia coli*, *Listeria monocytogenes*, and *Candida albicans*, clavanins are broadly effective against gram-positive bacteria, including *Staphylococcus aureus* (Menzel et al. 2002).

Two phenylalanine-rich antimicrobial peptides from *Styela clava*, styelin A and B, are effective against gram-negative and gram-positive bacterial pathogens of humans (Lee et al. 1997c). The styelins are highly basic polypeptides, encoded as prepropeptides, with a signal sequence and with cationic sequences in the mature protein counterbalanced by a polyanionic C-terminal extension in its precursor. Styelins also kill the marine bacteria *Psychrobacter immobilis* and *Planococcus citreus* in 0.4 M NaCl which approximates seawater salt concentrations (Lee et al. 1997c).

Plicatamide is a potently antimicrobial octapeptide from the blood cells of *Styela plicata* (Tincu et al. 2000, 2003). Wild-type and methicillin-resistant *Staphylococcus aureus* respond to plicatamide exposure with a massive potassium efflux that begins within seconds. Soon thereafter, treated bacteria largely cease consuming oxygen, and most become non-viable. Plicatamide forms cation-selective channels in model lipid bilayers composed of bacterial lipids.

Halocyamine A and B are two antimicrobial tetrapeptides isolated from the hemocytes of the ascidian *Halocynthia roretzi* (Azumi et al. 1990a). Halocyamine A was reported to inhibit the growth of yeast, *Escherichia coli* (Azumi et al. 1990a), and the marine bacteria *Achromobacter aquamarinus* and *Pseudomonas perfectomarinus* (Azumi et al. 1990b, c).

In addition to the halocyamines, antimicrobial peptides of 6.2 kDa and 3.4 kDa have been isolated from *Halocynthia roretzi*. The first, dicynthaurin, is composed of two 30-residue monomers without any sequence homology to previously identified peptides. Dicynthaurin's broad-spectrum activity includes *Micrococcus luteus*, *Staphylococcus aureus*, *Listeria monocytogenes*, *Escherichia coli*, and *Pseudomonas aeruginosa* but not *Candida albicans* (Lee et al. 2001a, b). The second, halocidin, has a mass of 3443 Da and is composed of two subunits containing 18 and 15 amino acid residues that are linked by a single disulfide bond (Jang et al. 2002). In antimicrobial assays halocidin was found to be active against methicillin-resistant *Staphylococcus aureus* and multidrug-resistant *Pseudomonas aeruginosa* (Jang et al. 2002).

Little is known about the signal transduction cascades involved in activating these antimicrobial peptides. Urochordates such as *Ciona intestinalis* (Azumi et al. 2003) and *Boltenia villosa* (Davidson and Swalla 2002), however, possess Toll-like receptors (TLRs) and share many components of the corresponding signal transduction cascades with vertebrates. Thus, similar to immune cells in higher organisms, cells in urochordates respond to microbes by the production and secretion of antimicrobial peptides and appear to use conserved signal transduction pathways.

4.2 Allorecognition in Urochordates

Despite their apparent "simplicity" and lack of obvious predecessors of an adaptive immune system, urochordates have several well documented forms of allorecognition. Two types of allogeneic recognition are observed. One is "colony specificity" in compound ascidians (see the review by Saito et al. 1994; Magor et al. 1999; Hirose 2003), and the other is "self-sterility" (blocked self-fertilization) which has been reported in colonial ascidians like Botryllids (see the review by Saito et al. 1994), solitary ascidians such as *Halocynthia roretzi* (see the review by Sawada 2002), and *Ciona intestinalis* (Murabe and Hoshi 2002).

Colony specificity was first observed by Bancroft, who found that, when two pieces of a single colony of *Botryllus schlosseri* come into contact with each other, they easily fuse to form a single colony (Bancroft 1903). Two pieces of different origin, however, never fuse after grafting. Oka and Watanabe (1957, 1960) analyzed this phenomenon in the Japanese botryllid ascidian *B. primigenus* and showed that this phenomenon was a type of self/non-self recognition. Genetic experiments in *B. primigenus* indicated that allorecognition is genetically controlled by a single highly polymorphic gene locus containing codominantly expressed alleles (Oka and Watanabe 1960; see the review by Saito et al. 1994). Publication of the hypothesis that the *Botryllus* fusibility/histocompatibility (FU/HC) locus may represent an ancestral form of vertebrate MHC (Scofield et al. 1982) attracted widespread attention, and *Botryllus* became one of the most widely studied models to unravel the evolutionary origin of MHC.

For an extended period of time it was expected that allorecognition in urochordates represented the ancestral state of the MHC-based histocompatibility reactions of vertebrates (Scoffield et al. 1982). However, with the sequencing of *Ciona intestinalis* and *C. savignyi* genomes it became clear that there are no MHC- or TCR-like molecules present at that level of animal evolution (Dehal et al. 2002; Azumi et al. 2003). This together with the characterization of putative allorecognition receptors in *Halocynthia* (Sawada et al. 2004; Ban et al. 2005) and *Botryllus* (De Tomaso and Weissman 2003; De Tomaso et al. 2005) provides convincing evidence that the allorecognition machinery in urochordates has nothing in common with the MHC-based histocompatibility reactions of vertebrates (De Tomaso et al. 2005; Klein 2006). The Fu/Hc receptor of *Botryllus schlosseri*, which appears to be responsible for the histocompatibility reactions in this colonial ascidian, has no direct homologs in vertebrates, nor in *Ciona intestinalis* (De Tomaso et al. 2005). Moreover, despite the fact that histocompatibility reactions in the form of allograft rejection were described in *Ciona* (Reddy et al. 1975), the whole *Botryllus* Fu/Hc locus itself has no clear syntenic region in the *Ciona* genome (De Tomaso et al. 2005), Thus, even within the urochordates, self/non-self discrimination systems may have branched off into a variety of unique and specialized systems during evolution.

Self-sterility is the second type of allogeneic recognition in urochordates. Fertilization is a precisely controlled process wherein the sperm binds to the vitelline coat (zona pellucida) of the egg in a species-specific manner and undergoes an acrosome reaction. Many species of solitary and colonial urochordates are hermaphrodites, and they release sperm and eggs simultaneously. They therefore have to evade self-fertilization. In *Halocynthia roretzi*, *Ciona intestinalis* and *Botryllus schlosseri*, studies of the genetic background of self-incompatibility have revealed several intriguing findings at the molecular level.

In ascidians which possess self-incompatibility, only a heterologous sperm is able to bind to the vitelline coat and the acrosome reaction proceeds only if the sperm recognizes the vitelline coat as non-self.

In *H. roretzi*, the main lytic target in the vitelline coat is HrVC120, a 120-kDa type I transmembrane protein with 12 EGF-like repeats at the N-terminus and a zona pellucida domain near the transmembrane region. HrVC120 has been shown to be the receptor in the vitelline coat which binds sperm (Sawada 2002). A comparative sequence analysis of HrVC120 disclosed (Sawada et al. 2004) that several amino acid residues in a restricted region were substituted at an individual level, with no identical sequences of HrVC120 among the ten individuals tested. Since the diversity in cDNA sequences was derived from genomic DNA polymorphism between different individuals (Sawada et al. 2004), HrVC120 is considered an important allorecognition molecule used during gamete interaction in urochordates. We recently obtained indirect support for that suggestion by discovering a group of related genes expressed in *Ciona* oocytes. In a screen designed to identify genes potentially involved in self-sterility in *C. intestinalis*, we identified a number of polymorphic genes encoding transmembrane proteins which are similar to HrVC120, with several EGF-like domains followed by a ZP domain (Kürn et al.

2007). All four *Ciona* genes are expressed in developing but not in mature oocytes. Thus, proteins with multiple EGF-like domains and ZP domains may play a crucial role in preventing self-fertilization in urochordates.

Interestingly, in *H. roretzi* the same monoclonal antibodies that prevent lysis between hemocytes from different individuals ("contact reaction") also prevent fertilization (Arai et al. 2001). The antigens recognized by these antibodies are present on the surface of hemocytes and the vitelline coat of the eggs. This is the first direct evidence that cell surface determinants responsible for allorecognition in urochordates may also be responsible for blocking self-fertilization (Arai et al. 2001).

4.3 Complement in Urochordates

The complement system is a cascade of serine proteases initiated by the detection of foreign agents, and it results in several different effector responses. In mammals, the cascade as a whole uses around 30 different proteins, which are all constituitively present in the blood serum.

Many components of the mammalian complement system can be traced back to urochordates. In the absence of antibodies, the activation pathway is of the lectin type and consists of mannose-binding lectins (MBLs), mannose associated serine proteases (MASPs), C3, and corresponding CR3/CR4 receptors present on macrophage-like cells (Kenjo et al. 2001; Sekine et al. 2001; Marino et al. 2002; Endo et al. 2003; for a review, see Fujita 2002). Thus, a lectin-based complement system is present in urochordates and functions in an opsonic manner. Obviously, this system has remained unchanged since its appearance at least 600 myr ago, well ahead of the emergence of adaptive immunity. From an evolutionary point of view, Fujita (2002) proposed that the primitive lectin pathway in innate immunity evolved into the classic pathway to serve as an effector system of adaptive immunity.

In mammals, the crusial step of complement function is the covalent attachement of serum protein C3 to the surface of autologous and bacterial cells and activation the complement cascade causing elimination of the "alien" cell by lysis and/or phagocytosis (Sahu and Lambris 2001). Attachment of C3 to normal autologous cells is prevented by the expression of "self" markers. The main markers of normal "self" in mammals are membrane cofactor protein (MCP/CD46), decay accelerating factor (DAF/CD55), and CD59, a member of the Ly6 family (Liszewski et al. 1991; Vanderpuye et al. 1992). These receptors function by interfering with complement components. CD46 and CD55 interact with C3 by preventing the formation of active C3 convertase. CD59 associates with C8 and/or C9 and thereby blocks the polymerization of C9 monomers required for the formation of lytic pores. As a final result, membrane attack complex (MAC) is not formed properly and cells are protected from complement-mediated lysis. Interestingly, the number of genes encoding proteins with multiple short consensus repeats (structurally similar to CD46, CD55, CR1, CR2) is much higher in number in ascidians than in mammals (Azumi et al. 2003). For poorly understood reasons (for a discussion, see Khalturin and Bosch 2007), this group of molecules has expanded considerably in the urochordate lineage.

The lytic pathway in urochordates consists of a large set of soluble proteins present in hemolymph, which have MAC/Perforin domains and an organization similar to that of the terminal vertebrate complement components, C6/C7/C8/C9 (Azumi et al. 2003).

Very recently, genomic views on the structure and evolution of the complement system in urochordates have been published (Nonaka and Yoshizaki 2004; Nonaka and Kimura 2006). Interestingly, most individual structural domains found in the urochordate complement components are specific to metazoans and are present in both invertebrates and vertebrates but are missing from yeast and plants. However, the unique domain architecture characteristic of vertebrate complement proteins is not present in protostomes, which led the authors (Nonaka and Yoshizaki 2004; Nonaka and Kimura 2006) to propose that the present-day complement system in vertebrates evolved by a unique combination of preexisting domains rather than by the invention of new domains.

In mammals the complement components are not variable between individuals, except for the C4 component in mice (Natsuume-Sakai et al. 1980). Surprisingly, in *Ciona* the number of genes encoding complement system components is greatly expanded compared with mammals (Azumi et al. 2003). In addition, some complement-related transmembrane proteins such as variable complement receptor-like 1 (vCRL1; Kürn et al. 2006) are highly variable between individuals. We have proposed elsewhere (Kürn et al. 2006; Khalturin and Bosch 2007) that these features allow us to explain the allorecognition reactions in *Ciona* by the involvement of complement-related receptors on the basis of the "missing self" concept: since cells in different individuals bear non-overlapping receptors and the corresponding ligands, cells within one individual are appropriately marked and are referred to as "self", while any cell of conspecific, but genetically different individuals are distinguished as "non-self".

In conclusion, there is accumulating evidence that early during chordate phylogeny the components of the complement system, in addition to their role in pathogen elimination, were involved in allorecognition. From an evolutionary point of view, the development of the adaptive immune system in the vertebrate lineage may have included the concomitant loss of involvement of complement molecules in allorecognition. This conclusion has an important implication. It indicates that urochordates use an allorecognition system which is unique and different from the one used by vertebrates (Khalturin and Bosch 2006).

4.4 Despite the Absence of MHC, Urochordate Blood Contains NK-Like Cells

In mammals, the immune system distinguishes self from non-self by three mechanisms: (a) molecules or metabolites typical for certain groups of pathogens are distinguished by pattern recognition molecules (e.g., Toll-like receptors for bacterial surface antigens), (b) T-cell receptors (TCR) that recognize non-self determinants presented by MHC, and (c) CTLD receptors and KIRs of natural killer (NK) cells that are used to screen for the presence of "self" determinates. Cells lacking these "self" markers are then eliminated.

When searching in the urochordate *Botryllus schlosseri* for genes which are altered in their expression in response to allogeneic contact, we discovered (Khalturin et al. 2003) a homolog of a mammalian NK cell receptor, CD94/NKR-P1. The *Botryllus* gene, termed bsCD94-1 (Khalturin et al. 2003) is down-regulated after allogenic contact. Immunostaining with a polyclonal antibody produced against the C-type lectin domain (CTLD) of bsCD94-1 indicates the presence of the receptor on the surface of a distinct group of *Botryllus* blood cells. Because the degree of structural conservation between the *Botryllus* bsCD94-1 protein and the vertebrate orthologs implies functional conservation, we proposed (Khalturin et al. 2003) that *B. schlosseri* blood cells carrying this receptor are mediators of allorecognition. Interestingly, close homologs of Botryllus bsCD94-1 are present in *Ciona intestinalis* (Khalturin et al. 2004). The *Ciona* protein has 45% identity to its *Botryllus* homolog and has the characteristic features of a type II transmembrane protein. It has even higher homology to mammalian CD94 and NKR-P1 than does the *Botryllus* homolog. Neither the ciCD94-1 protein function nor localization in *Ciona* is known. However, taking into consideration the high structural similarity to the family of NK receptors, one may speculate that ciCD94-1 participates in the NK cell-like cytotoxic activity described previously for *Ciona* blood cells (Parrinello et al. 1996).

Thus, key molecules of vertebrate NK cells are expressed in a subpopulation of urochordate blood cells. These cells, therefore, may be considered as ancestral NK cells.

5 Conclusion

The complexity of host defenses and recognition strategies in urochordates represents intricate processes that provide the backdrop for both the innate and adaptive immune systems of species that appeared much later. Understanding the strategies employed by our ancestors will undoubtedly provide novel insights leading to new approaches to modulate immune function in plants and animals.

Acknowledgements We are grateful to the Biological Station at Helgoland for supplying us with *Ciona intestinalis*. Supported by the Deutsche Forschungsgemeinschaft (grants BO 848/12,3 and KH 38/1–2 to T.C.G.B. and K.K.).

References

Ammerman JW, Fuhrman JA, Hagstrom A, Azam F (1984) Bacterioplankton growth in seawater. 1. Growth kinetics and cellular characteristics in seawater cultures. Mar Ecol Prog Ser 18:31–39

Arai M, Suzuki-Koike M, Ohtake S, Ohba H, Tanaka K, Chiba J (2001) Common cell-surface antigens functioning in self-recognition reactions by both somatic cells and gametes in the solitary ascidian *Halocynthia roretzi*. Microbiol Immunol 45:857–866

Azumi K, Yokosawa H, Ishi S (1990a) Haolcyamines: novel antimicrobial tetrapeptide-like substances isolated from the hemocytes of the solitary ascidian *Halocynthia roretzi*. Biochemistry 29:156–165

Azumi K, Yokosawa H, Ishii S (1990b) Presence of 3,4-dihydroxyphenylalanine containing peptides in hemocytes of the ascidian *Halocynthia roretzi*. Experientia 46:1020–1023

Azumi K, Yoshimizu M, Suzuki S, Ezura Y, Yokosawa H (1990c) Inhibitory effect of halocyamine, an antimicrobial substance from ascidian hemocytes, on the growth of fish viruses and marine-bacteria. Experientia 46:1066–1068

Azumi K, De Santis R, De Tomaso A, Rigoutsos I, Yoshizaki F, Pinto MR, et al (2003) Genomic analysis of immunity in a Urochordate and the emergence of the vertebrate immune system: 'waiting for Godot'. Immunogenetics 55:570–581

Ban S, Harada Y, Yokosawa H, Sawada H (2005) Highly polymorphic vitelline-coat protein HaVC80 from the ascidian, *Halocynthia aurantium*: structural analysis and involvement in self/nonself recognition during fertilization. Dev Biol 286:440–451

Bancroft FW (1903) Variation and fusion of colonies in compound ascidians. Proc Calif Acad Sci 3:137–186

Davidson B, Swalla BJ (2002) A molecular analysis of ascidian metamorphosis reveals activation of an innate immune response. Development 129:4739–4751

De Tomaso AW, Weissman IL (2003) Initial characterization of a protochordate histocompatibility locus. Immunogenetics 55:480–490

De Tomaso AW, Nyholm SV, Palmeri KJ, Ishizuka KJ, Ludington WB, Mitchel K, Weissman IL (2005) Isolation and characterization of a protochordate histocompatibility locus. Nature 438:454–459

Dehal P, Satou Y, Campbell RK, Chapman J, Degnan B, DeTomaso A, Davidson B, Di Gregorio A, Gelpke M, Goodstein DM, et al (2002) The draft genome of *Ciona intestinalis*: insights into chordate and vertebrate origins. Science 298:2157–2167

Delsuc F, Brinkmann H, Chourrout D, Philippe H (2006) Tunicates and not cephalochordates are the closest living relatives of vertebrates. Nature 439:965–968

Endo Y, Nonaka M, Saiga H, Kakinuma Y, Matsushita A, Takahashi M, Matsushita M, Fujita F (2003) Origin of mannose-binding lectin-associated serine protease (MASP)-1 and MASP-3 involved in the lectin complement pathway traced back to the invertebrate, *Amphioxus*. J Immunol 170:4701–4707

Fujita T (2002) Evolution of the lectin-complement pathway and its role in innate immunity. Nat Rev Immunol 2:346–353

Hirose E (2003) Colonial allorecognition, hemolytic rejection, and viviparity in botryllid ascidians. Zool Sci 20:387–394

Janeway CA Jr, et al (2005) Immunobiology: the immune system in health and disease, 6th edn. Garland, New York

Jang W, Kim K, Lee Y, Nam M, Lee I (2002) Halocidin: a new antimicrobial peptide from hemocytes of the solitary tunicate, *Halocynthia aurantium*. FEBS Lett 521:81–86

Kawamura K, Fujita H, Nakauchi M (1989) Concanavalin A modifies allo-specific sperm-egg interactions in the ascidian, *Ciona intestinalis*. Develop Growth Differ 31:493–501

Kenjo A, Takahashi M, Matsushita M, Endo Y, Nakata M, Mizuochi T, Fujita T (2001) Cloning and characterization of novel ficolins from the solitary ascidian, *Halocynthia roretzi*. J Biol Chem 276:19959–19965

Khalturin K, Bosch TCG (2007) Self/nonself discrimination at the basis of chordate evolution: limits on molecular conservation. Curr Opin Immunol 19:4–9

Khalturin K, Becker M, Rinkevich B, Bosch TCG (2003) Urochordates and the origin of natural killer cells: identification of a CD94/NKR-P1 related receptor in blood cells of *Botryllus*. Proc Natl Acad Sci USA 100:622–627

Khalturin K, Pancer Z, Cooper MD, Bosch TCG (2004) Recognition strategies in the innate immune system of ancestral chordates. Mol Immunol 41:1077–1087

Khalturin K, Kurn U, Pinnow N, Bosch TCG (2005) Towards a molecular code for individuality in the absence of MHC: screening for individually variable genes in the urochordate *Ciona intestinalis*. Dev Comp Immunol 29:759–773

Klein J (2006) The grapes of incompatibility. Dev Cell 10:2–4

Kürn U, Sommer F, Hemmrich G, Bosch TCG, Khalturin K (2006) Allorecognition in urochordates: identification of a highly variable complement receptor-like protein expressed in follicle cells of *Ciona*. Dev Comp Immunol 31:360–371

Kürn U, Sommer F, Bosch TC, Khalturin K (2007) In the urochordate Ciona intestinalis zona pellucida domain proteins vary among individuals. Dev Comp Immunol doi:10.1016/j.dci.2007.03.011

Lambert G (2005) Ecology and natural history of protochordates. Can J Zool 83:34–50

Lee I, Cho Y, Lehrer R (1997a) Effects of pH and salinity on the antimicrobial properties of clavanins. Infect Immun 65:2898–2903

Lee I, Zhao C, Cho Y, Harwig S, Cooper E, Lehrer R (1997b) Clavanins, helical antimicrobial peptides from tunicate hemocytes. FEBS Lett 400:158–162

Lee IH, Cho Y, Lehrer RI (1997c) Styelins, broad spectrum antimicrobial peptides from the solitary tunicate, *Styela clava*. Comp Biochem Physiol 118B:515–521

Lee I, Lee Y, Kim C, Hong C, Menzel LT, Boo L, Pohl J, Sherman M, Waring A, Lehrer R (2001a) Dicynthaurin: an antimicrobial peptide from hemocytes of the solitary tunicate, *Halocynthia aurantium*. Biochim Biophys Acta 1527:141–148

Lee I, Zhao C, Nguyen T, Menzel L, Waring A, Sherman M, Lehrer R (2001b) Clavaspirin, an antimicrobial and hemolytic peptide from *Styela clava*. J Peptide Res 58:445–456

Liszewski MK, Post TW, Atkinson JP (1991) Membrane cofactor protein (MCP or CD46): newest member of the regulators of complement activation gene cluster. Annu Rev Immunol 9:431–455

Magor BG, De Tomaso A, Rinkevich B, Weissman IL (1999) Allorecognition in colonial tunicates: protection against predatory cell lineages? Immunol Rev 167:69–79

Marino R, Kimura Y, De Santis R, Lambris JD, Pinto MR (2002) Complement in urochordates: cloning and characterization of two C3-like genes in the ascidian *Ciona intestinalis*. Immunogenetics, 53:1055–1064

Menzel L, Lee I, Sjostrand B, Lehrer R (2002) Immunolocalization of clavanins in *Styela clava* hemocytes. Dev Comp Immunol 26:505–515

Murabe N, Hoshi M (2002) Re-examination of sibling cross-sterility in the ascidian, *Ciona intestinalis*: genetic background of the self-sterility. Zool Sci 19:527–538

Naranjo SA, Carballo JL, Garcia-Gomez G (1996) Effects of environmental stress on ascidian populations in Algeciras Bay (southern Spain). Possible marine bioindicators? Mar Ecol Prog Ser 144:119–131

Natsuume-Sakai S, Kaidoh T, Nonaka M, Takahashi M (1980) Structural polymorphism of murine C4 and its linkage to H-2. J Immunol 1980 124:2714–2720

Nonaka M, Kimura A (2006) Genomic view of the evolution of the complement system. Immunogenetics 58:701–713

Nonaka M, Yoshizaki F (2004) Primitive complement system of invertebrates. Immunol Rev 198:203–215

Oka H, Watanabe H (1957) Colony specificity in compound ascidians as tested by fusion experiments (a preliminary report). Proc Jpn Acad 33:657–659

Oka K, Watanabe H (1960) Problems of colony specificity in compound ascidians. Bull Mar Biol Stn Asamushi 10:153–155

Parrinello N, Cammarata M, Arizza V (1996) Univacuolar refractile hemocytes from the tunicate Ciona intestinalis are cytotoxic for mammalian erythrocytes in vitro. Biol Bull 190:418–425

Reddy AL, Bryan B, Hildemann WH (1975) Integumentary allograft versus autograft reactions in Ciona intestinalis: a protochordate species of solitary tunicate. Immunogenetics 1:584–590

Sahu A, Lambris JD (2001) Structure and biology of complement protein C3, a connecting link between innate and acquired immunity. Immunol Rev 180:35–48

Saito Y, Hirose E, Watanabe H (1994) Allorecognition in compound ascidians. Int J Dev Biol 38:237–247

Satou Y, Yamada L, Mochizuki Y, Takatori N, Kawashima T, Sasaki A, Hamaguchi M, Awazu S, Yagi K, Sasakura Y, Nakayama A, Ishikawa H, Inaba K, Satoh N (2002) A cDNA resource from the basal chordate *Ciona intestinalis*. Genesis 33:153–154

Sawada H (2002) Ascidian sperm lysin system. Zool Sci 19:139–151

Sawada H, Tanaka E, Ban S, Yamasaki C, Fujino J, Ooura K, et al (2004) Self/nonself recognition in ascidian fertilization: vitelline coat protein HrVC70 is a candidate allorecognition molecule. Proc Natl Acad Sci USA 101:15615–15620

Scofield VL, Schlumpberger JM, West LA, Weissman IL (1982) Protochordate allorecognition is controlled by a MHC-like gene system. Nature 295:499–502

Sekine H, Kenjo A, Azumi K, Ohi G, Takahashi M, Kasukawa R, Ichikawa N, Nakata M, Mizuochi T, Matsushita M, Endo Y, Fujita T (2001) An ancient lectin-dependent complement system in an ascidian: novel lectin isolated from the plasma of the solitary ascidian, *Halocynthia roretzi*. J Immunol 167:4504–4510

Taylor SW, Kammerer B, Bayer E (1997) New perspectives in the chemistry and biochemistry of the tunichromes and related compounds. Chem Rev 97:333–346

Tincu JA, Taylor SW (2004) Antimicrobial peptides from marine invertebrates. Antimicrob Agents Chemother 48:3645–3654

Tincu JA, Craig AG, Taylor SW (2000) Plicatamide: a lead to the biosynthetic origins of the tunichromes? Biochem Biophys Res Commun 270:421–424

Tincu JA, Menzel LP, Azimov R, Sands J, Hong T, Waring AJ, Taylor SW, Lehrer RI (2003) Plicatamide, an antimicrobial octapeptide from *Styela plicata* hemocytes. J Biol Chem 278:13546–13553

Vanderpuye OA, Labarrere CA, McIntyre JA (1992) The complement system in human reproduction. Am J Reprod Immunol 27:145–155

Zasloff M (2002) Antimicrobial peptides of multicellular organisms. Nature 415:389–395

Chapter 6
Innate Immune System of the Zebrafish, *Danio rerio*

Con Sullivan and Carol H. Kim(✉)

1	Overview	114
2	Components of Innate Immunity	115
	2.1 General Description	115
	2.2 *Drosophila* Toll: Identification and Recognition of a Dually Functioning Pathway	116
	2.3 TLRs and TIR-Bearing Adaptor Proteins	117
3	Zebrafish as a Model for Infectious Disease and Innate Immune Responses	118
	3.1 Overview	118
	3.2 Forward and Reverse Genetics	119
	3.3 An Infectious Disease and Innate Immunity Model	120
4	NK-Like Cells	122
5	Additional Innate Immunity Receptors in Zebrafish	124
6	Zebrafish Phagocytes	126
7	Conclusion	126
	References	127

Abstract There has been a revolution in immunology in recent years that has transformed the paradigmatic underpinnings of vertebrate immunology to include the innate immune response. The utilization of basally diverging model systems, like the zebrafish, provides particular insight into the origins and evolution of vertebrate immunity. Investigations aimed at exposing the breadth and complexity of innate immunity using the zebrafish model system have uncovered a broad spectrum of mechanisms, both novel and conserved, that add depth to our understanding of how the immune system functions. Of particular significance is the fact that, during the first 4–6 weeks of development, the zebrafish relies upon innate immunity as its sole mechanism of defense. This unique characteristic, combined with the zebrafish model's inherent advantages including high fecundity, external development, and optical transparency during early development, make the zebrafish a particularly attractive model of study. The establishment

Department of Biochemistry, Microbiology, and Molecular Biology, University of Maine, Orono, ME 04469, USA, *carolkim@maine.edu* of bacterial and viral infectious disease models such as *Edwardsiella tarda* and snakehead rhabdovirus, respectively, as well as the addition of a wide range of reagents and techniques, including robust forward and reverse genetics approaches, have facilitated the zebrafish model's usage to study of a variety of innate immunity questions. Close examination of the zebrafish's innate immune system reveals a strong degree of sequence conservation in many of areas of study, including but not limited to pattern recognition receptors like the Toll-like receptors, their pathway components, and a variety of cytokines. Studies are currently underway to determine whether such sequence homology equates to functional homology. In addition, a variety of zebrafish genes encoding proteins of unique function are currently under study, including assorted lectins and novel immune type receptors. Close examination of these genes may provide needed insight into the evolutionary history of immunity in vertebrates. In its totality, the zebrafish is a vibrant and useful model system with the potential to offer immunologists alternative perspectives into the underlying mechanisms of vertebrate innate immunity, particularly during its early life, when innate immune responses are all that can be mustered.

1 Overview

The interplay between pathogenesis and immunity has shaped the course of natural history, influencing the population dynamics of organisms of all types and complexities. Our appreciation of these concepts has been influenced, in an historical sense, by the effects of disease on humankind. The linkage between pathogenesis and immunity can be traced in the written record to the Islamic physician Rhazes (Abu Bekr Mohammed ibn Zakariya al-Razi; 880–932 A.D.). In his *Treatise on the Smallpox and Measles*, Rhazes, the first to describe the clinical manifestations of smallpox, also noted that survivors of an initial infection acquired resistance to repeat infections (Silverstein 1989). Similar observations were recorded in the literature subsequently, but the mechanisms underlying this "acquired immunity" would not begin to be described accurately until the late 1800s when the modern field of immunology arose (deKruif 1926; Silverstein 1989, 2002; Tauber and Chernyak 1991). At that time, two competing schools of thought emerged. Ilya Mechnikov proposed that immunity was cell-mediated and dependent upon phagocytosis or cell engulfment (deKruif 1926; Silverstein 1989; Tauber and Chernyak 1991). This theory of cellular immunity was relegated for over 50 years in favor of the humoralist school, led by Paul Ehrlich, which argued that immunity was mediated by blood-borne factors, particularly antibodies and complement (deKruif 1926; Silverstein 1989; 2002; Tauber and Chernyak 1991). The humoralist school of immunology dominated the field, and theories abounded about the origins of antibody production. In 1959, MacFarlane Burnet offered his clonal selection theory, in which humoral responses were proposed to be mediated by cellular mechanisms (Burnet 1959). This theory reestablished the importance of cells in mediating immunity. The theory introduced the notion that, in a heterogeneous lymphocyte population, the activation of one antibody-bearing cell in that population by an antigen leads to its subsequent cloning so that additional cells bearing that particular antibody are made. Burnet's theory offered additional insight into how the immune system worked and became the framework upon which immunology proceeded. In his landmark essay *Approaching the asymptote? Evolution and revolution in immunology*, Charles Janeway argued that nonclonal cell-mediated immune responses, based upon what he termed "pattern recognition receptors," may be important for host defense. He recognized that these responses may constitute an essential aspect of innate immunity which, until that point in 1989, had not gained significant recognition or appreciation as a crucial component of the comprehensive immune response (Janeway 1989). His subsequent identification of such a pattern recognition receptor (Medzhitov et al. 1997) validated his earlier prediction. This finding of a human homolog of the fruitfly Toll receptor (known as hToll or Toll-like receptor-4) has indeed revolutionized immunology. As a result, the components underlying innate immune responses in mammals have garnered increasing attention and scrutiny.

The rapidly accelerating field of innate immunity is illuminating areas of immunology that heretofore were unknown or incomplete. As a result, the paradigm of mammalian host defense has been dramatically transformed to include concepts related to innate immunity. The advent and maturation of robust invertebrate and basally diverging vertebrate model systems, including the zebrafish, provide unique opportunities to examine the evolution of innate immunity. A fuller understanding of immunity in basally diverging vertebrates provides additional perspective into how the immune system functions in mammals.

2 Components of Innate Immunity

2.1 General Description

Innate immune responses are primordial and characteristic of all eukaryotic organisms, providing nonspecific, broad-spectrum host defense against a wide range of pathogens. Innate immunity constitutes the only mechanism of defense for eukaryotes at and "below" the agnathans, or jawless fishes: organisms ranging from the amoeba (Leippe 1995) to the lamprey (Alder et al. 2005) rely solely upon innate immune responses. Approximately 450 million years ago, antigen-specific, acquired immunity is thought to have originated with the introduction of a single transposable element encoding two recombination-activating genes (*RAG-1*, *RAG-2*), into the germ line of an ancestral, jawed, cartilaginous fish (Agrawal et al. 1998). It is theorized that this integration occurred in a primitive receptor exon, where the RAG proteins could induce the gene rearrangements necessary to spark the diversity and specificity associated with adaptive immune responses (Agrawal et al. 1998). These rearrangements manifested themselves in the immunoglobulins and T cell receptors now associated with B and T cells, respectively. For all organisms arising after the ancestral, jawed, cartilaginous fish, innate immunity serves a dual role: controlling and destroying pathogens and, if necessary, triggering a comprehensive immune response involving components of adaptive immunity.

Innate immunity describes a suite of broad-spectrum host defense mechanisms comprised of physical, chemical, cellular, and molecular elements. In recent years, the components that underlie innate immune responses have garnered increasing attention and appreciation. Foremost among these components are a family of evolutionarily conserved pattern recognition molecules collectively known as the Toll-like receptors (TLRs). TLRs, whose antecedent was originally defined in *Drosophila melanogaster* as Toll, possess the capacity to recognize conserved pathogen features like the lipopolysaccharides associated with bacteria or the double-stranded RNAs associated with viruses (Akira and Takeda 2004; Akira et al. 2001; Beutler 2004; Kopp and Medzhitov 2003; Takeda and Akira 2004; Ulevitch 2004). Upon recognition, these TLRs trigger a pro-inflammatory signal transduction cascade characterized by the induction of cytokines, antimicrobial peptides, and other immune factors. Each individual TLR has the unique capacity to recognize specific pathogen characteristics, thereby contributing to the diversity of responses that are generated. TLRs transduce the signals triggered by external stimuli through adaptor proteins characterized by the presence of Toll/interleukin-1 homologous region (TIR) domains (Akira et al. 2003). Functions have been assigned to four such adaptors: myeloid differentiation factor 88 (MyD88), TIR domain-containing adaptor protein (TIRAP), Toll-receptor-associated molecule (TRAM), and Toll-receptor-associated activator of interferon (TRIF; Akira and Takeda 2004; Akira et al. 2003; Beutler 2004; Kopp and Medzhitov 2003; Takeda and Akira 2004; Ulevitch 2004). These adaptor molecules propagate signals through specific interactions with other cytosolic proteins, which ultimately lead to the translocation of various transcription factors into the nucleus where they affect the transcription of specific genes.

2.2 Drosophila *Toll: Identification and Recognition of a Dually Functioning Pathway*

The dominant maternal effect mutation *Toll* was discovered as part of a screen to identify genes important for the establishment of *Drosophila* dorsal/ventral polarity (Anderson and Nusslein-Volhard 1984a). *Toll* was determined to be one of several genes present as maternal mRNA in *Drosophila* (Anderson and Nusslein-Volhard 1984b), serving to promote ventralization as the developmental polarity progressed (Anderson and Nusslein-Volhard 1984b; Anderson et al. 1985a, b). Its effects were shown to be the result of NF-κB-like transcription factors like Dorsal (Steward 1987). Close examination of the cytoplasmic domains revealed strong amino acid similarities between *Drosophila* Toll and mammalian interleukin 1 receptors (IL-1R; Heguy et al. 1992; Schneider et al. 1991). The strong similarities between Toll and

IL-1R as well as between NF-κB and Dorsal led to the realization that Toll may play a role in immunity (Lemaitre 2004). Lemaitre et al. (1996) established the link between Toll protein activation and upregulation of the antifungal peptide gene *drosomycin* via Dorsal translocation. Another protein, IMD, which had been shown to be important for the expression of antibacterial peptides (Lemaitre et al. 1995), was determined to regulate upregulation of the antibacterial peptide *diptericin*. Further experimentation revealed a capacity to discriminate among microorganisms and an ability to tailor specific immune responses to specific types of infections (Lemaitre et al. 1997). These findings provided important insight into innate immunity and became the template upon which investigations into mammalian TLR signal transduction were based.

2.3 TLRs and TIR-Bearing Adaptor Proteins

The TLRs are a family of evolutionarily-conserved, pattern-recognition molecules that function as essential components of innate immunity (Akira and Takeda 2004; Akira et al. 2001, 2003; Beutler 2004; Kopp and Medzhitov 2003; Takeda and Akira 2004; Ulevitch 2004). Vertebrate TLRs have been identified from zebrafish (Jault et al. 2004; Meijer et al. 2004; Phelan et al. 2005c) to humans (Akira and Takeda 2004; Akira et al. 2001, 2003; Beutler 2004; Kopp and Medzhitov 2003; Takeda and Akira 2004; Ulevitch 2004), with each possessing unique capacities to bind specific ligands ranging from the lipopolysaccharides (LPSs) characteristic of Gram-negative bacterial infections to the double-stranded (ds)RNAs characteristic of viral infections. Although divergent in their capacity to bind specific ligands, all TLRs described to date have in common a leucine-rich extracellular domain, a transmembrane domain, and a TIR (Toll-IL1 receptor) cytoplasmic domain. Functional TIR domains are necessary for the signal transduction events triggered by ligand binding. The absence of functional TLR TIR domains, as shown with the Lps^d point mutation (P712H; Hoshino et al. 1999; Poltorak et al. 1998), results in a failed initiation of a signal transduction cascade.

TLRs transduce extracellular signals through their TIR domains by homotypic protein–protein interactions with cytoplasmic adapter molecules also bearing TIR domains (Kopp and Medzhitov 2003). Perhaps the best described TIR-bearing adapter molecule is MyD88. The binding of a ligand to a TLR leads to the recruitment of MyD88, which is capable of interacting with a TLR's TIR domain with its own carboxy-terminus TIR domain (Medzhitov et al. 1998). From this position, MyD88 can interact through its amino-terminus death domain with the death domains of interleukin-1-receptor-associated kinases (IRAKs; Akira and Takeda 2004). IRAKs are a family of serine-threonine kinases that typically drive the signal through tumor-necrosis-factor-receptor-associated-factor-6 (TRAF6). According to one model (Kanayama et al. 2004), TRAF6 oligomerizes and becomes polyubiquitinated through its interactions with a dimeric ubiquitin-conjugating enzyme complex consisting of Ubc13 and Uev1A (Deng et al. 2000). Polyubiquitin chains are synthesized through lysine 63 of ubiquitin (Deng et al. 2000), and these chains are recognized and bound by the zinc finger domains of TAB2 or TAB3 (Kanayama et al. 2004). TAB2 or TAB3 recruits TAK1 to this large signaling complex, where TAK1 is activated. Activated TAK1 phosphorylates IKKb, which is complexed with IKKα and NEMO. This complex then phosphorylates an IkB/NF-κB complex that leads to its polyubiquitination and degradation. NF-κB is then liberated and can translocate into the nucleus to activate proinflammatory cytokines and antimicrobial peptides, and other immune factors are upregulated (Kanayama et al. 2004). This TRAF6–TAK1 complex is also responsible for the activation of MAP kinases and AP1 (Akira and Takeda 2004).

In addition to MyD88, other TIR-bearing adapter molecules have been identified and at least partially described, including TIRAP (or Mal), TRAM (or TICAM-2 or TIRP), SARM and TRIF (or TICAM-1; Beutler 2004). Such MyD88-independent signal pathways have recently gained greater appreciation. TIR domain-containing adaptor protein (TIRAP), for instance, has been implicated in the mediation of TLR4-to-IFN-β signaling (Fitzgerald et al. 2001; Henneke and Golenbock 2001; Horng et al. 2002; Horng et al. 2001; Shinobu et al. 2002; Takeda and Akira 2004; Vogel and Fenton 2003; Yamamoto et al. 2002a) as well as TLR4 to IL-6 signaling (Schilling et al. 2002). TIR domain-containing adaptor inducing IFN-β (TRIF), an adaptor molecule described in greater detail in following sections, plays an essential role in the TLR3-to-IFN-β signaling pathway (Oshiumi et al. 2003a; Yamamoto et al. 2002b) and may also be involved in TLR2 (Yamamoto et al. 2002b) and TLR4 signaling (Oshiumi et al. 2003b). The TRIF-related adaptor molecule (TRAM) has been shown to function within the TLR4 signaling pathway (Yamamoto et al. 2003) and the interleukin-1 signaling pathway (Bin et al. 2003). In addition, TRAM interacts with TRIF to facilitate the TLR4 to IFN-β pathway (Fitzgerald et al. 2003; Oshiumi et al. 2003b). The adaptor molecule SARM has not been characterized in mammals. However, its *Caenorhabditis elegans* ortholog, TIR-1, has been implicated as being part of a novel signaling pathway involved in the upregulation of antimicrobial genes following exposure to the fungus *Drechmeria coniospora* (Couillault et al. 2004).

3 Zebrafish as a Model for Infectious Disease and Innate Immune Responses

3.1 Overview

The zebrafish model system continues to evolve as an effective tool within the fields of biomedicine, as researchers continue to exploit its unique advantages to address their specific questions. Over time, the zebrafish system has migrated from the exclusive province of embryology to other fields, including genetics, toxicology, and immunology. The obvious advantages of external fertilization, large clutches, optical clarity during development, short generation times, and rapid development of organ primordia that embryologists exploited have been augmented with additional tools and techniques. These include: a variety of isogenic lines that have been bred; molecular, forward, and reverse genetics techniques that have been developed; a genome project that is nearly complete; and a broad range of reagents that are now available. As a result, numerous human disease models have been developed using the zebrafish, and these have proven extremely beneficial to understanding complex problems and processes (Fishman 2001; Traver et al. 2003; Trede et al. 2004). These are particularly useful because of the conserved synteny shown between the human and zebrafish genomes (Barbazuk et al. 2000; Dehal and Boore 2005).

3.2 Forward and Reverse Genetics

One of the most useful methods for determining gene function involves "knocking out" or disrupting the unknown coding sequence in an animal model and subsequently observing which structure or function is deleted. Knockout mice with distinct gene deletions are valuable tools for in vivo analysis of the immune response against infectious agents. These studies make possible the elucidation of the importance of different arms of the immune system to fight infection. Although there are numerous advantages to using the zebrafish system, including clear embryos, external fertilization, short generation times, and large clutches per mating, chromosomal integration is rare, and there are currently no widely available methods for generating gene knockouts. An alternative strategy for loss-of-function studies is large-scale *N*-ethyl-*N*-nitrosourea (ENU) chemical mutagenesis. In these screens, however, mutants of unknown molecular nature are isolated. Screening this large collection for a mutant corresponding to the gene under study, through mapping and linkage analysis, would require excessive effort and more than a small amount of luck.

An alternative approach is to inactivate the mRNA products of the gene by applying antisense methods. The drawback to this method is that the traditional antisense oligonucleotides are nuclease-sensitive, have low aqueous solubility, and require RNase H to function. RNA interference methods have given mixed results, but reports of disruptive effects have dissuaded many researchers from this approach (Liu et al. 2005; Oates et al. 2000; Zhao et al. 2001). To overcome these issues, alternative techniques have been devised. For example, a ribozyme directed against a zebrafish mRNA expression product of a targeted gene, *no tail*, has demonstrated the feasibility of downregulation or "knockdown" of the expression of targeted genes (Xie et al. 1997).

In a new technique developed for zebrafish, an alternative type of antisense oligonucleotide – a morpholino – provides substantially improved results, with efficiencies approaching 100% (Nasevicius and Ekker 2000). Morpholino oligonucleotides (MO) are non-ionic and function by an RNase H-independent mechanism. MO are assembled by joining up to 25 morpholino bases in a specific order by non-ionic phosphorodiamidate linkages. The cloning of zebrafish coding sequences, followed by their knockdown with a MO, may provide a rapid method for identifying their function. Because only knowledge of short sequences of the target mRNA is necessary for the design of a MO, only minimal sequence information needs to be determined for zebrafish.

With the advent of MO technology, zebrafish knockdowns for specific gene expression can now be created. In addition, combinations of MO may be microinjected simultaneously into a single developing embryo to determine the effects of double or even triple knockdowns. MO can be introduced into cells in vitro or into the developing embryos of zebrafish, where they block the translation of targeted mRNA transcripts. Innovative antisense RNA technologies have been developed to block splicing at specific exon–intron junctions, thereby altering protein products. MO have proven efficient at splice blocking in the zebrafish (Draper et al. 2003; Yan et al. 2002). Draper (2003) demonstrated that an MO directed against the exon 3 splice donor junction of zebrafish *fgf8* could block its splicing and result in the exclusion of exon 3 from the transcript. An MO directed against the splice donor junction of exon 1 of zebrafish *sox9* demonstrated a similar capacity to block splicing (Yan et al. 2002). These studies demonstrate the utility of MO antisense technology for blocking pre-mRNA splicing and altering gene products and functions.

A novel gene knockdown technique known as *t*argeting *i*nduced *l*ocal *l*esions *in g*enomes (TILLING), which was originally used to create gene knockouts in the mustard plant *Arabidopsis* (McCallum et al. 2000), was recently used in zebrafish (Wienholds et al. 2003). Zebrafish males are mutagenized with ENU and outcrossed with wild-type females to create a library of first filial generation fish (F_1). Single nucleotide polymorphisms from PCR products of specific genes are detected through cleavage by the celery enzyme CEL-1, which recognizes and cuts heteroduplexes 3′ to the mutation. Products are subsequently separated by denaturing polyacrylamide gel electrophoresis and those individuals possessing point mutations are identified and confirmed by sequencing.

3.3 An Infectious Disease and Innate Immunity Model

Recently, the zebrafish was established as a model for infection and immunity (Davis et al. 2002; Menudier et al. 1996; Neely et al. 2002; Phelan et al. 2005b; Pressley et al. 2005; Prouty et al. 2003; van der Sar et al. 2003, 2006). The zebrafish is particularly useful because each point in its development can be easily exploited to learn important information about the immune system. During the first four days of development, the zebrafish exhibits no adaptive immunity markers (Traver et al. 2003). Four days post-fertilization, *rag1* and *rag2* expression is noted (Willett et al. 1997), and zebrafish begin to develop the T and B cells necessary to mount adaptive immune responses (Danilova and Steiner 2002; Danilova et al. 2004). It has been shown that, while some adaptive immunity markers are present early in the developing larvae, the zebrafish requires 4–6 weeks to achieve a fully functional adaptive immune response (Lam et al. 2004). With the development of viral and bacterial infection models (Davis et al. 2002; Menudier et al. 1996; Neely et al. 2002; Phelan et al. 2005b; Pressley et al. 2005; Prouty et al. 2003; van der Sar et al. 2003), it is now possible to study innate immune responses exclusively in these first weeks of development and thereby examine the role of innate immunity without the complications of the adaptive immune response.

3.3.1 Bacterial Models

Several bacterial models of infection for zebrafish have been described and will prove useful in the study of zebrafish immunity (Davis et al. 2002; Menudier et al. 1996; Neely et al. 2002; Pressley et al. 2005; Prouty et al. 2003; van der Sar et al. 2003). For example, our laboratory established an effective, bacterial immersion infection model using *Edwardsiella tarda*, a Gram-negative rod (Pressley et al. 2005). Acute infection of both embryos and adults was noted through histopathology and cumulative percent mortalities. Additionally, upregulation of transcripts for the proinflammatory cytokines IL-1β and TNF-α was observed. Infection studies with strains of *Streptococcus* (Neely et al. 2002), *Mycobacterium* (Davis et al. 2002; Prouty et al. 2003), *Salmonella* (van der Sar et al. 2003), and *Listeria* (Menudier et al. 1996) have also been reported, but these infection schemes are limited to injection rather than immersion.

3.3.2 Viral Models

Our laboratory was also the first to demonstrate that zebrafish at varying stages of development were susceptible to lethal infection upon immersion challenge by snakehead rhabdovirus (SHRV), and that such infections could elicit potent antiviral responses throughout development, as measured by zebrafish type I IFN and Mx transcript levels (Phelan et al. 2005b). Other laboratories have shown that zebrafish can be infected with infectious hematopoietic necrosis virus (IHNV), infectious pancreatic necrosis virus (IPNV), and spring viremia of carp virus (SVCV; LaPatra et al. 2000; Sanders et al. 2003). Each of these infection models was limited to adult fishes. Unlike the SHRV model, infection of developing zebrafish from embryonic through juvenile stages was not described. IHNV and IPNV were able to replicate within the zebrafish, but these fish exhibited no mortalities (LaPatra et al. 2000). SVCV was shown to induce pathology consistent with viral infection in zebrafish adults, but these infections occurred at temperatures between 15 °C and 20 °C, well below the optimal temperature for maintaining zebrafish (28 °C; Sanders et al. 2003). As a result, zebrafish needed to be acclimated to lower temperatures and infections took longer to occur.

The establishment of infectious disease models in zebrafish has made it possible to assay mechanisms of host defense. Zebrafish homologs of mammalian TLRs and their pathway components have been identified by *in silico* analysis, and many have been partially cloned (Jault et al. 2004; Meijer et al. 2004; Phelan et al. 2005a). Phylogenetic analyses infer strong conservation amongst TLRs and their pathway components, from fishes to mammals (Iliev et al. 2005; Roach et al. 2005). Very little, however, has been done to characterize their function. Our laboratory characterized full-length zebrafish TLR3, IRAK4, and TRAF6 orthologs and assayed the effects of infection on their expression at different stages of development (Phelan et al. 2005a). All of the genes were upregulated in response to infection by *E. tarda*, but only TLR3 and TRAF6 expression were activated upon SHRV infection. The results demonstrated that a robust TLR-mediated antibacterial and antiviral response could be triggered upon infection. These findings were bolstered by a recent investigation into the role of zebrafish MyD88 in mediating response to infection. A MO-mediated gene knockdown of MyD88 was shown to disrupt clearance of *Salmonella enterica* serovar *Typhimurium* Ra bacteria (van der Sar et al. 2006), demonstrating an important role for MyD88-dependent signaling in zebrafish.

Homologs for a variety of mammalian cytokines have been identified in zebrafish, and some have begun to be characterized. These include IL-1b (Pressley et al. 2005), TNF-α (Praveen et al. 2006a; Pressley et al. 2005), and type I IFN (Altmann et al. 2003; Robertsen 2006), as have been described, as well as IL-10 (Zhang et al. 2005), IL-11 (Huising et al. 2005), IL-15 (Bei et al. 2006), IL-22 (Igawa et al. 2006), IL-26 (Igawa et al. 2006) and IFN-γ (Igawa et al. 2006; Robertsen 2006). In addition, a variety of chemokines have been identified (Baoprasertkul et al. 2005) and, based upon an extensive genome analysis, up to 46 CC chemokines in zebrafish may exist (Peatman and Liu 2006). In other fishes, homologs for lymphotoxin-β (Kono et al. 2006), granulocyte colony-stimulating factors (Santos et al. 2006), and IL-18 (Huising et al. 2004) have been identified.

4 NK-Like Cells

Nonspecific cytotoxic cells (NCCs) are considered to be the evolutionary precursors of mammalian natural killer (NK) cells (Evans et al. 2001; Jaso-Friedmann et al. 2001; Shen et al. 2002). They are capable of lysing many of the same target cell lines as human NK cells (Carlson et al. 1985; Evans et al. 1984a, b, c, 1987; Graves et al. 1984). The monoclonal antibodies (mAbs) 5C6, 6D32, and 6D34, directed against NCCs, exhibited cross-reactivity with human NK cells (Harris et al. 1991). In addition, mAbs 5C6 and 6D32 inhibited the lysis of K-562 cells by human NK cells. Furthermore, the binding of mAbs 5C6 and 6D34 to NK cell targets stimulated the release of tumor necrosis factor (TNF) and T cell growth factor (TCGF), indicating that these antibodies were capable of simulating induction by ligands of known signal transduction events (Harris et al. 1991; Jaso-Friedmann et al. 1993). Overall, these findings indicated evolutionary conservation and led to the conclusion that NCCs were the evolutionary precursors of NK cells.

Despite these similarities, there are still unresolved questions about the evolutionary relationship between NCCs and NK cells. One distinction between NCCs and NK cells is the inability of NCCs to recycle at the same rate as mammalian cells (Evans et al. 1984a). Another limitation is the failure to identify the NCC receptor protein-1 (NCCRP-1) homolog on NK cells (Shen et al. 2002). The evidence that such a receptor exists is the aforementioned cross-reactivity of NK cells with mAb 5C6, a monoclonal antibody directed specifically against NCCRP-1 (Evans et al. 1988; Jaso-Friedmann et al. 1997). Perhaps most interesting from our perspective is the seemingly agranular characteristic of NCCs (Evans et al. 1984a). One of the defining properties of mammalian NK cells, and cytotoxic T lymphocytes (CTLs) is the presence of cytolytic granules containing perforin, granulysin or NK-lysin, and granzymes. These granules are essential to the exocytosis pathway that is now considered to be the primary mechanism of defense against intracellular pathogens, viruses, and tumors (Lieberman 2003). The conspicuous absence of detectable granules from NCCs and their presence in other fish NK-like cells and CTLs (Shen et al. 2002) draws into question NCC ontogeny and lymphoid development in fishes. Recently, NCCs from channel catfish were shown to possess a variety of granzyme-like proteases (Praveen et al. 2006b), including the granzyme-like serine protease CFGR-1 (Praveen et al. 2004). In addition, the expression of transcripts for perforin, granulysin, and serglycin in catfish NCCs were detected by RT-PCR (Praveen et al. 2006b). The presence of components essential to granule-mediated exocytosis in supposedly agranular NCCs led to speculation that NCCs may in fact contain very small granules below the detection threshold by standard microscopy (Praveen et al. 2004, 2006b). Indeed, it has been shown that CFGR-1, based upon its tryptase activity, is released from NCCs upon targeting HL-60 cells and that this targeted release can be correlated with an increase in cytotoxicity, as measured by chromium release from radiolabeled HL-60 cells (Praveen et al. 2006b).

Mammalian NK and CTL cells, and presumably NCCs (Praveen et al. 2006b), mediate their killing responses through "immunological synapses," or the interfaces that develop between the cytotoxic effector cell and the potential target (Davis 2002; Lieberman 2003). When an NK cell comes in contact with another cell, it becomes activated or inhibited, based on recognition of the other cell as "self" or "nonself" (McCann et al. 2002). Activated NK cells destroy their targets in one of two ways: by secretion of cytotoxic granule contents into the immunological synapse or by signal transduction through death receptor pathways. The former mechanism predominates over the latter in response to virally infected cells, intracellular pathogens, and tumor cells (Lieberman 2003). Secretion of cytolytic granule contents leads to the introduction of perforin and granzymes into the synaptic space and subsequently into the target cell where they can induce apoptosis. The mechanisms by which this occurs remain controversial, but perforin's essential role in apoptosis induced by granzymes is clear: perforin is required for the activity of the granzymes (Walsh et al. 1994), serine proteases characterized by their ability to disrupt plasma, mitochondria, and nuclear membranes and to induce chromatin condensation (Lieberman 2003). Granule exocytosis also releases members of the granulysin/NK-lysin family into the immunological synapse.

Five distinct cytotoxic cell lines from channel catfish peripheral blood leukocytes (PBLs) all failed to react with the NCCRP-1 monoclonal antibody 5C6 (Stuge et al. 2000). It is argued that the cytotoxic effector cells they identified represent populations distinct from NCCs and that these effectors reveal the possibility that there is a multitude of CTL and NK-like effectors in the fishes. Of the clonal lines they isolated, group IV is considered to be the most NK-like. These cells kill targets nonspecifically, are TCRα/β negative, and contain granules. These characteristics raise the possibility of additional fish NK-like cells with cytolytic functions also linked to a granule exocytosis pathway.

The inherent advantages of the zebrafish model system, and the development of viable viral and bacterial infection models, including those with expressing fluorescent markers, make its use in studying NK-like and other cytotoxic cells a very attractive proposition. Zebrafish homologs for genes encoding NCCRP-1, along with perforin, granzymes, and NK-lysin-like protein, have been identified. With these genes, it will be possible to identify cytotoxic cells in real-time and to characterize their involvement in host response to pathogen exposure and infection. In particular, studies instigated early in development when the zebrafish rely solely upon their innate immunity will be instructive.

5 Additional Innate Immunity Receptors in Zebrafish

In mammals, NK cells mediate their activities through inhibitory and activating receptors which distinguish between self and nonself based upon recognition of MHC class I and/or activating ligands (Lanier 2005). Inhibitory NK receptors contain cytoplasmic immunoreceptor tyrosine-based inhibitory motifs (ITIMs) that trigger a signal transduction cascade that results in the prevention or reduction of NK cell activation (Lanier 2005). Activating NK receptors utilize adaptor proteins containing immunoreceptor tyrosine-based activation motifs (ITAMs; Lanier 2005). Phosphorylation of ITAMs triggers an activation of the NK cell, leading to degranulation and upregulation of cytokines and chemokines and attack on the target cell (Lanier 2005). As described, fishes do not possess NK cells per se; rather, they possess NCCs and other NK-like cells that function similarly in mediating cytotoxicity. The presence of such cells in fishes present interesting questions related to origins of vertebrate immune system and lymphocytes in particular. Novel immune type receptors (NITRs) have been identified in fishes and are speculated to play a role in mediated cytotoxicity, perhaps through NK-like cells (Hawke et al. 2001; Strong et al. 1999; Yoder and Litman 2000; Yoder et al. 2001, 2004). Indeed, abundant expression of NITRs has been noted in NK-like cell line 3H9, which was derived from channel catfish (*Ictalurus punctatus*; Hawke et al. 2001). NITRs were originally described in the southern pufferfish (*Spheroides nephelus*; Strong et al. 1999) and have since been identified in the zebrafish (Yoder et al. 2001, 2004). They possess two extracellular Ig domains, a transmembrane region, and a cytoplasmic tail containing ITIMs (Hawke et al. 2001; Strong et al. 1999; Yoder and

Litman 2000; Yoder et al. 2001, 2004). At this stage, the role of NITRs as NK receptors in a comprehensive immune response to pathogens is largely speculative and based upon the presence of ITIM and ITAM domains. Through extensive in vivo assays, particularly those targeting and disrupting specific individuals or groups of NITRs, it will possible to ascertain the specific role each NITR plays in the overall innate immune response.

C-type lectins, through specific recognition domains and in a calcium-dependent manner, function as carbohydrate receptors capable of binding pathogenic sugars (Meyer-Wentrup et al. 2005). In mammals, C-type lectins are known to be expressed on a variety of cells including dendritic cells and antigen presenting cells and thus have been shown to participate in host defense (Meyer-Wentrup et al. 2005). C-type lectins, like NITRs, may possess ITAMs or ITIMs and are therefore assumed to participate in signal transduction cascades (Meyer-Wentrup et al. 2005). In fact, it has recently been shown that lectins like Dectin-1 (Rogers et al. 2005) and DC-SIGN (Geijtenbeek et al. 2003) can synergize with TLRs and function as co-receptors to modulate host responses to infections. Several C-type lectins have been identified in the zebrafish, including the mannose-binding lectins (Vitved et al. 2000) and the Group II immune-related, lectin-like receptors (Illrs; Panagos et al. 2006). Each of the Illrs is predicted to possess a C-type lectin-like domain, a coiled-coil domain, a transmembrane domain, and one or two ITIM or ITIM-like cytoplasmic domains (Panagos et al. 2006). The presence of these ITIM or ITIM-like sequences indicate a likely role in inhibiting downstream signaling; however, it has been observed that in two of these lectins, Illr3a and Illr3b, there is an arginine residue in the transmembrane domain (Panagos et al. 2006). The presence of this residue indicates that it may interact with adaptor proteins like DAP10 and, as a result, have an activating role (Panagos et al. 2006). It has been noted that these Group II Illrs contain ITIM and ITIM-like sequences and that this characteristic makes them similar to Group V Illrs, which have not been shown to exist in fishes. Group V Illrs are typically associated with NK cells. It is speculated these Group Illrs may function as receptors in fish NK-like cells as well as in other cell types (Panagos et al. 2006).

The galectins (formerly known as S-type lectins) constitute a family of evolutionarily conserved, β-galactosidase-binding lectins that are nearly ubiquitous in nature, having been isolated from a broad range of organisms and viruses (Ahmed et al. 2004; Rabinovich et al. 2002; Vasta et al. 2004). In mammals, galectins have been shown to play several roles in immune defense, including apoptosis, cell adhesion, inflammation, cytokine induction, and chemotaxis (Rabinovich et al. 2002). Three classes of galectins have been shown to exist: proto-type, chimera-type, and tandem-repeat-type (Ahmed et al. 2004; Barondes et al. 1994; Rabinovich et al. 2002; Vasta et al. 2004). Proto-type galectins possess a single carbohydrate recognition domain per subunit. Chimera-type galectins possess a single carbohydrate recognition domain linked to an N-terminal peptide. Tandem-repeat galectins possess two carbohydrate recognition domains connected by a linker peptide. To date, five galectins have been identified in zebrafish: three proto-type, one chimera-type, and one tandem-repeat (Ahmed et al. 2004; Vasta et al. 2004). While their roles have not been defined, the role of zebrafish galectins in host defense is under active investigation (Vasta et al. 2004).

A novel lectin, possessing a fucose recognition domain, has recently been identified in the striped bass (*Morone saxatilis*; Odom and Vasta 2006). Known as MsaFBP32, the gene encoding this F-type lectin was upregulated approximately three-fold 72 h post-injection with turpentine, as determined by Northern blotting (Odom and Vasta 2006). This response to an inflammatory challenge indicates a potential role for immune defense (Odom and Vasta 2006). Additional homologs have been identified from broad representation of organisms, and it is believed that these constitute a novel family of F-type lectins (Odom and Vasta 2006). They are noticeably absent from the genomes of amniotes (Odom and Vasta 2006).

6 Zebrafish Phagocytes

The zebrafish has been an extremely successful model organism for the study of hematopoiesis (Bennett et al. 2001; de Jong and Zon 2005; Thisse and Zon 2002). Despite the advances in our understanding of blood cell origins and formation, less is known about how these blood cells participate in innate immunity. In particular, macrophages, neutrophils, and dendritic cells are indispensable to innate immunity in higher vertebrates and presumably play a similar role in zebrafish. While macrophages and neutrophils have been characterized (Bennett et al. 2001; de Jong and Zon 2005; Thisse and Zon 2002), dendritic cells have not.

The importance of phagocytes to zebrafish innate immunity is bolstered by data showing that both adult and embryonic macrophages infected by *Mycobacterium marinum* phagocytosed the bacteria and formed granuloma-like aggregates during in vivo studies (Davis et al. 2002). Additionally, it was shown that *M. marinum* infection could disrupt the hematopoietic program, causing early microglia destined for the brain to be redirected. In *panther* mutants, which lack functional macrophage colony stimulating factor receptor (M-CSF-R), infection did not disrupt macrophage recruitment, indicating that such activity is M-CSF-R-independent (Davis et al. 2002). Further evidence of phagocytic function in zebrafish has been established by our laboratory through an in vivo, fluorescence-based respiratory burst assay designed to detect reactive oxygen species (Hermann et al. 2004). The data showed areas of respiratory burst colocalized with phagocytes, as detected by the uptake of the phagocyte-specific dye neutral red.

7 Conclusion

Basally diverging vertebrate model systems, such the zebrafish, provide excellent insight into the origins of innate immunity in vertebrates. The conservation of syntenies across species simplifies comparative genome analyses, enabling the identification and characterization of true zebrafish orthologs to known mammalian genes; however, it has become clear through our comparative analyses that syntenies are not always conserved and that novel innate immune mechanisms, involving additional TLRs, NITRs, and lectins, have evolved. Thus, it is becoming apparent that sequence conservation does not always equate to functional conservation. A fuller appreciation of the zebrafish model system, and its utility to mammalian innate immune research, will only be attained through a greater understanding of its underlying mechanisms. Studies directed at deciphering these mechanisms will provide broad evolutionary insight into which pathway components have retained a functional similarity to their mammalian homologs and which have diverged and/or neo-functionalized.

The bulk of the work done so far in zebrafish innate immunity has concentrated on the identification and characterization of orthologs and the development of tools and reagents, including forward and reverse genetics techniques like ENU mutagenesis, TILLING, and MO-mediated gene knockdowns. For the zebrafish model system to achieve its fullest potential, and thus serve as a worthy complement to the mammalian innate immunity models, these tools must be used in a way that exploits the zebrafish model system's inherent advantages: specifically high fecundity, external development, transparent chorions, and innate immunity as the sole mechanism of defense through the first 4–6 weeks of development. These techniques are now becoming routine in the zebrafish community and can be used in combination with immunity and pathogenesis tools like fluorescently labeled pathogens and infectious disease models to create unique approaches to the study of pathogenesis and innate immunity.

References

Agrawal A, Eastman QM, Schatz DG (1998) Transposition mediated by RAG1 and RAG2 and its implications for the evolution of the immune system. Nature 394:744–751

Ahmed H, Du SJ, O'Leary N, Vasta GR (2004) Biochemical and molecular characterization of galectins from zebrafish (*Danio rerio*): notochord-specific expression of a prototype galectin during early embryogenesis. Glycobiology 14:219–232

Akira S, Takeda K (2004) Toll-like receptor signalling. Nat Rev Immunol 4:499–511

Akira S, Takeda K, Kaisho T (2001) Toll-like receptors: critical proteins linking innate and acquired immunity. Nat Immunol 2:675–680

Akira S, Yamamoto M, Takeda K (2003) Role of adapters in Toll-like receptor signalling. Biochem Soc Trans 31:637–642

Alder MN, Rogozin IB, Iyer LM, Glazko GV, Cooper MD, Pancer Z (2005) Diversity and function of adaptive immune receptors in a jawless vertebrate. Science 310:1970–1973

Altmann SM, Mellon MT, Distel DL, Kim CH (2003) Molecular and functional analysis of an interferon gene from the zebrafish, *Danio rerio*. J Virol 77:1992–2002

Anderson KV, Nusslein-Volhard C (1984a) Genetic analysis of dorsal–ventral embryonic pattern in *Drosophila*. In: Malacinski GM, Bryant SV (eds) Pattern formation: a primer in developmental biology. Macmillan, New York, pp 269–289

Anderson KV, Nusslein-Volhard C (1984b) Information for the dorsal–ventral pattern of the *Drosophila* embryo is stored as maternal mRNA. Nature 311:223–227

Anderson KV, Bokla L, Nusslein-Volhard C (1985a) Establishment of dorsal–ventral polarity in the *Drosophila* embryo: the induction of polarity by the Toll gene product. Cell 42: 791–798

Anderson KV, Jurgens G, Nusslein-Volhard C (1985b) Establishment of dorsal–ventral polarity in the *Drosophila* embryo: genetic studies on the role of the Toll gene product. Cell 42:779–789

Baoprasertkul P, He C, Peatman E, Zhang S, Li P, Liu Z (2005) Constitutive expression of three novel catfish CXC chemokines: homeostatic chemokines in teleost fish. Mol Immunol 42:1355–1366

Barbazuk WB, Korf I, Kadavi C, Heyen J, Tate S, Wun E, Bedell JA, McPherson JD, Johnson SL (2000) The syntenic relationship of the zebrafish and human genomes. Genome Res 10:1351–1358

Barondes SH, Cooper DN, Gitt MA, Leffler H (1994) Galectins. Structure and function of a large family of animal lectins. J Biol Chem 269:20807–20810

Bei JX, Suetake H, Araki K, Kikuchi K, Yoshiura Y, Lin HR, Suzuki Y (2006) Two interleukin (IL)-15 homologues in fish from two distinct origins. Mol Immunol 43:860–869

Bennett CM, Kanki JP, Rhodes J, Liu TX, Paw BH, Kieran MW, Langenau DM, Delahaye-Brown A, Zon LI, Fleming MD, Look AT (2001) Myelopoiesis in the zebrafish, *Danio rerio*. Blood 98:643–651

Beutler B (2004) Inferences, questions and possibilities in Toll-like receptor signalling. Nature 430:257–263

Bin LH, Xu LG, Shu HB (2003) TIRP, a novel Toll/interleukin-1 receptor (TIR) domain-containing adapter protein involved in TIR signaling. J Biol Chem 278:24526–24532

Burnet FM (1959) The clonal selection theory of acquired immunity. Vanderbilt University Press, Nashville

Carlson RL, Evans DL, Graves SS (1985) Nonspecific cytotoxic cells in fish (Ictalurus punctatus). V. Metabolic requirements of lysis. Dev Comp Immunol 9:271–280

Couillault C, Pujol N, Reboul J, Sabatier L, Guichou JF, Kohara Y, Ewbank JJ (2004) TLR-independent control of innate immunity in *Caenorhabditis elegans* by the TIR domain adaptor protein TIR-1, an ortholog of human SARM. Nat Immunol 5:488–494

Danilova N, Steiner LA (2002) B cells develop in the zebrafish pancreas. Proc Natl Acad Sci USA 99:13711–13716

Danilova N, Hohman VS, Sacher F, Ota T, Willett CE, Steiner LA (2004) T cells and the thymus in developing zebrafish. Dev Comp Immunol 28:755–767

Davis DM (2002) Assembly of the immunological synapse for T cells and NK cells. Trends Immunol 23:356–363

Davis JM, Clay H, Lewis JL, Ghori N, Herbomel P, Ramakrishnan L (2002) Real-time visualization of mycobacterium–macrophage interactions leading to initiation of granuloma formation in zebrafish embryos. Immunity 17:693–702

Dehal P, Boore JL (2005) Two rounds of whole genome duplication in the ancestral vertebrate. PLoS Biol 3:e314

deKruif P (1926) Microbe hunters. Harcourt Brace, Orlando

Deng L, Wang C, Spencer E, Yang L, Braun A, You J, Slaughter C, Pickart C, Chen ZJ (2000) Activation of the IkappaB kinase complex by TRAF6 requires a dimeric ubiquitin-conjugating enzyme complex and a unique polyubiquitin chain. Cell 103:351–361

Draper BW, Stock DW, Kimmel CB (2003) Zebrafish fgf24 functions with fgf8 to promote posterior mesodermal development. Development 130:4639–4654

Evans DL, Carlson RL, Graves SS, Hogan KT (1984a) Nonspecific cytotoxic cells in fish (*Ictalurus punctatus*). IV. Target cell binding and recycling capacity. Dev Comp Immunol 8:823–833

Evans DL, Graves SS, Cobb D, Dawe DL (1984b) Nonspecific cytotoxic cells in fish (*Ictalurus punctatus*). II. Parameters of target cell lysis and specificity. Dev Comp Immunol 8:303–312

Evans DL, Hogan KT, Graves SS, Carlson RL Jr, Floyd E, Dawe DL (1984c) Nonspecific cytotoxic cells in fish (*Ictalurus punctatus*). III. Biophysical and biochemical properties affecting cytolysis. Dev Comp Immunol 8:599–610

Evans DL, Smith EE Jr, Brown FE (1987) Nonspecific cytotoxic cells in fish (Ictalurus punctatus). VI. Flow cytometric analysis. Dev Comp Immunol 11:95–104

Evans DL, Jaso-Friedmann L, Smith EE Jr, St John A, Koren HS, Harris DT (1988) Identification of a putative antigen receptor on fish nonspecific cytotoxic cells with monoclonal antibodies. J Immunol 141:324–332

Evans DL, Leary JH 3rd, Jaso-Friedmann L (2001) Nonspecific cytotoxic cells and innate immunity: regulation by programmed cell death. Dev Comp Immunol 25:791–805

Fishman MC (2001) Genomics. Zebrafish–the canonical vertebrate. Science 294:1290–1291

Fitzgerald KA, Palsson-McDermott EM, Bowie AG, Jefferies CA, Mansell AS, Brady G, Brint E, Dunne A, Gray P, Harte MT, McMurray D, Smith DE, Sims JE, Bird TA, O'Neill LA (2001) Mal (MyD88-adapter-like) is required for Toll-like receptor-4 signal transduction. Nature 413:78–83

Fitzgerald KA, Rowe DC, Barnes BJ, Caffrey DR, Visintin A, Latz E, Monks B, Pitha PM, Golenbock DT (2003) LPS-TLR4 signaling to IRF-3/7 and NF-kappaB involves the toll adapters TRAM and TRIF. J Exp Med 198:1043–1055

Geijtenbeek TB, Van Vliet SJ, Koppel EA, Sanchez-Hernandez M, Vandenbroucke-Grauls CM, Appelmelk B, Van Kooyk Y (2003) Mycobacteria target DC-SIGN to suppress dendritic cell function. J Exp Med 197:7–17

Graves SS, Evans DL, Cobb D, Dawe DL (1984) Nonspecific cytotoxic cells in fish (*Ictalurus punctatus*). I. Optimum requirements for target cell lysis. Dev Comp Immunol 8: 293–302

Harris DT, Jaso-Friedmann L, Devlin RB, Koren HS, Evans DL (1991) Identification of an evolutionarily conserved, function-associated molecule on human natural killer cells. Proc Natl Acad Sci USA 88:3009–3013

Hawke NA, Yoder JA, Haire RN, Mueller MG, Litman RT, Miracle AL, Stuge T, Shen L, Miller N, Litman GW (2001) Extraordinary variation in a diversified family of immune-type receptor genes. Proc Natl Acad Sci USA 98:13832–13837

Heguy A, Baldari CT, Macchia G, Telford JL, Melli M (1992) Amino acids conserved in interleukin-1 receptors (IL-1Rs) and the Drosophila toll protein are essential for IL-1R signal transduction. J Biol Chem 267:2605–2609

Henneke P, Golenbock DT (2001) TIRAP: how Toll receptors fraternize. Nat Immunol 2:828–830

Hermann AC, Millard PJ, Blake SL, Kim CH (2004) Development of a respiratory burst assay using zebrafish kidneys and embryos. J Immunol Methods 292:119–129

Horng T, Barton GM, Medzhitov R (2001) TIRAP: an adapter molecule in the Toll signaling pathway. Nat Immunol 2:835–841

Horng T, Barton GM, Flavell RA, Medzhitov R (2002) The adaptor molecule TIRAP provides signalling specificity for Toll-like receptors. Nature 420:329–333

Hoshino K, Takeuchi O, Kawai T, Sanjo H, Ogawa T, Takeda Y, Takeda K, Akira S (1999) Cutting edge: Toll-like receptor 4 (TLR4)-deficient mice are hyporesponsive to lipopolysaccharide: evidence for TLR4 as the Lps gene product. J Immunol 162:3749–3752

Huising MO, Stet RJ, Savelkoul HF, Verburg-van Kemenade BM (2004) The molecular evolution of the interleukin-1 family of cytokines; IL-18 in teleost fish. Dev Comp Immunol 28:395–413

Huising MO, Kruiswijk CP, Schijndel JE van, Savelkoul HF, Flik G, Verburg-van Kemenade BM (2005) Multiple and highly divergent IL-11 genes in teleost fish. Immunogenetics 57:432–443

Igawa D, Sakai M, Savan R (2006) An unexpected discovery of two interferon gamma-like genes along with interleukin (IL)-22 and −26 from teleost: IL-22 and −26 genes have been described for the first time outside mammals. Mol Immunol 43:999–1009

Iliev DB, Roach JC, Mackenzie S, Planas JV, Goetz FW (2005) Endotoxin recognition: in fish or not in fish? FEBS Lett 579:6519–6528

Janeway CA Jr (1989) Approaching the asymptote? Evolution and revolution in immunology. Cold Spring Harb Symp Quant Biol 54:1–13

Jaso-Friedmann L, Leary JH 3rd, Evans DL (1993) Role of function-associated molecules in target cell lysis: analysis of rat adherent lymphokine-activated killer cells. Nat Immun 12:316–325

Jaso-Friedmann L, Leary JH 3rd, Evans DL (1997) NCCRP-1: a novel receptor protein sequenced from teleost nonspecific cytotoxic cells. Mol Immunol 34:955–965

Jaso-Friedmann L, Leary JH 3rd, Evans DL (2001) The non-specific cytotoxic cell receptor (NCCRP-1): molecular organization and signaling properties. Dev Comp Immunol 25:701–711

Jault C, Pichon L, Chluba J (2004) Toll-like receptor gene family and TIR-domain adapters in *Danio rerio*. Mol Immunol 40:759–771

Jong JL de, Zon LI (2005) Use of the zebrafish system to study primitive and definitive hematopoiesis. Annu Rev Genet 39:481–501

Kanayama A, Seth RB, Sun L, Ea CK, Hong M, Shaito A, Chiu YH, Deng L, Chen ZJ (2004) TAB2 and TAB3 activate the NF-kappaB pathway through binding to polyubiquitin chains. Mol Cell 15:535–548

Kono T, Zou J, Bird S, Savan R, Sakai M, Secombes CJ (2006) Identification and expression analysis of lymphotoxin-beta like homologues in rainbow trout *Oncorhynchus mykiss*. Mol Immunol 43:1390–1401

Kopp E, Medzhitov R (2003) Recognition of microbial infection by Toll-like receptors. Curr Opin Immunol 15:396–401

Lam SH, Chua HL, Gong Z, Lam TJ, Sin YM (2004) Development and maturation of the immune system in zebrafish, *Danio rerio*: a gene expression profiling, in situ hybridization and immunological study. Dev Comp Immunol 28:9–28

Lanier LL (2005) NK cell recognition. Annu Rev Immunol 23:225–274

LaPatra SE, Barone L, Jones GR, Zon LI (2000) Effects of infectious hematopoietic necrosis virus and infectious pancreatic necrosis virus infection on hematopoietic precursors of the zebrafish. Blood Cells Mol Dis 26:445–452

Leippe M (1995) Ancient weapons: NK-lysin, is a mammalian homolog to pore-forming peptides of a protozoan parasite. Cell 83:17–18

Lemaitre B (2004) The road to Toll. Nat Rev Immunol 4:521–527

Lemaitre B, Kromer-Metzger E, Michaut L, Nicolas E, Meister M, Georgel P, Reichhart JM, Hoffmann JA (1995) A recessive mutation, immune deficiency (imd), defines two distinct control pathways in the Drosophila host defense. Proc Natl Acad Sci USA 92: 9465–9469

Lemaitre B, Nicolas E, Michaut L, Reichhart JM, Hoffmann JA (1996) The dorsoventral regulatory gene cassette spatzle/Toll/cactus controls the potent antifungal response in *Drosophila* adults. Cell 86:973–983

Lemaitre B, Reichhart JM, Hoffmann JA (1997) Drosophila host defense: differential induction of antimicrobial peptide genes after infection by various classes of microorganisms. Proc Natl Acad Sci USA 94:14614–14619

Lieberman J (2003) The ABCs of granule-mediated cytotoxicity: new weapons in the arsenal. Nat Rev Immunol 3:361–370

Liu WY, Wang Y, Sun YH, Wang YP, Chen SP, Zhu ZY (2005) Efficient RNA interference in zebrafish embryos using siRNA synthesized with SP6 RNA polymerase. Dev Growth Differ 47:323–331

McCallum CM, Comai L, Greene EA, Henikoff S (2000) Targeted screening for induced mutations. Nat Biotechnol 18:455–457

McCann FE, Suhling K, Carlin LM, Eleme K, Taner SB, Yanagi K, Vanherberghen B, French PM, Davis DM (2002) Imaging immune surveillance by T cells and NK cells. Immunol Rev 189:179–192

Medzhitov R, Preston-Hurlburt P, Janeway CA Jr (1997) A human homologue of the *Drosophila* Toll protein signals activation of adaptive immunity. Nature 388:394–397

Medzhitov R, Preston-Hurlburt P, Kopp E, Stadlen A, Chen C, Ghosh S, Janeway CA Jr (1998) MyD88 is an adaptor protein in the hToll/IL-1 receptor family signaling pathways. Mol Cell 2:253–258

Meijer AH, Gabby Krens SF, Medina Rodriguez IA, He S, Bitter W, Ewa Snaar-Jagalska B, Spaink HP (2004) Expression analysis of the Toll-like receptor and TIR domain adaptor families of zebrafish. Mol Immunol 40:773–783

Menudier A, Rougier FP, Bosgiraud C (1996) Comparative virulence between different strains of *Listeria* in zebrafish (*Brachydanio rerio*) and mice. Pathol Biol (Paris) 44:783–789

Meyer-Wentrup F, Cambi A, Adema GJ, Figdor CG (2005) "Sweet talk": closing in on C type lectin signaling. Immunity 22:399–400

Nasevicius A, Ekker SC (2000) Effective targeted gene 'knockdown' in zebrafish. Nat Genet 26:216–220

Neely MN, Pfeifer JD, Caparon M (2002) *Streptococcus*–zebrafish model of bacterial pathogenesis. Infect Immun 70:3904–3914

Oates AC, Bruce AE, Ho RK (2000) Too much interference: injection of double-stranded RNA has nonspecific effects in the zebrafish embryo. Dev Biol 224:20–28

Odom EW, Vasta GR (2006) Characterization of a binary tandem domain F-type lectin from striped bass (*Morone saxatilis*). J Biol Chem 281:1698–1713

Oshiumi H, Matsumoto M, Funami K, Akazawa T, Seya T (2003a) TICAM-1, an adaptor molecule that participates in Toll-like receptor 3-mediated interferon-beta induction. Nat Immunol 4:161–167

Oshiumi H, Sasai M, Shida K, Fujita T, Matsumoto M, Seya T (2003b) TIR-containing adapter molecule (TICAM)-2, a bridging adapter recruiting to toll-like receptor 4 TICAM-1 that induces interferon-beta. J Biol Chem 278:49751–49762

Panagos PG, Dobrinski KP, Chen X, Grant AW, Traver D, Djeu JY, Wei S, Yoder JA (2006) Immune-related, lectin-like receptors are differentially expressed in the myeloid and lymphoid lineages of zebrafish. Immunogenetics 58:31–40

Peatman E, Liu Z (2006) CC chemokines in zebrafish: evidence for extensive intrachromosomal gene duplications. Genomics 88:381–385

Phelan PE, Mellon MT, Kim CH (2005a) Functional characterization of full-length TLR3, IRAK-4, and TRAF6 in zebrafish (*Danio rerio*). Mol Immunol 42:1057–1071

Phelan PE, Pressley ME, Witten PE, Mellon MT, Blake S, Kim CH (2005b) Characterization of snakehead rhabdovirus infection in zebrafish (*Danio rerio*). J Virol 79:1842–1852

Phelan PE III, Mellon MT, Kim CH (2005c) Functional characterization of full-length TLR3, IRAK-4, and TRAF6 in zebrafish (*Danio rerio*). Mol Immunol 42:1057–1071

Poltorak A, He X, Smirnova I, Liu MY, Van Huffel C, Du X, Birdwell D, Alejos E, Silva M, Galanos C, Freudenberg M, Ricciardi-Castagnoli P, Layton B, Beutler B (1998) Defective LPS signaling in C3H/HeJ and C57BL/10ScCr mice: mutations in Tlr4 gene. Science 282:2085–2088

Praveen K, Evans DL, Jaso-Friedmann L (2004) Evidence for the existence of granzyme-like serine proteases in teleost cytotoxic cells. J Mol Evol 58:449–459

Praveen K, Evans DL, Jaso-Friedmann L (2006a) Constitutive expression of tumor necrosis factor-alpha in cytotoxic cells of teleosts and its role in regulation of cell-mediated cytotoxicity. Mol Immunol 43:279–291

Praveen K, Leary JH 3rd, Evans DL, Jaso-Friedmann L (2006b) Nonspecific cytotoxic cells of teleosts are armed with multiple granzymes and other components of the granule exocytosis pathway. Mol Immunol 43:1152–1162

Pressley ME, Phelan PE 3rd, Witten PE, Mellon MT, Kim CH (2005) Pathogenesis and inflammatory response to *Edwardsiella tarda* infection in the zebrafish. Dev Comp Immunol 29:501–513

Prouty MG, Correa NE, Barker LP, Jagadeeswaran P, Klose KE (2003) Zebrafish–*Mycobacterium marinum* model for mycobacterial pathogenesis. FEMS Microbiol Lett 225:177–182

Rabinovich GA, Rubinstein N, Toscano MA (2002) Role of galectins in inflammatory and immunomodulatory processes. Biochim Biophys Acta 1572:274–284

Roach JC, Glusman G, Rowen L, Kaur A, Purcell MK, Smith KD, Hood LE, Aderem A (2005) The evolution of vertebrate Toll-like receptors. Proc Natl Acad Sci USA 102: 9577–9582

Robertsen B (2006) The interferon system of teleost fish. Fish Shellfish Immunol 20:172–191

Rogers NC, Slack EC, Edwards AD, Nolte MA, Schulz O, Schweighoffer E, Williams DL, Gordon S, Tybulewicz VL, Brown GD, Reis e Sousa C (2005) Syk-dependent cytokine induction by Dectin-1 reveals a novel pattern recognition pathway for C type lectins. Immunity 22:507–517

Sanders GE, Batts WN, Winton JR (2003) Susceptibility of zebrafish (*Danio rerio*) to a model pathogen, spring viremia of carp virus. Comp Med 53:514–521

Santos MD, Yasuike M, Hirono I, Aoki T (2006) The granulocyte colony-stimulating factors (CSF3s) of fish and chicken. Immunogenetics 58:422–432

Sar AM van der, Musters RJ, van Eeden FJ, Appelmelk BJ, Vandenbroucke-Grauls CM, Bitter W (2003) Zebrafish embryos as a model host for the real time analysis of *Salmonella typhimurium* infections. Cell Microbiol 5:601–611

Sar AM van der, Stockhammer OW, van der Laan C, Spaink HP, Bitter W, Meijer AH (2006) MyD88 innate immune function in a Zebrafish embryo infection model. Infect Immun 74:2436–2441

Schilling D, Thomas K, Nixdorff K, Vogel SN, Fenton MJ (2002) Toll-like receptor 4 and Toll-IL-1 receptor domain-containing adapter protein (TIRAP)/myeloid differentiation protein 88 adapter-like (Mal) contribute to maximal IL-6 expression in macrophages. J Immunol 169:5874–5880

Schneider DS, Hudson KL, Lin TY, Anderson KV (1991) Dominant and recessive mutations define functional domains of Toll, a transmembrane protein required for dorsal-ventral polarity in the *Drosophila* embryo. Genes Dev 5:797–807

Shen L, Stuge TB, Zhou H, Khayat M, Barker KS, Quiniou SM, Wilson M, Bengten E, Chinchar VG, Clem LW, Miller NW (2002) Channel catfish cytotoxic cells: a mini-review. Dev Comp Immunol 26:141–149

Shinobu N, Iwamura T, Yoneyama M, Yamaguchi K, Suhara W, Fukuhara Y, Amano F, Fujita T (2002) Involvement of TIRAP/MAL in signaling for the activation of interferon regulatory factor 3 by lipopolysaccharide. FEBS Lett 517:251–256

Silverstein AM (1989) A history of immunology. Academic, San Diego

Silverstein AM (2002) Paul Ehrlich's receptor immunology: the magnificent obsession. Academic, San Diego

Steward R (1987) Dorsal, an embryonic polarity gene in *Drosophila*, is homologous to the vertebrate proto-oncogene, c-rel. Science 238:692–694

Strong SJ, Mueller MG, Litman RT, Hawke NA, Haire RN, Miracle AL, Rast JP, Amemiya CT, Litman GW (1999) A novel multigene family encodes diversified variable regions. Proc Natl Acad Sci USA 96:15080–15085

Stuge TB, Wilson MR, Zhou H, Barker KS, Bengten E, Chinchar G, Miller NW, Clem LW (2000) Development and analysis of various clonal alloantigen-dependent cytotoxic cell lines from channel catfish. J Immunol 164:2971–2977

Takeda K, Akira S (2004) TLR signaling pathways. Semin Immunol 16:3–9

Tauber AI, Chernyak L (1991) Metchnikoff and the origins of immunology: from metaphor to theory. Oxford University Press, New York

Thisse C, Zon LI (2002) Organogenesis – heart and blood formation from the zebrafish point of view. Science 295:457–462

Traver D, Herbomel P, Patton EE, Murphey RD, Yoder JA, Litman GW, Catic A, Amemiya CT, Zon LI, Trede NS (2003) The zebrafish as a model organism to study development of the immune system. Adv Immunol 81:253–330

Trede NS, Langenau DM, Traver D, Look AT, Zon LI (2004) The use of zebrafish to understand immunity. Immunity 20:367–379

Ulevitch RJ (2004) Therapeutics targeting the innate immune system. Nat Rev Immunol 4:512–520

Vasta GR, Ahmed H, Du S, Henrikson D (2004) Galectins in teleost fish: zebrafish (*Danio rerio*) as a model species to address their biological roles in development and innate immunity. Glycoconj J 21:503–521

Vitved L, Holmskov U, Koch C, Teisner B, Hansen S, Salomonsen J, Skjodt K (2000) The homologue of mannose-binding lectin in the carp family Cyprinidae is expressed at high level in spleen, and the deduced primary structure predicts affinity for galactose. Immunogenetics 51:955–964

Vogel SN, Fenton M (2003) Toll-like receptor 4 signalling: new perspectives on a complex signal-transduction problem. Biochem Soc Trans 31:664–668

Walsh CM, Matloubian M, Liu CC, Ueda R, Kurahara CG, Christensen JL, Huang MT, Young JD, Ahmed R, Clark WR (1994) Immune function in mice lacking the perforin gene. Proc Natl Acad Sci USA 91:10854–10858

Wienholds E, Eeden F van, Kosters M, Muddc J, Plasterk RH, Cuppen E (2003) Efficient target-selected mutagenesis in zebrafish. Genome Res 13:2700–2707

Willett CE, Cherry JJ, Steiner LA (1997) Characterization and expression of the recombination activating genes (rag1 and rag2) of zebrafish. Immunogenetics 45:394–404

Xie Y, Chen X, Wagner TE (1997) A ribozyme-mediated, gene "knockdown" strategy for the identification of gene function in zebrafish. Proc Natl Acad Sci USA 94:13777–13781

Yamamoto M, Sato S, Hemmi H, Sanjo H, Uematsu S, Kaisho T, Hoshino K, Takeuchi O, Kobayashi M, Fujita T, Takeda K, Akira S (2002a) Essential role for TIRAP in activation of the signalling cascade shared by TLR2 and TLR4. Nature 420:324–329

Yamamoto M, Sato S, Mori K, Hoshino K, Takeuchi O, Takeda K, Akira S (2002b) Cutting edge: a novel Toll/IL-1 receptor domain-containing adapter that preferentially activates the IFN-beta promoter in the Toll-like receptor signaling. J Immunol 169:6668–6672

Yamamoto M, Sato S, Hemmi H, Uematsu S, Hoshino K, Kaisho T, Takeuchi O, Takeda K, Akira S (2003) TRAM is specifically involved in the Toll-like receptor 4-mediated MyD88-independent signaling pathway. Nat Immunol 4:1144–1150

Yan YL, Miller CT, Nissen RM, Singer A, Liu D, Kirn A, Draper B, Willoughby J, Morcos PA, Amsterdam A, Chung BC, Westerfield M, Haffter P, Hopkins N, Kimmel C, Postlethwait JH (2002) A zebrafish sox9 gene required for cartilage morphogenesis. Development 129:5065–5079

Yoder JA, Litman GW (2000) Immune-type diversity in the absence of somatic rearrangement. Curr Top Microbiol Immunol 248:271–282

Yoder JA, Mueller MG, Wei S, Corliss BC, Prather DM, Willis T, Litman RT, Djeu JY, Litman GW (2001) Immune-type receptor genes in zebrafish share genetic and functional properties with genes encoded by the mammalian leukocyte receptor cluster. Proc Natl Acad Sci USA 98:6771–6776

Yoder JA, Litman RT, Mueller MG, Desai S, Dobrinski KP, Montgomery JS, Buzzeo MP, Ota T, Amemiya CT, Trede NS, Wei S, Djeu JY, Humphray S, Jekosch K, Hernandez Prada JA, Ostrov DA, Litman GW (2004) Resolution of the novel immune-type receptor gene cluster in zebrafish. Proc Natl Acad Sci USA 101:15706–15711

Zhang DC, Shao YQ, Huang YQ, Jiang SG (2005) Cloning, characterization and expression analysis of interleukin-10 from the zebrafish (*Danio rerion*). J Biochem Mol Biol 38:571–576

Zhao Z, Cao Y, Li M, Meng A (2001) Double-stranded RNA injection produces nonspecific defects in zebrafish. Dev Biol 229:215–223

Chapter 7
Toll-Like Receptors in the Mammalian Innate Immune System

Andrei E. Medvedev(✉) and Stefanie N. Vogel

1	Introduction..	136
2	TLRs as Primary Sensors of Pathogenic PAMPS and Endogenous "Danger" Molecules	137
3	TLR Signaling Pathways	141
	3.1 Interaction of TLRs with PAMPs and Co-Receptors Initiates Signaling.................	141
	3.2 Role of TIR-Containing Adapter Molecules in TLR Signaling	142
	3.3 TLR Specificity for PAMPs in the Ectodomain and Adapters in the TIR Domain Underlie a Dual Recognition/Response System........................	146
	3.4 The IRAK Family: Key Regulators of TLR Signaling	147
4	Mutations in TLRs and IRAK-4: Implications for Disease...	149
5	Conclusions...	155
References..		156

Abstract Toll-like receptors (TLRs) are evolutionary conserved, germ-line encoded molecules that express an ectodomain with leucine-rich regions, a single transmembrane domain, and a cytoplasmic region that contains the Toll-IL-1R resistance (TIR) signaling domain. TLRs recognize bacterial and viral pathogen-associated molecular patterns (PAMPs), as well as certain endogenous mammalian "danger signals," i.e., proteins, oligosaacharides, and nucleic acids released from damaged cells as a consequence of stress, inflammation, and wounding. TLR2, TLR4, TLR5, TLR9, and TLR11 preferentially respond to bacterial, yeast, and protozoan PAMPs, while TLR3, TLR7, and TLR8 sense viral nucleic acids. TLRs exhibit different cellular localization: TLR2, TLR4, and TLR5 are expressed on the cell surface, while TLR3, TLR7, TLR8, and TLR9 are localized intracellularly in endosomal compartments. Recognition of PAMPs by TLRs triggers TLR oligomerization and conformational changes within TIR domains, leading to recruitment of various adapter proteins and kinases. This, in turn, activates MAP kinases, transcription factors, and, ultimately, production of nitric oxide (NO) and reactive oxygen intermediates, up-regulation of expression of accessory and co-stimulatory molecules, and secretion of cytokines

---

Department of Microbiology and Immunology, School of Medicine, University of Maryland, Baltimore (UMB), 660 W. Redwood Street, Rm. HH 324, Baltimore, MD 21201, USA, *amedvedev@som.umaryland.edu*

and chemokines. TLR-mediated up-regulation of co-stimulatory and MHC molecules on dendritic cells (DC), coupled with cytokine production and DC maturation, facilitates the adaptive immune response, providing a link between the innate and adaptive immunity. This review focuses on mechanisms of TLR signaling, known mutations/polymorphisms in genes encoding TLRs and IRAK-4 and their implication for susceptibility to infectious and autoimmune diseases and asthma.

1 Introduction

Molecular heterogeneity and rapid evolution of pathogens led to the development of a recognition strategy that enables the host to sense conserved pathogen-associated molecular patterns (PAMPs) derived from bacteria, viruses, fungi, and protozoa by "pattern-recognition receptors" (PRRs; Medzhitov and Janeway 2000). The non-clonal distribution of PRRs and their ubiquitous expression on various cell types enables a rapid immune response upon encounter with pathogens, providing a first line of host immune defense (Medzhitov and Janeway 2000, 2002; Janeway and Medzhitov 2002). PAMP recognition by PRRs is an ancient mechanism of host immune defense present in plants, zebra fish, crabs, *Drosophila*, and mammals (Janeway and Medzhitov 2002; Inamori et al. 2004; Meijer et al. 2004). One of the most important families of PRRs, the Toll-like Receptor (TLR) family, was initially identified in *Drosophila*, where a single protein, Toll, mediates dorsal–ventral partitioning during embryogenesis and anti-fungal immune defense in adult flies (Anderson et al. 1985; Lemaitre et al. 1996). In 1997, Janeway and colleagues reported that overexpression of a constitutively active CD4–TLR4 fusion protein resulted in transcription factor activation, cytokine secretion, and up-regulation of co-stimulatory and accessory molecules (Medzhitov et al. 1997), suggesting that TLR4 triggers innate immune responses. Confirming this conclusion, positional cloning studies led to the identification of mouse *tlr4* as the elusive *Lps* gene responsible for LPS recognition and sensing Gram negative bacteria (Poltorak et al. 1998; Qureshi et al. 1999). To date, 12 mammalian TLRs have been identified and most of their ligand recognition patterns characterized (for a review, see Kaisho and Akira 2006). Concurrently, a complex picture of signal transduction pathways triggered by various TLRs has emerged, with many questions regarding agonist recognition, as well as adapter and kinase utilization, still unanswered. In addition, another family of intracellular PRRs, the nucleotide-binding oligomerization domain (NOD) proteins, NOD1 and NOD2, have been identified as important intracellular sensors of peptidoglycan (PGN)-derived components [D-glutamyl-meso-diaminopimelic acid and muramyl dipeptide (MDP), respectively; Idohara and Nunez 2003; Kawai and Akira 2006]. Although bacterial PGN was initially reported as a TLR2 agonist (Dziarski et al. 2001), lipoprotein and lipoteichoic acid (LTA) contaminants of PGN were the likely source of TLR2 activity, as highly purified PGN does not activate TLR2 (Travassos et al. 2004). Very recently, other non-TLR-mediated intracellular sensors of viral nucleic acids and bacterial flagellin have been described (e.g., RIG-I, MDA-5, and Ipaf-1, respectively), adding another host safeguard system against viral invasion (Kato et al. 2006; Miao et al. 2006). This review summarizes current knowledge about TLR structure and functions, mechanisms of TLR signaling, and mutations/polymorphisms within TLRs and IRAK-4 associated with autoimmune and infectious diseases and asthma.

2 TLRs as Primary Sensors of Pathogenic PAMPS and Endogenous "Danger" Molecules

TLRs are evolutionarily conserved, non-clonally distributed, type I signaling receptors comprised of an ectodomain with leucine-rich repeats, a single transmembrane region, and a cytoplasmic tail that contains a conserved signaling TIR domain (Martin and Wesche 2002; Underhill and Ozinski 2002; Kaisho and Akira 2006). TLRs are expressed on epithelial and endothelial cells, monocytes, macrophages, neutrophils, dendritic cells, and lymphocytes (Zarember and Godowski 2002; Iwasaki and Medzhitov 2005; Kaisho and Akira 2006), providing the host a means to react rapidly to pathogens in various locations in the body. TLR4-deficent and knockout (KO) mice fail to respond to LPS and whole Gram negative bacteria (Poltorak et al. 1998; Qureshi et al. 1999; Hoshino et al. 1999), whereas overexpression of TLR4 in LPS-unresponsive cell lines imparts LPS sensitivity (Chow et al. 1999; Hirschfeld et al. 2000; Yang et al. 2000), indicating TLR4 as the principal sensor of Gram negative bacteria and LPS. In addition to LPS, TLR4 also recognizes other, structurally unrelated PAMPs, *e.g.*, the fusion (F) protein of respiratory syncytial virus (RSV; Kurt-Jones et al. 2000), chlamydial heat shock protein (HSP) 60 (Bulut et al. 2002), and pneumolysin (Malley et al. 2003), although the molecular basis for such broad ligand specificity is unknown. An extracellular protein, MD-2, binds LPS and presents it to TLR4 (Shimazu et al. 1999; Yang et al. 2000; Viriyakosol et al. 2001) and is essential for conferring LPS sensitivity to TLR4, as evidenced by a complete failure of LPS to induce responses in MD-2$^{-/-}$ cells *in vitro* and MD-2 KO mice *in vivo* (Nagai et al. 2002). Cross-linking studies in HEK293T cells that overexpress TLR4, CD14, and MD-2 revealed the interaction of a photoactivatable LPS probe with CD14, MD-2, and TLR4 (Da Silva Correia et al. 2001). Using a biotinylated LPS probe, Visintin et al. (2003) found that CD14 enhances LPS binding to MD-2, whereas MD-2 enables TLR4 binding to LPS, resulting in the formation of stable receptor complexes with TLR4 and recruitment of the adapter protein MyD88. It is important to note that the additional accessory molecules, e.g., CD14 and CD11b/CD18, bind LPS and facilitate LPS signaling via the TLR4-MD-2 complex (Wright et al. 1990; Golenbock et al. 1993; Pugin et al. 1994; Haziot et al. 1996; Perera et al. 1997; Ingalls et al. 1998; Moore et al. 2000; Perera et al. 2001; Visintin et al. 2003). Macrophages from CD14$^{-/-}$ mice or mice with a deficiency in CD11b/CD18 expression exhibit selective inhibition of gene expression when compared with wild-type macrophages (Perera et al. 1997, 2001). CD14 has been shown to be more than simply an LPS-binding protein, as it is required for TLR4/MD-2-mediated activation of both MyD88-dependent and MyD88-independent signaling pathways in response to smooth LPS, whereas in the absence of CD14, only MyD88-dependent signaling occurs in response to rough LPS and lipid A (Jiang et al. 2005). Taken together, these data support a model in which co-receptors such as CD14 and CD11b/CD18 contribute not only to sensing of microbial structures, but also to the signaling that is completely TLR4-dependent (Perera et al. 2001; Vogel et al. 2001).

TLR2 triggers host responses to components of Gram positive bacteria [*e.g.*, lipoteichoic acid (LTA), lipopeptides; Lien et al. 1999; Schwandner et al. 1999], mycobacterium [*e.g.*, lipoarabinomannan (LAM); Means et al. 1999], mycoplasma lipopeptides (Lien et al. 1999; Takeuchi et al. 2000), and mammalian HSP70 (Asea et al. 2002). TLR1 and TLR6 do not elicit signaling on their own, but rather form complexes with TLR2, amplifying TLR2-mediated responses (Ozinsky et al. 2000; Bulut et al. 2001; Hajjar et al. 2001). Thus, it has been shown that triacylated lipopeptides are recognized by TLR2/TLR1 heterodimers, whereas diacylated lipopeptides use TLR2/TLR6 for signaling (Takeuchi et al. 2002). However, di- and triacylated lipopeptides were recently reported to mediate TLR2-dependent responses in TLR1- and in TLR6-deficient mice (Buwitt-Beckmann et al. 2006). These data indicate that lipopeptides with distinct acylation patterns can be recognized by TLR2 in a TLR1- and TLR6-independent manner. Further studies will be needed to address the question of whether a tri- or diacylation patterns confer specificity upon a lipopeptide for TLR1- or TLR6-dependency, respectively, or if an additional, compensatory recognition system exists. Other molecules, e.g., CD36 and Dectin-1, also contribute to TLR2-mediated signal transduction. CD36 has been characterized as a selective and non-redundant sensor of microbial diacylglycerides that signal via the TLR2/6 heterodimer (e.g., MALP-2, LTA; Hoebe et al. 2005). Indeed, CD36 KO– mice exhibit defective clearance of *Staphylococcus aureus in vivo*, develop profound bacteremia, and $CD36^{-/-}$ macrophages fail to internalize *S. aureus* and secrete TNF-α or IL-12 (Hoebe et al. 2005; Stuart et al. 2005). The macrophage β-glucan receptor, Dectin-1, has been characterized as a co-receptor that facilitates TLR2-mediated recognition of the pathogenic fungus *Candida posadasii* (Viriyakosol et al. 2005). Engagement of Dectin-1 initiates an independent signaling cascade via tyrosine phosphorylation of the Dectin-1 cytoplasmic domain and recruitment of Syk kinase that promotes IL-2 and IL-10 production, thus contributing to the proinflammatory response induced by TLR2 engagement (Rogers et al. 2005). Thus, these data support the existence of multi-component, integrated TLR complexes whose engagement by PAMPSs or endogenous "danger" molecules brings TIR domains into close proximity, altering their conformation, and creating novel scaffolds within the cytoplasmic domain, initiating signaling (Vogel et al. 2003).

TLR3 and TLR5 agonists include viral double-stranded RNA and various flagellin species, respectively (Alexopoulou et al. 2001; Gewirtz et al. 2001; Hayashi et al. 2001), while TLR9 responds to unmethylated CpG motifs present in bacterial

DNA (Hemmi et al. 2000), as well as DNA from herpes simplex virus, baculovirus, and the fungal pathogens *Candida albicans* and *Aspergillus fumigatus* (Bellocchio et al. 2004; Hochrein et al. 2004; Abe et al. 2005). The antiviral imidazoquinoline compounds, imiquimod and resiquimod, activate cells via human TLR7 and TLR8 (Hemmi et al. 2002; Jurk et al. 2002), respectively, and TLR7 senses another compound, loxoribine (Heil et al. 2003). Imidazoqinoline and loxoribine are structurally similar to guanosine nucleoside; therefore it came as no surprise when TLR7 and TLR8 were reported to recognize guanosine- or uridine-rich single-stranded RNA from human immunodeficiency virus, vesicular stomatitis virus, and influenza virus (Heil et al. 2004; Lund et al. 2004). TLR3 and TLR7-9 sense their respective agonists intracellularly inside membrane-bound compartments and, for TLR7-9, these processes require endosomal maturation (Ahmad-Nejad et al. 2002; Takeda and Akira 2005), whereas TLR2, 4, and 5 are localized at the cell surface (for a review, see Takeda and Akira 2005; Fig. 1). However, once TLR2 or TLR4 recognition occurs at the surface, signaling can further continue within the phagosome, and phagocytosis of bacteria is impaired in the absence of TLR signaling (Blander and Medzhitov 2004). Thus, TLRs 1, 2, 4, 5, and 6 are more involved in recognition of bacterial products and host proteins, while TLR3 and TLR7-9 preferentially detect viral or bacterial nucleic acids (Fig. 1).

Human TLR10 is localized to chromosome 4p14 and closely related to TLR1 and TLR6 (Chuang and Ulevitch 2001). TLR10 mRNA is highly expressed in lymphoid tissues such as spleen, lymph node, thymus, and tonsil (Chuang and Ulevitch 2001), and TLR10 protein has been detected in B cell lines, B cells from peripheral blood, and plasmacytoid dendritic cells from tonsils (Hasan et al. 2005). While TLR10 has been reported to homodimerize, as well as heterodimerize with TLR1 and TLR6, the ligands and functions of TLR10 are still unknown. Mouse TLR11 was found to sense uropathogenic *Escherichia coli* (Zhang et al. 2004) and a profilin-like molecule from the protozoan parasite *Toxoplasma gondii* (Yarovinsky et al. 2005). Activation of mouse TLR11 results in potent, MyD88-dependent IL-12 production in mouse dendritic cells and is required for parasite-induced IL-12 production *in vivo* and optimal resistance to infection (Yarovinsky et al. 2005). In contrast to mice, human TLR11 is a pseudogene (Lauw et al. 2005), suggesting a possible divergence of TLR11 functions in these two species. Readers should be aware that the protein sequences of mouse TLR11 and TLR12 (accession numbers AAS83531.1 AAS37673, respectively, in the NCBI, NIH) show 100% identities upon alignment, indicating that these represent the same proteins, despite different designations.

Interestingly, a number of endogenous molecules are recognized by TLRs (Fig. 1). For instance, TLR4 has been reported to respond to Hsp 60 and Hsp 70 (Ohashi et al. 2000; Vabulas et al. 2002b), fibronectin (Okamura et al. 2001), fibrinogen (Smiley et al. 2001), surfactant protein A (Guillot et al. 2002), mouse β-defensin 2 (Biragyn et al. 2002), high mobility group box (HMGB)-1 protein (Wang et al. 1999, Park et al. 2004), and oligosaccharides of hyaluronic acid (Termeer et al. 2002). TLR2 was also implicated in recognition of Hsp60, Hsp70, Hsp Gp96 (Vabulas et al. 2001, 2002a, b), and HMGB-1 protein (Wang et al. 1999; Park et al.

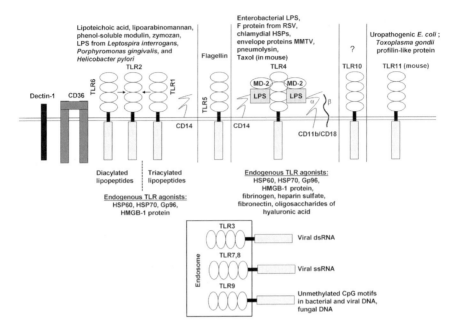

Fig. 1 TLRs and their cognate PAMPs and endogenous ligands. TLR2 recognize triacylated and diacylated lipopeptides either directly or in co-operation with TLR1 or TLR6 and CD14, CD36, and Dectin-1. TLR4 senses LPS and many other structurally unrelated agonists. MD-2, an accessory protein, binds to the extracellular domain of TLR4 and is necessary for TLR4-mediated LPS recognition and signaling. CD14 is a high-affinity co-receptor for LPS and other TLR4 agonists, which fails to mediate signaling due to a lack of an intracytoplasmic signaling domain, but enables activation of the MyD88-independent pathway by the TLR4/MD-2 complex. TLR3 induces cell activation in response to viral dsRNA, TLR5 recognizes flagellin, and TLR7/8 sense single-stranded RNA from viruses. TLR9 responds to the unmethylated CpG motifs present in bacterial DNA, as well as in DNA from herpes simplex virus, baculovirus, and the fungal pathogens *Candida albicans* and *Aspergillus fumigatus*. Mouse TLR11 is activated by uropathogenic *Escherichia coli* and a profilin-like molecule from *Toxoplasma gondii*. Endogenous proteins, oligosaccharides, and nucleic acids released from necrotic cells as a consequence of stress, wounding, or inflammation can activate TLRs and represent "danger" signals. TLRs recognize PAMPs and endogenous ligands either at the cell surface (TLR2, TLR4, TLR5), or in intracellular compartments such as endosomes (TLR3, TLR7–9)

2004), while TLR3 and TLR9 sense endogenous mRNA and chromatin-IgG complexes, respectively (Leadbetter et al. 2002; Kariko et al. 2004). TLR sensing of endogenous agonists released upon cellular damage has been proposed to alert the host to "danger" signals associated with stress, inflammation, and wounding. This mechanism may also provide "danger signals" to the host during an acute or chronic inflammatory incident, and may exacerbate an ongoing inflammatory reaction. For example, inflammation mediated by monosodium urate monohydrate crystals, the hallmark of gout, was found to be attenuated in TLR2$^{-/-}$ and TLR4$^{-/-}$ mice (Liu-Bryan et al. 2005); and hyaluronic acid fragments, that are plentiful during an ongoing inflammatory response, are TLR4 agonists (Termeer et al. 2002). It should be noted, however, that many original findings of TLR4 activation by such endogenous "danger" ligands were later associated with microbial PAMP contamination, emphasizing the need to work with highly purified preparations of recombinant proteins and to rule out a possible contribution of microbial PAMPs by using several approaches, *e.g.*, heat inactivation, polymyxin B, and others (for a review, see Tsan and Gao 2004).

Taken together, these data indicate that TLRs discriminate diverse microbial and viral structures, as well as endogenous host "danger" molecules released from necrotizing or apoptotic cells due to wounding and/or infection, initiating the innate immune response and facilitating the development of efficient adaptive immunity by means of DC differentiation, up-regulation of accessory and MHC molecules, and cytokine production.

3 TLR Signaling Pathways

3.1 Interaction of TLRs with PAMPs and Co-Receptors Initiates Signaling

Despite extensive studies, many aspects of initiation of TLR signaling by PAMPs remain poorly understood. It has been postulated that binding of PAMPs to co-receptors (e.g., CD14, CD11b/CD18, Dectin-1, CD36) facilitates their subsequent presentation to TLRs, resulting in TLR oligomerization (*e.g.*, TLR4) or heterodimerization (*e.g.*, TLR2 with TLR1 or TLR6; Martin and Wesche 2002; Akira and Takeda 2004; Bell et al. 2006). Alternatively, others postulate that TLRs exist as pre-formed dimers, and ligand binding triggers a conformational change within the TIR domain, initiating signaling (Tapping and Tobias 2003; Lee and Tobias 2004; Choe et al. 2005). Regardless of which theory proves to be true, the engagement of multi-component, integrated TLR complexes by PAMPs is thought to bring TIR domains into close proximity, altering their conformation and creating novel scaffolds within the cytoplasmic domain (Vogel et al. 2003). One consequence of TLR engagement by PAMPs described for TLR2, TLR3, and TLR4 is tyrosine phosphorylation. Arbibe et al. (2000) first demonstrated that stimulation of TLR2-expressing HEK293 cells with heat-killed *S. aureus* causes tyrosine phosphorylation of TLR2 and the recruitment of active Rac1 and PI3K to the TLR2 cytosolic domain, leading to activation of Akt and p65 transactivation. We recently confirmed and extended these data by showing that TLR2 becomes phosphorylated on tyrosine following stimulation with other TLR2 agonists, mycobacterial soluble tuberculosis factor (STF) and synthetic lipopeptide Pam3Cys (Medvedev et al. 2007). TLR4 also undergoes LPS-inducible tyrosine phosphorylation in human monocytes (Chen et al. 2003), and tyrosine-deletion mutants of a constitutively-active CD4-TLR4 exhibit a diminished ability to activate NF-κB-, AP-1-, and

C/EBP-dependent reporter plasmids (Ronni et al. 2003). Our recent results indicate that deficient signal transduction capacities of constitutively active mouse ectodomain-deficient TLR4 and human CD4-TLR4 carrying P712H or P714A mutations in their TIR domains correlate with impaired tyrosine phosphorylation (Medvedev et al. 2007). Moreover, LPS-mediated TLR4 tyrosine phosphorylation is blunted in endotoxin-tolerant HEK293/TLR4/MD-2 transfectants and primary human monocytes, again suggesting an important role for TLR4 tyrosine phosphorylation in signaling and endotoxin tolerance. Sarkar et al. (2004) reported that dsRNA-activated phosphorylation of two specific tyrosine residues of TLR3 is essential for signaling activation of TBK-1, PI3K, and Akt, leading to full phosphorylation and activation of IRF-3. Future studies will be required to elucidate whether agonist-mediated TLR tyrosine phosphorylation is a general phenomenon for activation of most TLRs and to identify the specific tyrosine protein kinase(s) involved. In summary, TLR recognition of PAMPs could initiate tyrosine phosphorylation and conformational changes within the TIR domain, leading to recruitment of adapter proteins and kinases, and activation of MAPKs and transcription factors (*e.g.*, NF-κB). This ultimately results in cellular activation manifested by phagocytosis, respiratory burst, production of nitric oxide and reactive oxygen species, as well as secretion of many cytokines and chemokines (Martin and Wesche 2002; Kaisho and Akira 2006). Both common and specific signaling pathways have been described as a consequence of TLR engagement by PAMPs that seem to depend on a complex interplay and utilization of various intracellular adapter proteins and kinases by various TLRs as described below.

3.2 Role of TIR-Containing Adapter Molecules in TLR Signaling

Five TIR-containing intracellular adapter molecules have been identified, and four are known to participate in TLR signaling. MyD88 was the first adapter protein found to mediate IL-1R and TLR signaling (Muzio et al. 1997; Wesche et al. 1997; Burns et al. 1998; Kawai et al. 1999, 2001). MyD88 is comprised of a C-terminal TIR domain involved in interaction with TLRs and an N-terminal death domain that associates with IRAK family members (Muzio et al. 1997; Wesche et al. 1997; Burns et al. 1998). MyD88 is required for responses by most TLRs except TLR3 (Kawai et al. 1999, 2001; Kaisho et al. 2001; Yamamoto et al. 2002a), and is recruited to TLRs via homotypic TIR-TIR domain interactions (Muzio et al. 1997; Wesche et al. 1997; Burns et al. 1998). This triggers the association of MyD88 with IL-1R-associated kinases (IRAK)-4 and IRAK-1, and subsequent phosphorylation reactions by IRAK-4 and IRAK-1 (Muzio et al. 1997; Wesche et al. 1997; Burns et al. 1998; Li et al. 2004; Kollewe et al. 2004), resulting in IRAK-1-mediated phosphorylation of Tollip (Burns et al. 2000), an inhibitory protein that sequesters IRAK-1 in unstimulated cells (Burns et al. 2000). IRAK-1 dissociates from the TLR complex and interacts with TNFR-associated factor 6 (TRAF-6) via a downstream adapter TIFA, leading to TRAF-6 activation (Takatsuna et al. 2003). TRAF-6 exists in a complex with two ubiquitin-conjugating enzymes, UEV1A and UBC13, that activate TGF-β-activated kinase (TAK) 1 through the assembly of a lysine 63-linked polyubiquitin chain and engagement of additional intracellular intermediates Pellino-1 and Pellino-2 (Deng et al. 2000; Jensen and Whitehead 2003; Ea et al. 2004). Activated TAK-1 triggers phosphorylation of IKK-α/β and MAPK kinase (MKK) 3 and 6, leading to activation of MAPKs (*e.g.*, JNK, p38) and transcription factors (*e.g.*, NF-κB, AP-1; Takaesu et al. 2000, 2001; Wang et al. 2001; Jiang et al. 2002, 2003; Jensen and Whitehead 2003; Yoshida et al. 2005). Another downstream intracellular intermediate, evolutionarily conserved signaling intermediate in Toll pathways (ECSIT) was reported to bridge TRAF-6 and MAPK/ERK kinase kinase-1 (MEKK-1) and to regulate MEKK-1 processing and activation of NF-κB and AP-1 (Kopp et al. 1999). IRAK-1 is subsequently ubiquitinated and targeted to the proteosome where it is degraded, which prevents hyperactivation of TLR-stimulated cells (Yamin and Miller 1997; Fig. 2). This MyD88 signaling pathway results in rapid NF-κB and MAP kinase activation, B cell proliferation, and expression of pro-inflammatory cytokines. Studies in MyD88 KO mice revealed the existence of an MyD88-independent pathway initiated by TLR3 and TLR4 that leads to dendritic cell maturation, phosphorylation of interferon (IFN)-regulatory factor (IRF)-3, type I IFN expression, and delayed NF-κB and MAPK activation (Kawai et al. 1999, 2001, Kaisho et al. 2001; Yamamoto et al. 2002b).

The TIR domain-containing adapter protein/MyD88 adapter-like protein (TIRAP/MAL) was initially identified as a second adapter protein involved in TLR4-, but not IL-1R-mediated signaling (Fitzgerald et al. 2001; Horng et al. 2001). Coupled with the observation that TIRAP/MAL$^{-/-}$ mice have an impaired response to TLR2- and TLR4-agonists, while responses to ligands for TLR3, 5, 7, and 9 are preserved (Horng et al. 2002; Yamamoto et al. 2002a, 2004), these findings suggest that TIRAP/MAL may contribute to TLR signaling specificity. Both MyD88$^{-/-}$ and TIRAP/MAL$^{-/-}$ mice exhibit very similar phenotypes in terms of TLR2 and TLR4 signaling, manifested by delayed activation of NF-κB and MAP kinases and the lack of induction of TNF-α, but normal activation of IRF-3 and IFN-β (Horng et al. 2002; Yamamoto et al. 2002a, 2004), indicating that TIRAP/MAL is restricted to the MyD88-dependent pathway. Interestingly, only TLR2 and TLR4 utilize both MyD88 and TIRAP/MAL, whereas TLR3 does not use either adapter (Horng et al. 2002; Yamamoto et al. 2002a, 2004). In contrast, TLR5, 7, 8, and 9 use only MyD88 for triggering both NF-κB activation, production of proinflammatory cytokines, and, surprisingly, induction of type I IFNs (Hemmi et al. 2002; Horng et al. 2002; Yamamoto et al. 2004). However, these responses are triggered by distinct signaling molecules than those involved in TLR3- and TLR4-mediated type I IFN production (Fig. 2, see below). Agonist-induced phosphorylation of two critical tyrosine residues (Y86, Y187) in TIRAP/MAL by Bruton's tyrosine kinase was shown to play a critical role in TLR2 and TLR4 signaling (Gray et al. 2006). In addition, suppressor of cytokine signaling (SOCS)-1 was found to associate with TIRAP/MAL and targets this adapter for polyubiquitination and subsequent

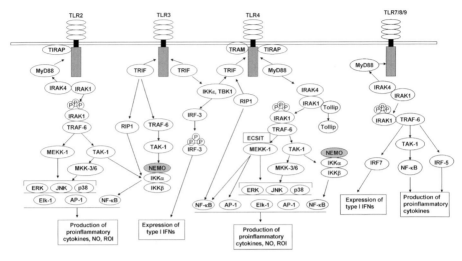

Fig. 2 MyD88-dependent and MyD88-independent TLR signaling pathways. TLRs respond to PAMPs or endogenous TLR ligands either directly or in co-operation with depicted co-receptors, leading to TLR oligomerization, conformational changes within the TIR domain, and recruitment of adapter proteins and kinases. TLR2 utilizes exclusively TIRAP/Mal and MyD88 adapter proteins and kinases IRAK-4 and IRAK-1 for triggering proinflammatory cytokine production via engagement of downstream adapter TRAF-6 and kinases TAK-1, IKKα/β, MEKK-1, and MKK3/6. The TLR4-mediated MyD88 signaling pathway is initiated by engagement of TIRAP/Mal and association of MyD88 with the cytoplasmic region of TLR4 via their TIR domain interactions, followed by recruitment of IRAK-4 and IRAK-1-Tollip complex. This triggers phosphorylation of IRAK-1, dissociation of Tollip, interaction of IRAK-1 with TRAF-6, and stimulation of MAP kinases and transcription factors via the engagement of the downstream adapters (e.g., ECSIT) and activation of IKK, TAK-1, MEKK-1, and MKK-3,4,6 kinases. TLR4 stimulates MyD88-independent pathway via adapter proteins TRAM and TRIF that signal activation of IRF-3 via stimulation of two non-typical IKKs, IKKε and TBK-1 and the induction of IFN-α and IFN-β-dependent genes. Engagement of the MyD88-independent pathway also mediates dendritic cell maturation and delayed activation of MAPK and NF-κB. TLR3 signaling uses the adapter protein TRIF that activates NF-κB and production of proinflammatory cytokines via TRAF6-TAK-1 and RIP-1 and induction of type I IFN expression via IKKε/TBK-1 and IRF-3. TLR7/9 signaling occurs by utilization of the MyD88/IRAK-4/IRAK-1/TRAF-6 module of adapter proteins and kinases whereby TRAF-6 (by still poorly understood mechanisms) activates proinflammatory cytokine production via NF-κB and IRF-5 and type I IFN expression via IRF-7

degradation (Mansell et al. 2006), representing a rapid and selective means of limiting the innate immune response. Whereas TIRAP/MAL has been postulated to serve as a bridging adapter between TLR2, TLR4 and MyD88, it cannot be ruled out that it also mediates specific signaling functions. In this respect, Horng et al. (2002) showed that PKR is a downstream target of TIRAP/Mal and Kagan and Medzhitov (2006) recently reported that TIRAP/Mal is recruited to the plasma membrane lipid rafts through a phosphoinositol-4,5-bisphosphate binding site where it facilitates MyD88 recruitment to TLR4.

Two other adapter proteins are involved in MyD88-independent signaling triggered by TLR3 and TLR4. TIR domain-containing adapter inducing IFN-β (TRIF), also called TICAM-1, was first identified as an adapter protein that mediates the MyD88-independent signaling pathway elicited by TLR3 (Yamamoto et al. 2002a, 2003a, b; Oshiumi et al. 2003a, b). The functional importance of TRIF was first demonstrated in TRIF KO mice that exhibited impaired IFN-β induction and IRF-3 activation in response to TLR3 and TLR4 agonists (Hoebe et al. 2003; Oshiumi et al. 2003a, b; Yamamoto et al. 2003a, b). TRIF mediates downstream signaling by interacting with IKKε and TANK-binding kinase 1 (TBK1), resulting in phosphorylation and nuclear translocation of IRF-3 (Sato et al. 2003; Oganesyan et al. 2006). In addition to IRF-3 activation, TRIF was also shown to cause NF-κB activation via its interactions with downstream adapters, TRAF6 and receptor-interacting protein (RIP) 1 (Yamamoto et al. 2002a, b; Meylan et al. 2004). Ultimately, a co-ordinated activation of NF-κB and IRF-3 leads to expression of type I IFNs (Nusinzon and Horvath 2006). TLR3 was shown to associate directly with TRIF (Oshiumi et al. 2003a, b), whereas another adapter, TRIF-related adapter molecule (TRAM), also known as TICAM-2, is required for TLR4 engagement of TRIF (Fitzgerald et al. 2003; Oshiumi et al. 2003a, b; Yamamoto et al. 2003a, b). TRAM interacts with TLR4 and TRIF, but not with TLR3, and is involved in TLR4-mediated MyD88-independent signaling (Fitzgerald et al. 2003; Oshiumi et al. 2003a, b; Yamamoto et al. 2003a, b). TRAM is myristoylated and localized to the plasma membrane in unstimulated cells (Rowe et al. 2006). LPS triggers transient phosphorylation of TRAM by PKCε on serine-16, which results in its translocation from the membrane in a PKCε-dependent manner (McGettrick et al. 2006). Overexpression of a TRAMS16A mutant failed to activate NF-κB- and ISRE-dependent reporters and showed an impaired ability to reconstitute signaling in TRAM-deficient cells (McGettrick et al. 2006), indicating a critical role for serine-16 phosphorylation in signaling. Moreover, TRAM-dependent activation of IRF-3 and induction of RANTES were attenuated in PKCε-deficient cells (McGettrick et al. 2006), suggesting that TRAM is the target for PKCε. It is tempting to speculate that PKCε-dependent TRAM phosphorylation and translocation from the membrane into the cytoplasm enables TRAM interactions with downstream adapters (e.g., TRIF) or kinases (e.g., TBK1/IKKε). Further studies will be needed to address this hypothesis.

Interestingly, TRIF, IKKε, and TBK1 are not required for TLR7/9-mediated induction of type I IFNs, although they are necessary for TLR4-induced IFN-β expression (for a review, see Kaishi and Akira 2006). While MyD88 and TIRAP/MAL are not involved in type I IFN induction by TLR3 or TLR4, TLR7- and TLR9-mediated activation of IFN-α requires MyD88 and involves IRAK-4/IRAK-1, TRAF-6, and IRF-7 (Heil et al. 2003; Honda et al. 2004; Uematsu et al. 2005). Recently, complex molecular pathways involved in MyD88-dependent production of inflammatory cytokines and expression of type I IFNs via TLR7/9 were revealed. MyD88 was shown to co-localize in endosomes with IRF-7, but not IRF-3, and IRF-7 can directly associate with IRAK-1 and TRAF6, resulting in type I IFN production in plasmacytoid DC (Honda et al. 2004). Interestingly, plasmacytoid DC obtained from IRF-7-KO mice exhibit a severely impaired ability to produce type I IFNs, but not other cytokines (Honda et al. 2005), indicating selective involvement of IRF-7 in type I IFN expression. Conversely, IRF-5 also associates with MyD88 and TRAF6 and is required for induction of inflammatory cytokines, but not type I IFN, in response to TLR7/9 engagement (Takaoka et al. 2005). Another IRF family member, IRF-4, also interacts with MyD88 and negatively regulates TLR7/TLR9-mediated, MyD88-dependent NF-κB and MAPK activation and production of inflammatory cytokine, while not affecting the ability of TLR7/9-stimulated pDCs to secrete IFN-α (Negishi et al. 2005). Thus, distinct members of the IRF families have important, yet distinct functions in regulation of TLR7/9-induced, MyD88-dependent signaling cascades leading to pro-inflammatory cytokine production and secretion of type I IFNs (Fig. 2). Further studies will determine how the same signaling components (*e.g.*, MyD88-IRAK-4/IRAK-1-TRAF-6) can impart activation of both proinflammatory cytokine release and type I IFNs expression (*e.g.*, TLR7/TLR9), or only proinflammatory cytokine production (*e.g.*, TLR4-mediated MyD88-dependent signaling).

Sterile α and Armadillo motifs (SARM) is the fifth putative adapter that contains a TIR domain. SARM is a 690-amino-acid protein that expresses two sterile α motif (SAM) domains and an Armadillo repeat motif (ARM) and exhibits a high degree of sequence similarity to proteins in *Drosophila melanogaster* and *Caenorhabditis elegans*. SAM domains can homo- and hetero-oligomerize and mediate protein–protein interactions. The Armadillo repeat mediates the interaction of β-catenin with its ligands and is involved in protein–protein interactions with the small GTPase Ras (for a review, see O'Neill et al. 2003). Thus, the structural organization and domain functions of SARM are consistent with its regulatory role in signaling. Of note, a nematode ortholog of SARM, TIR-1, was found to be critical for TLR-independent innate immunity (Couillault et al. 2004). Recently, SARM was shown to negatively regulate TRIF-mediated, MyD88-independent signaling (Carty et al. 2006).

3.3 TLR Specificity for PAMPs in the Ectodomain and Adapters in the TIR Domain Underlie a Dual Recognition/Response System

Molecular mechanisms that control TLR signaling specificities are obscure at present and need to be elucidated further. Indeed, it is difficult to explain how putative ligand-recognition motifs within the leucine-rich, N-terminal domains enable surprisingly broad TLRs ligand specificities, especially evident for TLR2 and TLR4. In addition, no compelling evidence is available for direct agonist binding by various TLRs. However, TLR-mediated signaling can be blocked by antibodies raised against the N-terminal region of the respective TLRs (Flo et al. 2002; Uehori et al. 2003), and mutations in the extracellular region impair TLR signaling and are associated with infectious and autoimmune diseases (see below). Thus, it is plausible that TLRs respond to PAMPs presented by various accessory molecules primarily by engagement of their ectodomain. Vogel et al. (2003) proposed that interaction of different combinations of adapter molecules and kinases with individual TLRs yields distinct signaling complexes capable of activating different signaling cascades (*e.g.*, distinct repertoires of cytokines and other proinflammatory genes). For instance, TIRAP/MAL and TRAM were reported to be constitutively associated with TLR4, and LPS stimulation triggers recruitment of TRIF to TRAM and MyD88 to TIRAP/MAL. This was proposed to create a signaling platform for downstream kinases and adapters, leading to the recruitment of IRAK-4 and IRAK-1 to MyD88, while TBK-1 and IKKε are recruited to TRIF, kinase activation, and signaling for the full spectrum of MyD88-dependent and MyD88-independent pathways. In contrast, PAMP-mediated engagement of TLR2 triggers its association with MyD88 and TIRAP/MAL, but not TRIF/TRAM. As a result, TLR2-mediated signaling activates the MyD88-dependent pathway only and fails to trigger activation of IRF-3, STAT-1 phosphorylation, and induction of IFN-β or IFN-β-dependent genes (Vogel et al. 2003). Thus, the response to specific TLR agonists is constrained by two sets of specificities: through the N-terminal domain, sensing of specific PAMPs (without or with co-receptors) initiates TLR oligomerization and activation, while engagement of different combinations of adapter molecules through receptor/adapter TIR-TIR interactions at the intracellular TLR region leads to recruitment of the distinct combinations of enzymes and substrates associated with the specific receptor/adapter complex.

3.4 The IRAK Family: Key Regulators of TLR Signaling

IRAKs belong to the Ser/Thr kinase family and include two active kinases, IRAK-4 and IRAK-1, as well as two enzymatically inactive proteins, IRAK-2 and IRAK-M (Muzio et al. 1997; Kobayashi et al. 2002; Li et al. 2002; Wietek and O'Neill 2002; Janssens and Beyaert 2003). Human IRAK-1, IRAK-2, and IRAK-4 mRNA exhibit a wide tissue distribution, whereas IRAK-M mRNA is expressed predominantly in peripheral blood leukocytes and monocytic cell lines (for reviews, see Wietek and O'Neill 2002; Janssens and Beyaert 2003). All IRAKs express a conserved N-terminal death domain (DD) that functions as a protein interaction motif implicated in binding of IRAK-1 to MyD88, and a central kinase domain (KD) with structural features common to other serine/threonine kinase domains (Muzio et al. 1997; Wesche et al. 1999; Li et al. 2002; Janssens and Beyaert 2003). A functional ATP binding pocket with an invariant lysine residue in kinase subdomain II is characteristic of all IRAK species, whereas only IRAK-1 and IRAK-4 contain a functional catalytic site with a critical aspartate residue (D340 for IRAK-1, D311 for IRAK-4) in kinase subdomain VIb (Wesche et al. 1999). In contrast to IRAK-1, IRAK-2, and IRAK-M, IRAK-4 does not express a unique 90- to 170-residue C-terminal domain (Li et al. 2002). Both IRAK-4 and IRAK-1 exhibit kinase activity (Knopp and Martin 1999; Li et al. 2002), and IRAK-4 directly phosphorylates Thr387 and Ser376 in the activation loop of IRAK-1 (Li et al. 2002). However, kinase activity of IRAK-1 is not obligatory for signal transduction, as evidenced by the ability of kinase-defective IRAK-1 variants to restore IL-1 signaling in IRAK-1$^{-/-}$ HEK293 cells (Knopp and Martin 1999; Li et al. 1999; Jensen and Whitehead 2001). In contrast,

IRAK-4 appears to represent the only IRAK that requires kinase activity for signaling. Indeed, Lye et al. (2004) revealed impaired LPS-induced IRAK-1 recruitment and activation, and the failure of kinase-deficient mouse IRAK-4 species to restore NF-κB activation, in IRAK-4$^{-/-}$ mouse embryonic fibroblasts. Interestingly, whereas Qin et al. (2004) demonstrated that the "kinase-dead" human IRAK-4 KK213AA variant restores IL-1 signaling in IRAK-4$^{-/-}$ human fibroblasts upon transfection, a truncated IRAK-4 species (1-191 amino acids) failed to elicit NF-κB reporter activation. Our results also showed significantly suppressed LPS- and IL-1-mediated NF-κB activation, MAPK phosphorylation, and cytokine production in a patient who expressed two IRAK-4 variants with predicted truncations in the kinase domain (discussed below; Medvedev et al. 2003). Overexpression of truncated IRAK-4 species inhibited IL-1-mediated phosphorylation of endogenous IRAK-1 (Medvedev et al. 2003). Moreover, overexpressed IRAK-4 mutants showed impaired association with IL-1R, TLR4, and IRAK-1 and increased sequestration of MyD88 in the cytoplasm (Medvedev et al. 2005), underscoring a possible mechanism for deficient signaling by IRAK-4 species truncated within the kinase domain. The central role for IRAK-4 in TLR/IL-1R signaling is strongly supported by the profound hyporesponsive phenotype observed in IRAK-4 KO mice (Suzuki et al. 2002) and in humans with inherited IRAK-4 deficiencies, who suffer from recurrent pyogenic infections (Kuhns et al. 1997; Medvedev et al. 2003; Picard et al. 2003; Day et al. 2004). Overexpression of IRAK-1, IRAK-2, or IRAK-M, but not IRAK-4, triggers activation of NF-κB in IRAK-1$^{-/-}$ cells (Muzio et al. 1997; Li et al. 1999, 2002). In contrast to the severe effect of IRAK-4 deficiency (Suzuki et al. 2002), only partial inhibition of TLR- or IL-1R-mediated responses were found in IRAK-1 KO mice (Thomas et al. 1999; Swantek et al. 2000). These results suggest that IRAK-4 functions upstream of IRAK-1 and that IRAK-1, IRAK-2, and IRAK-M are functionally redundant. To confirm the position of IRAK-4 in the TLR signaling cascade, the effect of overexpression of upstream and downstream adapter proteins in IRAK-4-deficient mouse embryonic fibroblasts on NF-κB activation was assessed. Since overexpression of TRAF-6, but not MyD88 or TIRAP/Mal, stimulated NF-κB reporter activation, IRAK-4 was suggested to lie downstream of MyD88/Mal, but upstream of TRAF-6 (Suzuki et al. 2002). An alternatively spliced, truncated variant of MyD88, MyD88s, that lacks the intermediate domain between the N-terminal DD and the C-terminal TIR domain, interacts with IRAK-1, but fails to recruit IRAK-4 and trigger IRAK-1 phosphorylation (Janssens et al. 2002). These data strongly suggest that the intermediate domain of MyD88 is required for IRAK-1 phosphorylation, most likely, via the recruitment of IRAK-4. Thus, binding of IRAK-4 and IRAK-1 to distinct domains within MyD88 appears to enable sufficient proximity to facilitate IRAK-4-mediated IRAK-1 phosphorylation. In addition to its critical role in innate immunity, IRAK-4 is essential for T cell activation (Suzuki et al. 2006).

Other full-length or alternatively spliced IRAK species, IRAK-M and IRAK-2, appear to be mainly involved in the regulation of TLR signaling. Despite the presence of a functional ATP binding pocket, IRAK-2 and IRAK-M express a substitution of a critical aspartate residue in kinase subdomain VIb to an asparagine or a serine, respectively, that renders them catalytically inactive (Wesche et al. 1999). In contrast to ubiquitous expression of other IRAKs, IRAK-M is expressed primarily in monocytes and macrophages and is up-regulated upon TLR stimulation (Wesche et al. 1999). Both IRAK-M and IRAK-2 trigger NF-κB-dependent gene expression (Muzio et al. 1997; Wesche et al. 1999) and restore IL-1-induced signaling in IRAK-1-deficient cells upon overexpression (Li et al. 1999), suggesting functional redundancy for IRAK-2 and IRAK-M. IRAK-M appears to negatively regulate IL-1R and TLR signaling, as evidenced by increased cytokine production and inflammatory responses to bacterial infection observed in IRAK-M KO mice (Kobayashi et al. 2002). Overexpression of IRAK-M was found to prevent dissociation of IRAK-1 and IRAK-4 from MyD88 and association of IRAK-1 with TRAF6 complexes (Kobayashi et al. 2002), revealing a possible molecular mechanism of negative regulation of TLR signaling. IRAK-2, but not IRAK-1, interacts with TIRAP/Mal, and an IRAK-2, but not IRAK-1, dominant-negative construct suppresses TIRAP/Mal-induced NF-κB activation (Fitzgerald et al. 2001). Although these data suggest specific involvement of IRAK-2 in TIRAP/MAL signaling, further studies with IRAK-2 KO mice will be required to delineate the exact physiological role of IRAK-2. Interestingly, the mouse, but not human, *IRAK2* gene encodes four alternatively spliced isoforms, two of which, Irak2c and Irak2d, exert inhibitory effects on LPS-mediated NF-κB activation (Hardy and O'Neill 2004). Irak2c expression was shown to be up-regulated by LPS (Hardy and O'Neill 2004), suggesting that, at least in the mouse, alternatively spliced IRAK-2 negatively regulates TLR signaling.

Thus, IRAK-4 and IRAK-1 are critical kinases required for efficient signaling from IL-1RI and many TLRs, whereas IRAK-M and IRAK-2 lack the kinase domain and are preferentially involved in negative regulation of TLR signaling. Mutations within TLRs and IRAK-4 affect the ability of the host to mount efficient antibacterial immune responses, and the possible molecular mechanisms by which they affect TLR signaling are summarized in the next section.

4 Mutations in TLRs and IRAK-4: Implications for Disease

Since TLRs are essential sensors of PAMPs and endogenous "danger" proteins, TLR mutations would be expected to affect host innate immune defense, changing susceptibility to infectious, allergic, and autoimmune diseases. The first identified TLR polymorphisms were Asp299Gly and Thr399Ile amino acid substitutions within the ectodomain of TLR4 (Arbour et al. 2000). Analysis of 25 different publications indicated that these two mutations are expressed at a carrier frequency of approximately 11% and 7%, respectively, and an allele frequency of 6% and 3%, respectively (Vogel et al. 2005; Awomoyi et al. 2007), with evidence of co-segregation published in several studies (Arbour et al. 2000; Schmitt et al. 2002; Awomoyi et al. 2007). Arbour et al. (2000) reported that overexpression of the mutant Asp299Gly TLR4 in THP-1 cells suppressed LPS-mediated NF-κB activation, and our own results showed suppressed LPS-inducible activation of an NF-κB reporter in HEK293T cells that overexpress either of the two TLR4 mutations (Rallabhandi et al. 2006). Using the HEK293T transient transfection system, we observed the most pronounced suppressive effect upon overexpression of both polymorphic TLR4 variants (to mimic co-segregation; Rallabhandi et al. 2006). This result correlates with an earlier finding of decreased LPS-induced IL-1 production exhibited by human airway epithelial cells obtained from Asp299Gly/Thr399Ile heterozygous individuals (Arbour et al. 2000). At present, the molecular mechanisms by which TLR4 polymorphisms affect the ability of TLR4 to respond to PAMPs remain unclear. In their original report, Arbour et al. (2000) observed that human airway epithelia obtained from Asp299Gly/Thr399Ile heterozygous subjects exhibited lower TLR4 cell surface expression compared to that detected in control wild-type (WT) cells. However, our Western blot and FACS analyses did not confirm that the TLR4 polymophisms affect protein expression of TLR4 in HEK293T transfected with the expression vectors encoding WT and mutant TLR4 variants (Rallabhandi et al. 2006). Interestingly, overexpression of Flag-tagged WT and mutant TLR4 species at input concentrations that are constitutively active (due to Flag-enforced TLR4 oligomerization) led to comparable activation of an NF-κB reporter. In contrast, when untagged TLR4 is overexpressed to achieve agonist- and CD14/MD-2-dependent signaling, mutant TLR4 species showed an impaired ability to mediate NF-κB activation compared to wild-type (WT) TLR4 (Rallabhandi et al. 2006). Most importantly, a similar suppressive effect of these polymorphisms on TLR4-mediated NF-κB activation was observed using two other structurally distinct TLR4 agonists, the F protein of RSV and chlamydial Hsp60 (Rallabhandi et al. 2006). Our data suggest that Asp299Gly and Thr399Ile mutations either impair TLR4 interaction with agonists, or the co-receptors, CD14 and/or MD-2, and/or affect the ability of TLR4 to associate with downstream adapter molecules and kinases. Further analysis will be required to determine the exact molecular basis for the diminished response of mutant TLR4 molecules to PAMPs.

Because of the essential role of TLR4 in sensing PAMPs, many studies have been carried out to elucidate possible association of the TLR4 polymorphisms with various infectious and inflammatory diseases. The Asp299Gly polymorphism is associated with increased susceptibility to Gram negative bacteremia and sepsis (Lorenz et al. 2002a), and another study linked this mutation to an increased incidence of systemic inflammatory response syndrome (Agnese et al. 2002). Interestingly, the incidence of polymicrobial sepsis did not correlate with the presence of the Asp299Gly polymorphism (Feterowski et al. 2003), suggesting that this polymorphism is likely to predispose individuals selectively to sepsis caused by Gram negative infections. Mockenhaupt et al. (2006a) reported that the Asp299Gly and Thr399Ile TLR4 polymorphisms conferred 1.5- and 2.6-fold increased risks of severe malaria, respectively. Tal et al. (2004) found an association of the Asp299Gly and Thr399Ile mutations with an increased risk of severe RSV bronchiolitis in full-term, healthy Israeli infants, suggesting that these mutations impair TLR4 sensing of the F protein of RSV. In line with this suggestion, we recently demonstrated an impaired ability of the Asp299Gly and Thr399Ile TLR4 mutants overexpressed in

HEK293T cells to trigger NF-κB activation in response to the RSV F protein (Rallabhandi et al. 2006). Since intrauterine infection is the major cause of prematurity, it is plausible that impaired PAMP sensing by polymorphic TLR4 variants would lead to pre-term delivery. Indeed, Lorenz et al. (2002b) showed a positive association between prematurity and the Asp299Gly mutation in Finnish cohort. Awomoyi et al. (2007) studied the inheritance of these polymorphisms in a highly select group of high-risk children (premature, without or with bronchopulmonary dysplasia) with confirmed RSV and found that nearly 90% of these children expressed both the Asp299Gly and Thr399Ile polymorphisms. Studies are ongoing to determine whether these same polymorphisms predispose children to be delivered prematurely or whether diminished lung development observed in many premature children predisposes them to increased risk of RSV and other TLR4-dependent infections.

Of importance, the Asp299Gly *Tlr4* allele was found to be significantly more prevalent in patients with Crohn's disease (Franchimont et al. 2004), suggesting this TLR4 polymorphism as a risk factor for the development of Crohn's disease. However, the Asp299Gly and Thr399Ile mutations are not associated with other diseases including candidiasis (Morre et al. 2002), tuberculosis (Newport et al. 2004), or infection by *Chlamydia trachomatis* (Morre et al. 2003). Controversial data have been published regarding whether these TLR4 polymorphisms are related to the development of meningococcal disease. Whereas Read et al. (2001) reported that they are not associated with increased susceptibility to or severity of meningococcal disease, Faber et al. (2006) found that both the Asp299Gly and Thr399Ile mutations represent a risk factor for meningococcal disease in infants. In contrast, the Asp299Gly and Thr399Ile polymorphisms seem to confer a reduced risk for certain infectious and inflammatory diseases, including carotid artery atherosclerosis (Kiechl et al. 2002), acute coronary events (Ameziane et al. 2003), and rheumatoid arthritis (Kilding et al. 2003). Interestingly, the Asp299Gly TLR4 polymorphism was linked to increased susceptibility to infections with *Gardnerella vaginalis* and anaerobic Gram negative rods (Genc et al. 2004), but confers resistance to Legionnaires' disease (Hawn et al. 2005). Rare heterozygous missense mutations of TLR4 were found to be associated with the development of systemic meningococcal disease (Smirnova et al. 2003). These data suggest that various TLR4 mutations may differentially affect TLR4-mediated sensing of distinct PAMPs. Thus, TLR4 mutations reduce innate immune defense against bacterial, viral, and possibly other infectious agents while protecting from developing over-exuberant inflammatory responses characteristic of atherogenesis or rheumatoid arthritis.

Other TLRs have also been analyzed for the presence of polymorphisms that correlate with other infectious diseases and asthma. Two mutations in TLR2, Arg677Trp and Arg753Gln, have been reported to correlate with an increased susceptibility to leprosy and tuberculosis (Kang et al. 2002; Ben-Ali et al. 2004; Ogus et al. 2004). The Arg677Trp TLR2 polymorphism impairs the ability human PBMC to secrete IL-2, TNF-α, IL-12, and IFN-γ in response to *Mycobacterium leprae* (Kang et al. 2002, Kang et al. 2004), indicating defective signal-transducing capacity of polymorphic TLR2 variants. The Arg753Gln TLR2 mutation was originally associated with susceptibility to staphylococcal infections, and cells obtained from individuals with this TLR2 mutation showed diminished responses to bacterial lipopeptides obtained from *Borellia burgdorferi* and *Treponema pallidum* (Lorenz et al. 2000). However, a later study failed to confirm a link between the presence of these TLR2 polymorphisms and the occurrence of severe staphylococcal infections (Moore et al. 2004). Interestingly, the Arg753Gln TLR2 polymorphism was linked to protection from the development of Lyme disease associated with profound inflammatory reactions (Schroder et al. 2005). This protection is likely to be due to impaired production of proinflammatory TNF-α and IFN-γ by *B. burgdorferi*-stimulated PBMC obtained from individuals who express this TLR2 polymorphism heterozygously (Schroder et al. 2005). A mutation in TLR5, a common stop codon 392STOP, leads to a decreased response to bacterial flagellin *in vitro* and is associated with increased susceptibility to a pneumonia caused by a flagellated bacterium, *Legionella pneumophila* (Hawn et al. 2003). Controversial results have been reported in the literature with respect to whether TLR polymorphisms are associated with asthma. For example, studies of different ethnic groups led to contradictory results, with the Asp299Gly and A896G TLR4 polymorphisms being associated with a predisposition to asthma in Swedish and Turkish children (Fageras Bottcher et al. 2004; Sackesen et al. 2005), but with no similar association in UK and Canadian Caucasian families, as well as in a Japanese population (Raby et al. 2002; Noguchi et al. 2004; Yang et al. 2004). Thus, the Asp299Gly polymorphism may be predictive of airway and atopic responses in specific subsets of certain populations. A promoter polymorphism for *TLR9*, -1237C, was identified as predisposing to the development of asthma in Europeans (Lazarus et al. 2003), whereas a *TLR2* promoter polymorphism (-16934 A→T) was associated with a lower risk of developing asthma (Eder et al. 2004). In addition, in *Plasmodium falciparum*-infected women, both the TLR4 Asp299Gly and the TLR9 T-1486C polymorphisms were linked to a 6-fold increased risk of low birth weight in term infants (Mockenhaupt et al. 2006b), suggesting their role in the manifestation of malaria during pregnancy. Thus, mutations in TLRs impair responses to various PAMPs and are linked to certain infectious diseases and asthma.

Mutations in *IRAK4* have been linked to the hyporesponsive phenotype in more than 25 patients within many unrelated families. These patients suffer from repeated, life-threatening Gram positive infections early in life that correlate with severely suppressed macrophage responses to LPS, IL-1, IL-18, and TLR agonists, but not to TNF-α or PMA (for reviews, see Puel et al. 2005; Vogel et al. 2005; Medvedev et al. 2006). Picard et al. (2003) first identified autosomal recessive mutations in exons 7 and 8 of the *IRAK-4* gene in three unrelated patients suffering from pyogenic infections and hyporesponsive to IL-1, IL-18, LPS, as well as to TLR2, TLR3, TLR5, TLR9 agonists. One patient had a homozygous deletion of thymidine in exon 7 (821delT in mRNA), whereas two others expressed the identical point mutation in exon 8 (C877T substitution in mRNA). These were predicted to yield premature stop codons and the lack of expression of functional full-length IRAK-4 mRNA and protein (Picard et al. 2003). Interestingly, Day et al. (2004) reported that the third patient, who exhibited sustained recurring bacterial and fungal infections, also showed an impaired antibody response upon booster immunizations with diphtheria/tetanus toxoid, pneumococcal polysaccharide, and bacteriophage øX174. These data suggest an association of IRAK-4 deficiency with the failure of B lymphocytes to be properly activated. An alternative interpretation would be that the mutation of IRAK-4 may lead to decreased co-stimulatory molecule expression on antigen-presenting cells, resulting in insufficient T-helper function (that, in turn, would limit B cell response to antigens). Further studies will be required to determine the exact molecular mechanism of impaired antibody production in IRAK-4-deficient patients.

Our own studies identified a patient with two distinct mutations in *IRAK4*, a point mutation C877T [identical to the one described by Picard et al. (2003)], and a two nucleotide deletion (620–621del), both of which were expressed heterozygously (Fig 3; Medvedev et al. 2003), in contrast to autosomal recessive genotypes expressed by all but one of the other patients genotyped to date (Puel et al. 2005; Vogel et al. 2005). The point mutation expressed by our patient has since been observed in at least five unrelated families from eight different countries that differ in their ethnicities (Picard et al. 2003; Puel et al. 2005; Vogel et al. 2005; D. Speert, personal communication). Although a founder effect has not been formally excluded, these patients were from widely differing geographic locations and of diverse ethnicities. In addition, the sequence surrounding the C877T point mutation does not appear to be a typical hypermutable sequence or "hot spot" (Jacobs and Bross 2001). The compound heterozygous genotype of our IRAK-4-deficient patient was associated with a history of repeated bacterial infections and the failure to develop febrile response or produce cytokines upon LPS challenge *in vivo*

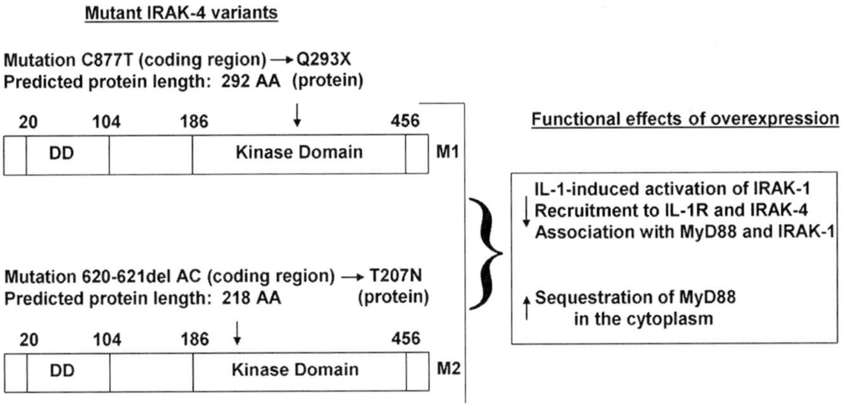

Fig. 3 IRAK-4 mutations in a patient with a compound heterozygous genotype and their effects on IL-1- and TLR4-mediated signaling. Two IRAK-4 mutations identified in our patient suffering from repeated bacterial infections and hyporesponsive to LPS and IL-1 are depicted: the C877T point mutation and the 620–621AC del. Shown are the points of truncations in IRAK-4 proteins (indicated by *arrows*) and predicted changes in their amino acid composition. The functional effects of overexpression of truncated, kinase-deficient IRAK-4 mutant variants on IL-1- and TLR4-signaling revealed by overexpression analysis are depicted on the *right*

(Kuhns et al. 1997). *In vitro*, our patient's PBMC and neutrophils were hyporesponsive to LPS and IL-1, as evidenced by markedly suppressed NF-κB activation and p38 MAPK phosphorylation, yet TNF-α responsiveness was normal (Medvedev et al. 2003). In addition, our patient's polymorphonuclear cells (PMNs) failed to upregulate expression of a number of co-stimulatory molecules, including CD18 and CD67, in response to LPS, while their responses to a non-TLR-specific stimulus, fMLF, were not inhibited (Medvedev et al. 2005). Interestingly, our patient showed an abnormal inflammatory response to a non-microbial stimulus *in vivo*, as evidenced by depressed influx of neutrophils and decreased cytokine production in the region of the blister, whereas complement activation was normal (Medvedev et al. 2003, 2005). These data indicate that the deficiency in IRAK-4 in this patient also underlies decreased responsiveness to non-microbial insults, suggesting impaired TLR signaling upon recognition of endogenous TLR ligands (e.g., HSPs, fibrinogen) released during tissue damage.

Translation of IRAK-4 mRNAs expressed by our patient (i.e., the C877T point mutation, the 620–621del two-nucleotide deletion), resulted in the predicted truncated forms of the IRAK-4 protein, with both truncations occurring within the kinase domain (Fig. 3; Medvedev et al. 2003). Despite extensive efforts to detect endogenous IRAK-4 species in PBMC cell lysates and IRAK-1 immune complexes obtained from our patient versus healthy volunteers, we were unable to detect IRAK-4 protein by Western blot analysis using a polyclonal antiserum obtained from Tularik. This, most likely, was the result of extremely low levels of expression of endogenous IRAK-4 protein. Indeed, reliable detection of endogenous IRAK-4 in the HEK293 cell line has been reported by others to require a very large number of cells (~200×10^6 cells per treatment; Li et al. 2002). Unfortunately, this large number of PBMCs could not be obtained from from our patient. Therefore, we used overexpression approaches to analyze the effect of mutations found in patient's IRAK-4 mRNA on IRAK-4 protein. HEK293T cells were transfected with the pRK7-IRAK-4 expression vector encoding Flag-tagged WT, the point mutation (mutant 1, or M1) or the deletion mutation (M2) IRAK-4, followed by immunoprecipitation and immunoblot analysis of overexpressed IRAK-4 species with anti-Flag antibody. As expected, transfection of HEK293T cells with expression vectors encoding mutant IRAK-4 species yielded truncated species of IRAK-4, in contrast to the full-length wild-type IRAK-4 (Fig. 3; Medvedev et al. 2003, 2005). Interestingly, both mutant IRAK-4 forms act as dominant negative inhibitors of wild-type, endogenous IRAK-4 upon their overexpression in HEK293 cells, as evidenced by suppression of IL-1-induced IRAK-1 kinase activity (Medvedev et al. 2003) and recruitment of endogenous IRAK-1 and MyD88 to IL-1RI (Medvedev et al. 2005). Subsequent co-immunoprecipitation analyses revealed that mutations in the kinase domain of IRAK-4 impairs its IL-1-inducible association with IL-1RI and IRAK-1 (Medvedev et al. 2005), and failed to be recruited to TLR4 upon stimulation of HEK293/TLR4/MD-2 cells with LPS (Medvedev et al. 2006). We also found that overexpression of truncated IRAK-4 variants results in constitutive association of kinase-defective IRAK-4 with endogenous or overexpressed cytoplasmic MyD88 (Medvedev et al. 2005). These results demonstrate that the hyporesponsive phenotype of our IRAK-4-defective patient may result from the failure of mutant IRAK-4 species to form functional signaling complexes with components of the IL-1R/TLR4 pathways in response to stimulation with IL-1 and LPS (summarized in Fig. 3). Since the heterozygous parents, grandparents, and sibling of this patient show normal LPS sensitivity and resistance to infection, it is plausible that normal IRAK-4 protein (present in quantities at least equivalent to mutant IRAK-4 in these individuals) compensates adequately for the failure of the "truncated" forms of IRAK-4 to form functional TLR/IL-1R signaling complexes. In contrast, upon overexpression, a large proportion of kinase-defective IRAK-4 variants are likely to be present in the cells relative to the amount of WT IRAK-4, leading to an inhibitory effect. Our results also suggest that small molecular weight mimetic compounds may be designed based on molecular structures of truncated IRAK-4 species to diminish IRAK-4-dependent signaling in people with hyperinflammatory syndromes (Medvedev et al. 2005).

5 Conclusions

Rapid progress has been made over the past decade in the identification and functional characterization of receptors responsible for induction and maintenance of antimicrobial and antiviral innate immune responses. TLRs have been shown to represent principal sensors of bacterial and viral infections, as well as receptors involved in recognition of "danger" endogenous proteins, oligosaccharides, and nucleic acids released as a consequence of cellular damage. The extreme complexity and potential for cross-talk within the MyD88-dependent and MyD88-independent TLR signaling pathways has been revealed, and several TLR polymorphisms have been associated with predisposition to a number of infectious diseases or reducing the risk for development of inflammatory diseases, *e.g.*, atherosclerosis. Because of the central role for IRAK-4 in TLR signaling, it is perhaps no surprise that mutations within the kinase domain of IRAK-4 have been shown to severely affect antibacterial immune defense mechanisms. This defect is associated with the failure of mutant IRAK-4 species to form functional signaling complexes with receptor and intracellular components of the IL-1R/TLR4 pathway. The capacity of overexpressed, truncated IRAK-4 species to inhibit assembly of signaling complexes among wild-type IRAK-1, TLR4, and MyD88 may be employed for possible future therapeutic intervention in hyperinflammatory states by using kinase-inactive IRAK-4 mimetics to inhibit signaling. Future studies will likely reveal the feasibility of such an approach, and delineate how mutations within TLRs and IRAK-4 affect other antibacterial immune defense mechanisms, including the development of adaptive immunity.

Acknowledgements This work was supported by NIH grants AI47233, AI057490, AI44936, AI18797 (to S.N.V.), and AI-059524 (to A.E.M.).

References

Abe T, Hemmi H, Miyamoto H, Moriishi K, Tamura S, Takaku H, Akira S, Matsuura Y (2005) Involvement of the Toll-like receptor 9 signaling pathway in the induction of innate immunity by baculovirus. J Virol 79:2847–2858

Agnese DM, Calvano JE, Hahm SJ, Coyle SM, Corbett SA, Calvano SE, Lowry SF (2002) Human toll-like receptor 4 mutations but not CD14 polymorphisms are associated with an increased risk of gram-negative infections. J Infect Dis 186:1522–1525

Ahmad-Nejad P, Kacker H, Rutz M, Bauer S, Vabulas RM, Wagner H (2002) Bacterial CpG-DNA and lipopolysaccharides activate Toll-like receptors at distinct cellular compartments. Eur J Immunol 32: 1958–1968

Akira S, Takeda K (2004) Toll-like receptor signaling. Nat Rev Immunol 4:499–511

Alexopoulou L, Holt AC, Medzhitov R, Flavell RA (2001) Recognition of double-stranded RNA and activation of NF-kappaB by Toll-like receptor 3. Nature 413:732–738

Ameziane N, Beillat T, Verpillat P, Chollet-Martin S, Aumont MC, Seknadji P, Lebret D, Ollivier V, Prost D de (2003) Association of the Toll-like receptor 4 gene Asp299Gly polymorphism with acute coronary events. Arterioscler Thromb Vasc Biol 23:e61–e64

Anderson KV, Bokla L, Nusslein-Volhard C (1985) Establishment of dorsal-ventral polarity in the *Drosophila* embryo: the induction of polarity by the Toll gene product. Cell 42:791–798

Arbibe L, Mira JP, Teusch N, Guha M, Macman N, Godowski PJ, Ulevitch RJ, Knaus UG (2000) Toll-like receptor 2-mediated NF-kappa B activation requires a Rac1-dependent pathway. Nat Immunol 1:533–540

Arbour NC, Lorenz E, Schutte BC, Zabner J, Kline JN, Jones M, Frees K, Watt JL, Schwartz DA (2000) TLR4 mutations are associated with endotoxin hyporesponsiveness in humans. Nat Genet 25:187–191

Asea A, Rehli M, Kabingu E, Boch JA, Bare O, Auron PE, Stevenson MA, Calderwood SK (2002) Novel signal transduction pathway utilized by extracellular HSP70: role of toll-like receptor (TLR) 2 and TLR4. J Biol Chem 277:15028–15034

Awomoyi AA, Rallabhandi P, Pollin TI, Lorenz E, Sztein MB, Boukhvalova MS, Hemming VG, Blanco JCG, Vogel SN (2007) Association of TLR4 polymorphisms with symptomatic respiratory syncytial virus infection in high-risk infants and young children. J Immunol (in press)

Bell JK, Askins J, Hall PR, Davies DR, Segal DM (2006) The dsRNA binding site of human Toll-like receptor 3. Proc Natl Acad Sci USA 103:8792–8797

Bellocchio S, Montagnoli C, Bozza S, Gaziano R, Rossi G, Mambula SS, Vecchi A, Mantovani A, Levits SM, Romani L (2004) The contribution of the Toll-like/IL-1 receptor superfamily to innate and adaptive immunity to fungal pathogens in vivo. J Immunol 172:3059–3069

Ben-Ali M, Barbouche MR, Bousnina S, Chabbou A, Dellagi K (2004) Toll-like receptor 2 Arg677Trp polymorphism is associated with susceptibility to tuberculosis in Tunisian patients. Clin Diagn Lab Immunol 11:625–626

Biragyn A, Ruffini PA, Leifer CA, Klyushnenkova E, Shakhov A, Chertov O, Shirakawa AK, Farber JM, Segal DM, Oppenheim JJ, Kwak LW (2002) Toll-like receptor 4-dependent activation of dendritic cells by beta-defensin 2. Science 298:1025–1029

Blander JM, Medzhitov R (2004) Regulation of phagosome maturation by signals from toll-like receptors. Science 304:1014–1018

Bulut Y, Faure E, Thomas L, Equilis O, Arditi M (2001) Cooperation of Toll-like receptor 2 and 6 for cellular activation by soluble tuberculosis factor and *Borrelia burgdorferi* outer surface protein A lipoprotein: role of Toll-interacting protein and IL-1 receptor signaling molecules in Toll-like receptor 2 signaling. J Immunol 167:987–994

Bulut Y, Faure E, Thomas L, Karahashi H, Michelsen KS, Equils O, Morrison SG, Arditi M (2002) Chlamydial heat shock protein 60 activates macrophages and endothelial cells through Toll-like receptor 4 and MD2 in a MyD88-dependent pathway. J Immunol 168:1435–1440

Burns K, Martinon F, Esslinger C, Pahl H, Schneider P, Bodmer J-L, Di Marco F, French L, Tschopp J (1998) MyD88, an adaptor protein involved in IL-1 signaling. J Biol Chem 273:12203–12209

Burns K, Cladworthy J, Martin L, Martinon F, Plumpton C, Maschera B, Lewis A, Ray K, Tschopp J, Volpe F (2000) Tollip, a new component of the IL-1RI pathway, links IRAK to the IL-1 receptor. Nat Cell Biol 2:346–351

Buwitt-Beckmann U, Heine H, Wiesmuller KH, Jung G, Brock R, Akira S, Ulmer AJ (2006) TLR1- and TLR6-independent recognition of bacterial lipopeptides. J Biol Chem 281:9049–9057

Carty M, Goodbody R, Schroder M, Stack J, Moynagh PN, Bowie AG (2006) The human adaptor SARM negatively regulates adaptor protein TRIF-dependent Toll-like receptor signaling. Nat Immunol 7:1074–1081.

Chen LY, Zuraw BL, Zhao M, Liu FT, Huang S, Pan ZK (2003) Involvement of protein tyrosine kinase in Toll-like receptor 4-mediated NF-kappa B activation in human peripheral blood monocytes. Am J Physiol Lung Cell Mol Physiol 284:L607–L613

Choe J, Kelker MS, Wilson IA (2005) Crystal structure of human toll-like receptor 3 (TLR3) ectodomain. Science 309:581–585

Chow JC, Young DW, Golenbock DT, Christ WJ, Gusovsky F (1999) Toll-like receptor 4 mediates lipopolysaccharide-induced signal transduction. J Biol Chem 274:10689–10692

Chuang T, Ulevitch RJ (2001) Identification of hTLR10: a novel human Toll-like receptor preferentially expressed in immune cells. Biochim Biophys Acta 1518:157–161

Couillault C, Pujol N, Reboul J, Sabatier L, Guichou JF, Kohara Y, Ewbank JJ (2004) TLR-independent control of innate immunity in Caenorhabditis elegans by the TIR domain adaptor protein TIR-1, an ortholog of human SARM. Nat Immunol 5:488–494

Da Silva Correia J, Soldau K, Christen U, Tobias PS, Ulevitch RJ (2001) Lipopolysaccharide is in close proximity to each of the proteins in its membrane receptor complex: transfer from CD14 to TLR4 and MD-2. J Biol Chem 276:21129–21135

Day N, Tangsinmankong N, Ochs H, Rucker R, Picard C, Casanova JL, Haraguchi S, Good R (2004) Interleukin receptor-associated kinase (IRAK-4) deficiency associated with bacterial infections and failure to sustain antibody responses. J Pediatr 144:524–526

Deng L, Wang C, Spencer E, Yang L, Braun A, You J, Slaughter C, Pickart C, Chen ZJ (2000) Activation of the IκB kinase complex by TRAF6 requires a dimeric ubiquitin-conjugating enzyme complex and a unique polyubiquitin chain. Cell 103:351–361

Dziarski R, Wang Q, Miyake K, Kirschning CJ, Gupta D (2001) MD-2 enables Toll-like receptor 2 (TLR2)-mediated responses to lipopolysaccharide and enhances TLR2-mediated responses to Gram-positive and Gram-negative bacteria and their cell wall components. J Immunol 166:1938–1944

Ea CK, Sun L, Inoue J, Chen ZJ (2004) TIFA activates IkappaB kinase (IKK) by promoting oligomerization and ubiquitination of TRAF6. Proc Natl Acad Sci USA 101:15318–15323

Eder W, Klimecki W, Yu L, Mutius E von, Riedler J, Braun-Fahrander C, Nowak D, Martinez FD (2004) Toll-like receptor 2 as a major gene for asthma in children of European farmers. J Allergy Clin Immunol 113:482–488

Faber J, Meyer CU, Gemmer C, Russo A, Finn A, Murdoch C, Zenz W, Mannhalter C, Zabel BU, Schmitt HJ, Habermehl P, Zepp F, Knuf M (2006) Human toll-like receptor 4 mutations are associated with susceptibility to invasive meningococcal disease in infancy. Pediatr Infect Dis J 25:80–81

Fageras Bottcher M, Hmani-Aifa M, Lindstrom A, Jenmalm MC, Mai XM, Nilsson L, Zdolek HA, Bjorksten B, Soderkvist P, Vaarala O (2004) A TLR4 polymorphism is associated with asthma and reduced lipopolysaccharide-induced interleukin-12(p70) responses in Swedish children. J Allergy Clin Immunol 114:561–567

Feterowski C, Emmanuilidis K, Miethke T, Gerauer K, Rump M, Ulm K, Holzmann B, Weighardt H (2003) Effects of functional Toll-like receptor-4 mutations on the immune response to human and experimental sepsis. Immunology 109:426–431

Fitzgerald KA, Palsson-McDermott EM, Bowie AG, Jefferies CA, Mansell AS, Brady G, Brint E, Dunne A, Gray P, Harte MT, McMurray D, Smith DE, Sims JE, Bird TA, O'Neill LA (2001) Mal (MyD88-adapter-like) is required for Toll-like receptor-4 signal transduction. Nature 413:78–83

Fitzgerald KA, Rowe DC, Barnes BJ, Caffrey DR, Visintin A, Latz E, Monks B, Pitha PM, Golenbock DT (2003) LPS-TLR4 signaling to IRF-3/7 and NF-κB involves the Toll adapters TRAM and TRIF. J Exp Med 198:1043–1055

Flo TH, Ryan L, Latz E, Takeuchi O, Monks BG, Lien E, Halaas O, Akira S, Skjak-Braek, Golenbock DT, Espevik T (2002) Involvement of toll-like receptor (TLR) 2 and TLR4 in cell activation by mannuronic acid polymers. J Biol Chem 277:35489–3595

Franchimont D, Vermeire S, El Housni H, Pierik M, Van Steen K, Gustot T, Quertinmont E, Abramowicz M, Van Gossum A, Deviere J, Rutgeerts P (2004) Deficient host-bacteria interactions in inflammatory bowel disease? The toll-like receptor (TLR)-4 Asp299gly polymorphism is associated with Crohn's disease and ulcerative colitis. Gut 53:987–992

Genc MR, Vardhana S, Delaney ML, Onderdonk A, Tuomala R, Norwitz E, Witkin SS (2004) Relationship between a toll-like receptor-4 gene polymorphism, bacterial vaginosis-related flora and vaginal cytokine responses in pregnant women. Eur J Obstet Gynecol Reprod Biol 116:152–156

Gewirtz AT, Navas TA, Lyons S, Godowski PJ, Madara JL (2001) Cutting edge: bacterial flagellin activates basolaterally expressed TLR5 to induce epithelial proinflammatory gene expression. J Immunol 167:1882–1885

Golenbock DT, Liu Y, Millham FH, Freeman MW, Zoeller RA (1993) Surface expression of human CD14 in Chinese hamster ovary fibroblasts imparts macrophage-like responsiveness to bacterial endotoxin. J Biol Chem 268:22055–22059

Gray P, Dunne A, Brikos C, Jefferies CA, Doyle SL, O'Neill LA (2006) MyD88 adapter-like (Mal) is phosphorylated by Bruton's tyrosine kinase during TLR2 and TLR4 signal transduction. J Biol Chem 281:10489–10495

Guillot L, Balloy V, McCormack FX, Golenbock DT, Chignard M, Si-Tahar M (2002) Cutting edge: the immunostimulatory activity of the lung surfactant protein-A involves Toll-like receptor 4. J Immunol 168:5989–5992

Hacker H, Redecke V, Blagoev B, Kratchmarova I, Hsu LC, Wang GG, Kamps MP, Raz E, Wagner H, Hacker G, Mann M, Karin M (2006) Specificity in Toll-like receptor signalling through distinct effector functions of TRAF3 and TRAF6. Nature 439:204–207

Hajjar AM, O'Mahony DS, Ozinsky A, Underhill DM, Aderem A, Klebanoff SJ, Wilson CB (2001) Cutting edge: functional interactions between toll-like receptor (TLR) 2 and TLR1 or TLR6 in response to phenol-soluble modulin. J Immunol 166:15–19

Hardy M, O'Neill L (2004) The murine Irak2 gene encodes four alternatively spliced isoforms, two of which are inhibitory. J Biol Chem 279:27669–27708

Hasan U, Chaffois C, Gaillard C, Saulnier V, Merck E, Tancredi S, Guiet C, Briere F, Vlach J, Lebecque S, Trinchieri G, Bates EE (2005) Human TLR10 is a functional receptor, expressed by B cells and plasmacytoid dendritic cells, which activates gene transcription through MyD88. J Immunol 174:2942–2950

Hawn TR, Verbon A, Lettinga KD, Zhao LP, Li SS, Laws RJ, Skerret SJ, Beutler B, Schroeder L, Nachman A, Ozinsky A, Smith KD, Aderem A (2003) A common dominant TLR5 stop codon polymorphism abolishes flagellin signaling and is associated with susceptibility to legionnaires' disease. J Exp Med 198:1563–1572

Hawn TR, Verbon A, Janer M, Zhao LP, Beutler B, Aderem A (2005) Toll-like receptor 4 polymorphisms are associated with resistance to Legionnaires' disease. Proc Natl Acad Sci USA 102:2487–2489

Hayashi F, Smith KD, Ozinsky A, Hawn TR, Yi EC, Goodlett DR, Eng JK, Akira S, Underhill DM, Aderem A (2001) The innate immune response to bacterial flagellin is mediated by Toll-like receptor 5. Nature 410:1099–10103

Haziot A, Ferrero E, Kontgen F, Hijiya N, Yamomoto S, Silver J, Stewart CL, Goyert SM (1996) Resistance to endotoxin shock and reduced dissemination of Gram-negative bacteria in CD14-deficient mice. Immunity 4:407–414

Heil F, Ahmad-Nejad P, Hemmi H, Hochrein H, Ampenberger F, Gellert T, Dietrich H, Lipford G, Takeda K, Akira S, Wagner H, Bauer S (2003) The Toll-like receptor 7 (TLR7)-specific stimulus loxoribine uncovers a strong relationship within the TLR7, 8 and 9 subfamily. Eur J Immunol 33:2987–2997

Heil F, Hemmi H, Hochrein H, Ampenberger F, Kirschning C, Akira S, Lipford G, Wagner H, Bauer S (2004) Species-specific recognition of single-stranded RNA via toll-like receptor 7 and 8. Science 303:1526–1529

Hemmi H, Takeuchi O, Kawai T, Kaisho T, Sato S, Sanjo H, Matsumoto M, Hoshino K, Wagner H, Takeda K, Akira S (2000) A Toll-like receptor recognizes bacterial DNA. Nature 408:740–745

Hemmi H, Kaisho T, Takeuchi O, Sato S, Sanjo H, Hoshino K, Horiuchi T, Tomizawa H, Takeda K, Akira S (2002) Small anti-viral compounds activate immune cells via the TLR7 MyD88-dependent signaling pathway. Nat Immunol 3:196–200

Hirschfeld M, Ma Y, Weis JH, Vogel SN, Weis JJ (2000) Cutting edge: repurification of lipopolysaccharide eliminates signaling through both human and mouse Toll-like receptor 2. J Immunol 165:618–622

Hochrein H, Schlatter B, O'Keeffe M, Wagner C, Schmitz F, Schiemann M, Bauer S, Suter M, Wagner H (2004) Herpes simplex virus type-1 induces IFN-alpha production via Toll-like receptor 9-dependent and -independent pathways. Proc Natl Acad Sci USA 101:11416–11421

Hoebe K, Du X, Georgel P, Janssen E, Tabeta K, Kim SO, Goode J, Lin P, Mann N, Mudd S, Crozat K, Sovath S, Han J, Beutler B (2003) Identification of Lps2 as a key transducer of MyD88-independent TLR signaling. Nature 424:743–748

Hoebe K, George P, Rutschmann S, Du X, Mudd S, Crozat K, Sovath S, Shamet L, Hartung T, Zahringer U, Beutler B (2005) CD36 is a sensor of diacylglycerides. Nature 433:523–527

Honda K, Yanai H, Mizutani T, Negishi H, Shimada N, Suzuki N, Ohba Y, Takaoka A, Yeh WC, Taniguchi T (2004) Role of a transductional-transcriptional processor complex involving MyD88 and IRF-7 in Toll-like receptor signaling. Proc Natl Acad Sci USA 101:15416–14521

Honda K, Ohba Y, Yanai H, Negishi H, Mizutani T, Takaoka A, Taya C, Taniguchi T (2005) Spatiotemporal regulation of MyD88-IRF-7 signalling for robust type-I interferon induction. Nature 434:1035–1040

Horng T, Barton GM, Medzhitov R (2001) TIRAP: an adapter molecule in the Toll signaling pathway. Nat Immunol 2:835–841

Horng T, Barton GM, Flavell RA, Medzhitov R (2002) The adaptor molecule TIRAP provides signalling specificity for Toll-like receptors. Nature 420:329–333

Hoshino K, Takeuchi O, Kawai T, Sanjo H, Ogawa T, Takeda Y, Takeda K, Akira S (1999) Cutting edge: Toll-like receptor 4 (TLR4)-deficient mice are hyporesponsive to lipopolysaccharide: evidence for TLR4 as the Lps gene product. J Immunol 162:3749–3752

Inamori K, Ariki S, Kawabata S (2004) A Toll-like receptor in horseshoe crabs. Immunol Rev 198:106–115

Ingalls RR, Monks BG, Savedra RJr, Christ WJ, Delude RL, Medvedev AE, Espevik T, Golenbock DT (1998) CD11/CD18 and CD14 share a common lipid A signaling pathway. J Immunol 161:5413–5420

Inohara N, Nunez G (2003) NODs: intracellular proteins involved in inflammation and apoptosis. Nat Rev Immunol 3:371–382

Iwasaki A, Medzhitov R (2004) Toll-like receptor control of the adaptive immune responses. Nat Immunol 5:987–995

Jacobs H, Bross L (2001) Towards an understanding of somatic hypermutation. Curr Opin Immunol 13:208–218

Janeway CA Jr, Medzhitov R (2002) Innate immune recognition. Annu Rev Immunol 20:197–216

Janssens S, Beyaert R (2003) Functional diversity and regulation of different interleukin-1 receptor-associated kinase (IRAK) family members. Mol Cell 11:293–302

Janssens S, Burns K, Tschopp J, Beyaert R (2002) Regulation of interleukin-1- and lipopolysaccharide-induced NF-κB activation by alternative splicing of MyD88. Curr Biol 12:467–471

Jensen LE, Whitehead AS (2001) IRAK1b, a novel alternative splice variant of interleukin-1 receptor-associated kinase (IRAK), mediates interleukin-1 signaling and has prolonged stability. J Biol Chem 276:29037–29044

Jensen LE, Whitehead AS (2003) Pellino2 activates the mitogen activated protein kinase pathway. FEBS Lett 545:199–202

Jiang Z, Ninomiya-Tsuji J, Qian Y, Matsumoto K, Li X (2002) Interleukin-1 (IL-1) receptor-associated kinase-dependent IL-1-induced signaling complexes phosphorylate TAK1 and TAB2 at the plasma membrane and activate TAK1 in the cytosol. Mol Cell Biol 22:7158–7167

Jiang Z, Johnson HJ, Nie H, Qin J, Bird TA, Li X (2003) Pellino 1 is required for Interleukin-1 (IL-1)-mediated signaling through its interaction with the IL-1 receptor-associated kinase 4 (IRAK4)-IRAK-tumor necrosis factor receptor-associated factor 6 (TRAF6) complex. J Biol Chem 278:10952–10956

Jiang Z, Georgel P, Du X, Shamel L, Sovath S, Mudd S, Huber M, Kalis C, Keck S, Galanos C, Freudenberg M, Beutler B (2005) CD14 is required for MyD88-independent LPS signaling. Nat Immunol 6:565–570

Jurk M, Heil F, Vollmer J, Schetter C, Krieg AM, Wagner H, Lipford G, Bauer S (2002) Human TLR7 or TLR8 independently confer responsiveness to the antiviral compound R-848. Nat Immunol 3:499

Kagan JC, Medzhitov R (2006) Phosphoinositide-mediated adaptor recruitment controls Toll-like receptor signaling. Cell 125:943–955

Kaisho T, Akira S (2006) Toll-like receptor function and signaling. J Allergy Clin Immunol 117:979–987

Kaisho T, Takeuchi O, Kawai T, Hoshino K, Akira S (2001) Endotoxin-induced maturation of MyD88-deficient dendritic cells. J Immunol 166:5688–5694

Kang TJ, Lee SB, Chae GT (2002) A polymorphism in the toll-like receptor 2 is associated with IL-12 production from monocyte in lepromatous leprosy. Cytokine 20:56–62

Kang TJ, Yeum CE, Kim BC, You EY, Chae GT (2004) Differential production of interleukin-10 and interleukin-12 in mononuclear cells from leprosy patients with a Toll-like receptor 2 mutation. Immunology 112:674–680

Kariko K, Ni H, Capodici J, Lamphier M, Weissman D (2004) mRNA is an endogenous ligand for Toll-like receptor 3. J Biol Chem 279:12542–12550

Kato H, Takeuchi O, Sato S, Yoneyama M, Yamamoto M, Matsui K, Uematsu S, Jung A, Kawai T, Ishii KJ, Yamaguchi O, Otsu K, Tsujimura T, Koh CS, Reis e Sousa C, Matsuura Y, Fujita T, Akira S (2006) Differential roles of MDA5 and RIG-I helicases in the recognition of RNA viruses. Nature 441:101–105

Kawai T, Akira S (2006) Innate immune recognition of viral infection. Nat Immunol 7:131–137

Kawai T, Adashi O, Ogawa T, Takeda K, Akira S (1999) Unresponsiveness of MyD88-deficient mice to endotoxin. Immunity 11:115–122

Kawai T, Takeuchi O, Fujita T, Inoue J, Muhlradt PF, Sato S, Hoshino K, Akira D (2001) Lipopolysaccharide stimulates the MyD88-independent pathway and results in activation of IFN-regulatory factor 3 and the expression of a subset of lipopolysaccharide-inducible genes. J Immunol 167:5887–94

Kiechl S, Lorenz E, Reindl M, Wiedermann CJ, Oberhollenzer F, Bonora E, Williet J, Schwartz DA (2002) Toll-like receptor 4 polymorphisms and atherogenesis. N Engl J Med 347:185–192

Kilding R, Akil M, Till S, Amos R, Winfield J, Iiles MM, Wilson AG (2003) A biologically important single nucleotide polymorphism within the toll-like receptor-4 gene is not associated with rheumatoid arthritis. Clin Exp Rheumatol 21:340–342

Kobayashi K, Hernandez LD, Galan JE, Janeway CA Jr, Medzhitov R, Flavell RA (2002) IRAK-M is a negative regulator of Toll-like receptor signaling. Cell 110:191–202

Kollewe C, Mackensen AC, Neumann D, Knop J, Cao P, Wesche H, Martin MU (2004) Sequential autophosphorylation steps in the interleukin-1 receptor-associated kinase-1 regulate its availability as an adapter in interleukin-1 signaling. J Biol Chem 279:5227–5236

Knopp J, Martin MU (1999) Effects of IL-1 receptor-associated kinase (IRAK) expression on IL-1 signaling are independent of its kinase activity. FEBS Lett 448:81–85

Kopp E, Medzhitov R, Carothers J, Xiao C, Douglas I, Janeway CA, Ghosh S (1999) ECSIT is an evolutionarily conserved intermediate in the Toll/IL-1 signal transduction pathway. Genes Dev 13:2059–2071

Kuhns DB, Long Priel DA, Gallin JI (1997) Endotoxin and IL-1 hyporesponsiveness in a patient with recurrent bacterial infections. J Immunol 158:3959–3964

Kurt-Jones EA, Popova L, Kwinn L, Haynes LM, Jones LP, Tripp RA, Walsh EE, Freeman MW, Golenbock DT, erson LJ, et al (2000) Pattern recognition receptors TLR4 and CD14 mediate response to respiratory synsytial virus. Nat Immunol 1:398–401

Lauw FN, Caffrey DR, Golenbock DT (2005) Of mice and man:TLR11 (finally) finds profilin. Trends Immunol 26:509–511

Lazarus R, Klimecki WT, Raby BA, Vercelli D, Palmer LJ, Kwiatkowski DJ, Silverman EK, Martinez F, Weiss ST (2003) Single-nucleotide polymorphisms in the Toll-like receptor 9 gene (TLR9):frequencies, pairwise linkage disequilibrium, and haplotypes in three U.S. ethnic groups and exploratory case-control disease association studies. Genomics 81:85–91

Leadbetter EA, Rifkin IR, Hohlbaum AM, Beadette BC, Shlomnik MJ, Marshak-Rothstein A (2002) Chromatin-IgG complexes activate B cells by dual engagement of IgM and Toll-like receptors. Nature 416:603–607

Lee H, Dunzendorfer S, Tobias PS (2004). Cytoplasmic domain-mediated dimerizations of Toll-like receptor 4 observed by β-lactamase enzyme fragment complementation J Biol Chem 279:10564

Lemaitre B, Nicolas E, Michaut L, Reichhart JM, Hoffmann JA (1996) The dorsoventral regulatory gene cassette spatzle/Toll/cactus controls the potent antifungal response in Drosophila adults. Cell 86:973–983

Li S, Strelow A, Fontana EJ, Wesche H (2002) IRAK-4:a novel member of the IRAK family with the properties of an IRAK-kinase. Proc Natl Acad Sci USA 99:5567–5572

Li X, Commane M, Burns C, Vithalani K, Cao Z, Stark GR (1999) Mutant cells that do not respond to interleukin-1 (IL-1) reveal a novel role for IL-1 receptor-associated kinase. Mol Cell Biol 19:4643–4652

Lien E, Sellati TJ, Yoshimura A, Flo TH, Rawadi G, Finberg RW, Carroll JD, Espevik T, Ingalls RR, Radolf JD, Golenbock DT (1999) Toll-like receptor 2 functions as a pattern recognition receptor for diverse bacterial products. J Biol Chem 274:33419–33425

Lorenz E, Mira JP, Cornish KL, Arbour NC, Schwartz DA (2000) A novel polymorphism in the toll-like receptor 2 gene and its potential association with staphylococcal infection. Infect Immun 68:6398–6401

Lorenz E, Mira JP, Frees KL, Schwartz DA (2002a) Relevance of mutations in the TLR4 receptor in patients with gram-negative septic shock. Arch Intern Med 162:1028–1032

Lorenz E, Hallman M, Marttlia R, Haataja R, Schwartz DA (2002b) Association between the Asp299Gly polymorphisms in the Toll-like receptor 4 and premature births in the Finnish population. Pediatr Res 52:373–376

Lund JM, Alexopoulou L, Sato A, Karow M, Adams NC, Gale NW, Iwasaki A, Flavell RA (2004) Recognition of single-stranded RNA viruses by Toll-like receptor 7. Proc Natl Acad Sci USA 101:5598–5603

Lye E, Mirtsos C, Suzuki N, Suzuki S, Yah WC (2004) The role of interleukin 1 receptor-associated kinase-4 (IRAK-4) kinase activity in IRAK-4-mediated signaling. J Biol Chem 279:40653–40658

Malley R, Henneke P, Morse SC, Cieslewicz MJ, Lipsitch M, Thompson CM, Kurt-Jones E, Paton JC, Wessels MR, Golenbock DT (2003) Recognition of pneumolysin by Toll-like receptor 4 confers resistance to pneumococcal infection. Proc Natl Acad Sci USA 100:1966–1971

Mansell A, Smith R, Doyle SL, Gray P, Fenner JE, Crack PJ, Nicholson SE, Hilton DJ, O'Neill LA, Hertzog PJ (2006) Suppressor of cytokine signaling 1 negatively regulates Toll-like receptor signaling by mediating Mal degradation. Nat Immunol 7:148–155

Martin MU, Wesche H (2002) Summary and comparison of the signaling mechanisms of the Toll/interleukin-1 receptor family. Biochim Biophys Acta 1592:265–280

McGettrick AF, Brint EK, Palsson-McDermott EM, Rowe DC, Golenbock DT, Gay NJ, Fitzgerald KA, O'Neill LA (2006) Trif-related adapter molecule is phosphorylated by PKCvarepsilon during Toll-like receptor 4 signaling. Proc Natl Acad Sci USA 103:9196–9201

Means TK, Lien E, Yoshimura A, Wang S, Golenbock DT, Fenton MJ (1999) The CD14 ligands lipoarabinomannan and lipopolysaccharide differ in their requirement for Toll-like receptors. J Immunol 163:6748–6755

Medvedev AE, Lentschat A, Kuhns DB, Blanco JC, Salkowski C, Zhang S, Arditi M, Gallin JI, Vogel SN (2003) Distinct mutations in IRAK-4 confer hyporesponsiveness to lipopolysaccharide and interleukin-1 in a patient with recurrent bacterial infections. J Exp Med 198:521–531

Medvedev AE, Thomas K, Awomoyi A, Kuhns DB, Gallin JI, Li X, Vogel SN (2005) Cutting edge: expression of IL-1 receptor-associated kinase-4 (IRAK-4) proteins with mutations identified in a patient with recurrent bacterial infections alters normal IRAK-4 interaction with components of the IL-1 receptor complex. J Immunol 174:6587–6591

Medvedev AE, Kuhns DB, Gallin JI, Vogel SN (2006) IRAK-4: a key kinase involved in toll-like receptor signaling and resistance to bacterial infection. In: O'Neill LAJ, Brint E (eds) Toll-like receptors in inflammation. Birkhäuser, Basel, pp 173–192

Medvedev AE, Piao W, Shoenfelt J, Rhee SH, Chen H, Basu S, Wahl LM, Fenton MJ, Vogel SN (2007) Role of TLR4 tyrosine phosphorylation in signal transduction and endotoxin tolerance. J Biol Chem 282:16042–16053

Medzhitov R, Janeway C Jr (2000) Innate immunity. N Engl J Med 343:338–344

Medzhitov R, Janeway C Jr (2002) Decoding the patterns of self and nonself by the innate immune system. Science 296:298–300

Medzhitov R, Preston-Hurlburt P, Janeway CA Jr (1997) A human homologue of the Drosophila Toll protein signals activation of adaptive immunity. Nature 388:394–397

Meijer AH, Gabby Krens SF, Medina Rodriguez IA, He S, Bitter W, Snaar-Jagalska B, Spaink HP (2004) Expression analysis of the Toll-like receptor and TIR domain adaptor families of zebrafish. Mol Immunol 40:773–83

Meylan E, Burns K, Hofmann K, Blancheteau V, Martinon F, Kelliher M, Tschopp J (2004) RIP1 is an essential mediator of Toll-like receptor 3-induced NF-kappa B activation. Nat Immunol 5:503–507

Miao EA, Alpuche-Aranda CM, Dors M, Clark AE, Bader MW, Miller SI, Aderem A (2006) Cytoplasmic flagellin activates caspase-1 and secretion of interleukin 1beta via Ipaf. Nat Immunol 7:569–575

Mockenhaupt FP, Cramer JP, Hamann L, Stegemann MS, Eckert J, Oh NR, Otchwemah RN, Dietz E, Ehrardt S, Schroder NW, Bienzle U, Schumann RR (2006a) Toll-like receptor (TLR) polymorphisms in African children: common TLR-4 variants predispose to severe malaria. Proc Natl Acad Sci USA 103:177–182

Mockenhaupt FP, Hamann L, Gaertner C von, Bedu-Addo G, Kleinsorgen C von, Schumann RR, Bienzle U (2006b) Common polymorphisms of toll-like receptors 4 and 9 are associated with the clinical manifestation of malaria during pregnancy. J Infect Dis 194:184–188

Moore CE, Segal S, Berendt AR, Hill AV, Day NP (2004) Lack of association between Toll-like receptor 2 polymorphisms and susceptibility to severe disease caused by *Staphylococcus aureus*. Clin Diagn Lab Immunol 11:1194–1197

Moore KJ, Andersson LP, Ingalls RR, Monks BG, Li R, Arnaout MA, Golenbock DR, Freeman MW (2000) Divergent response to LPS and bacteria in CD14-deficient murine macrophages. J Immunol 165:4272–4280

Morre SA, Murillo LS, Spaargaren J, Fennema HS, Pena AS (2002) Role of the toll-like receptor 4 Asp299Gly polymorphism in susceptibility to *Candida albicans* infection. J Infect Dis 186:1377–1379

Morre SA, Murillo LS, Bruggeman CA, Pena AS (2003) The role that the functional Asp299Gly polymorphism in the toll-like receptor-4 gene plays in susceptibility to *Chlamydia trachomatis*-associated tubal infertility. J Infect Dis 187:341–342

Muzio M, Ni J, Feng P, Dixit V (1997) IRAK (Pelle) family member IRAK-2 and MyD88 as proximal mediators of IL-1 signaling. Science 278:1612–1615

Nagai Y, Akashi S, Nagafuku M, Ogata M, Twakura Y, Akira S, Kitamura T, Kosugi A, Kimoto M, Miyake K (2002) Essential role of MD-2 in LPS responsiveness and TLR4 distribution. Nat Immunol 3:667–672

Negishi H, Ohba Y, Yanai H, Takaoka A, Honma K, Yui K, Matsuyama T, Taniguchi T, Honda K (2005) Negative regulation of Toll-like-receptor signaling by IRF-4. Proc Natl Acad Sci USA 102:15989–15994

Newport MJ, Allen A, Awomoyi AA, Dunstan SJ, McKinney E, Marchant A, Sirugo G (2004) The toll-like receptor 4 Asp299Gly variant: no influence on LPS responsiveness or susceptibility to pulmonary tuberculosis in the Gambia. Tuberculosis 84:347–352

Noguchi E, Nishimura F, Fukai H, Kim J, Ichikawa K, Shibasaki M, Arinami T (2004) An association study of asthma and total serum immunoglobin E levels for Toll-like receptor polymorphisms in a Japanese population. Clin Exp Allergy 34:177–183

Nusinzon I, Horvath CM (2006) Positive and negative regulation of the innate antiviral response and beta interferon gene expression by deacetylation. Mol Cell Biol 26:3106–3113

Oganesyan G, Saha SK, Guo B, He JQ, Shahangian A, Zarnegar B, Perry A, Cheng G (2006) Critical role of TRAF3 in the Toll-like receptor-dependent and -independent antiviral response. Nature 439:208–211

Ogus AC, Yoldas B, Ozdemir T, Uguz A, Olcen S, Keser I, Coskun M, Cilli A, Yegin O (2004) The Arg753GLn polymorphism of the human toll-like receptor 2 gene in tuberculosis disease. Eur Respir J 23:219–223

Ohashi K, Burkart V, Flohe S, Kolb H (2000) Cutting edge: heat shock protein 60 is a putative endogenous ligand of the toll-like receptor-4 complex. J Immunol 164:558–561

Okamura Y, Watari M, Jerud ES, Young DW, Ishizaka ST, Rose J, Chow JC, Strauss JF (2001) The extra domain A of fibronectin activates Toll-like receptor 4. J Biol Chem 276:10229–10233

O'Neill LA, Fitzgerald KA, Bowie AG (2003) The Toll-IL-1 receptor adaptor family grows to five members. Trends Immunol 24:286–290

Oshiumi H, Matsumoto M, Funami K, Akazawa T, Seya T (2003a) TICAM-1, an adaptor molecule that participates in Toll-like receptor 3-mediated interferon-β induction. Nat Immunol 4:161–167

Oshiumi H, Sasai M, Shida K, Fujita T, Matsumoto M, Seya T (2003b) TICAM-2: a bridging adapter recruiting to Toll-like receptor 4 TICAM-1 that induces interferon-β. J Biol Chem 278:49751–49762

Ozinsky A, Underhill DM, Fontenot JD, Hajjar AM, Smith KD, Wilson CB, Schroeder L, Aderem A (2000) The repertoire for pattern recognition of pathogens by the innate immune system is defined by cooperation between toll-like receptors. Proc Natl Acad Sci USA 97:13766–13771

Park JS, Svetkauskaite D, He Q, Kim JY, Strassheim D, Ishizaka A, Abraham E (2004) Involvement of toll-like receptors 2 and 4 in cellular activation by high mobility group box 1 protein. J Biol Chem 279:7370–7377

Perera P-Y, Vogel SN, Detore GR, Haziot A, Goyert SM (1997) CD14-dependent and CD14-independent signaling pathways in murine macrophages from normal and CD14 knockout mice stimulated with lipopolysaccharide or taxol. J Immunol 158:4422–4429

Perera PY, Mayadas TN, Takeuchi O, Akira S, Zaks-Zilberman M, Goyert SM, Vogel SN (2001) CD11b/CD18 acts in concert with CD14 and Toll-like receptor (TLR) 4 to elicit full lipopolysaccharide and taxol-inducible gene expression. J Immunol 166:574–581

Picard C, Puel A, Bonnet M, Ku CL, Bustamante J, Yang K, Soudais C, Dupuis S, Feinberg J, Fieschi C, Elbim C, Hitchcock R, Lammas D, Davies G, Al-Ghonaium A, Al-Rayes H, Al-Jumaah S, Al-Hajjar S, Al-Mohsen IZ, Frayha HH, Rucker R, Hawn TR, Aderem A, Tufenkeji H, Haraguchi S, Day NK, Good RA, Gougerot-Pocidalo MA, Ozinsky A, Casanova JL (2003) Pyogenic bacterial infections in humans with IRAK-4 deficiency. Science 299:2076–2079

Poltorak A, He X, Smirnova I, Liu MY, Van Huffel C, Du X, Birdwell D, Alejos E, Silva M, Galanos C, Freudenberg M, Ricciardi-Castagnoli P, Layton B, Beutler B (1998) Defective LPS signaling in C3H/HeJ and C57BL/10ScCr mice: mutations in Tlr4 gene. Science 282:2085–2088

Puel A, Yamg K, Ku CL, von Bernuth H, Bustamante J, Santos OF, Lawrence T, Chang HH, Al-Mousa H, Picard C, Casanova JL (2005) Heritable defects of the human TLR signalling pathways. J Endotoxin Res 11:220–224

Pugin J, Heumann D, Tomasz A, Kravchenko VV, Akamatsu Y, Nishijima M, Glauser MP, Tobias PS, Ulevitch RJ (1994) CD14 as a pattern recognition receptor. Immunity 1:509–516

Qin J, Jiang Z, Qian Y, Casanova J-L, Li X (2004) IRAK4 kinase activity is redundant for interleukin-1 (IL-1) receptor-associated kinase phosphorylation and IL-1 responsiveness. J Biol Chem 279:26748–26753

Qureshi ST, Lariviere L, Leveque G, Clermont S, Moore KJ, Gros P, Malo D (1999) Endotoxin-tolerant mice have mutations in Toll-like receptor 4 (Tlr4). J Exp Med 189:615-2-65

Raby BA, Klimecki WT, Laprise C, Renaud Y, Faith J, Lemire M, Greenwood C, Weiland KM, Lange C, Palmer LJ, Lazarus R, Vercelli D, Kwiatkowski DJ, Silverman EK, Martinez FD, Hudson TJ, Weiss ST (2002) Polymorphisms in toll-like receptor 4 are not associated with asthma or atopy-related phenotypes. Am J Respir Crit Care Med 166:1449–1456

Rallabhandi P, Bell J, Boukhvalova MS, Medvedev A, Lorenz E, Arditi M, Hemming VG, Blanco JC, Segal DM, Vogel SN (2006) Analysis of TLR4 polymorphic variants: new insights into TLR4/MD-2/CD14 stoichiometry, structure, and signaling. J Immunol 177:322–332

Read RC, Pullin J, Gregory S, Borrow R, Kaczmarski EB, di Giovine FS, Dower SK, Cannings C, Wilson AG (2001) A functional polymorphism of toll-like receptor 4 is not associated with likelihood or severity of meningococcal disease. J Infect Dis 184:640–642

Rogers NC, Slack EC, Edwards AD, Nolte MA, Schulz O, Schweighoffer E, Williams DL, Gordon S, Tybulewicz VL, Brown GD, Reis e Sousa C (2005) Syk-dependent cytokine induction by Dectin-1 reveals a novel pattern recognition pathway for C type lectins. Immunity 22:507–517

Ronni T, Agarwal V, Haykinson M, Haberland ME, Cheng G, Smale ST (2003) Common interaction surfaces of the toll-like receptor 4 cytoplasmic domain stimulate multiple nuclear targets. Mol Cell Biol 23:2543–2555

Rowe DC, McGettrick AF, Latz E, Monks BG, Gay NJ, Yamamoto M, Akira S, O'Neill LA, Fitzgerals KA, Golenbock DT (2006) The myristoylation of TRIF-related adaptor molecule is essential for Toll-like receptor 4 signal transduction. Proc Natl Acad Sci USA 103:6299–6304

Sackesen C, Karaaslan C, Keskin O, Tokol N, Tahan F, Civelek E, Sover OU, Adalioglu G, Tuncer A, Birben E, Oner C, Kalayci O (2005) The effect of polymorphisms at the CD14 promoter and the TLR4 gene on asthma phenotypes in Turkish children with asthma. Allergy 60:1485–1492

Sarkar SN, Peters KL, Elco CP, Sakamoto S, Pal S, Sen GC (2004) Novel roles of TLR3 tyrosine phosphorylation and PI3 kinase in double-stranded RNA signaling. Nat Struct Mol Biol 11:1060–1067

Sato S, Sugiyama M, Yamamoto M, Watanabe Y, Kawai T, Takeda K, Akira S (2003) Toll/IL-1 receptor domain-containing adaptor inducing IFN-beta (TRIF) associates with TNF receptor-associated factor 6 and TANK-binding kinase 1, and activates two distinct transcription factors, NF-kappa B and IFN-regulatory factor-3, in the Toll-like receptor signaling. J Immunol 171:4304–4310

Schmitt C, Humeny A, M. Becker C, Brune K, Pahl A (2002) Polymorphisms of TLR4: rapid genotyping and reduced response to lipopolysaccharide of TLR4 mutant alleles. Clin Chem 48:1661–1667

Schroder NW, Diterich I, Zinke A, Eckert J, Draing C, von Baehr V, Hassler D, Priem S, Hahn K, Michelsen KS, Hartung T, Burmester GR, Gobel UB, Hermann C, Schumann RR (2005) Heterozygous Arg753Gln polymorphism of human TLR-2 impairs immune activation by Borrelia burgdorferi and protects from late stage Lyme disease. J Immunol 175:2534–2540

Schwandner R, Dziarski R, Wesche H, Rothe M, Kirschning CJ (1999) Peptidoglycan- and lipoteichoic acid-induced cell activation is mediated by toll-like receptor 2. J Biol Chem 274:17406–17409

Shimazu R, Akashi S, Ogata H, Nagai Y, Fukudome K, Miyake K, Kimoto M (1999) MD-2, a molecule that confers lipopolysaccharide responsiveness on Toll-like receptor 4. J Exp Med 189:1777–1782

Smiley ST, King JA, Hancock VW (2001) Fibrinogen stimulates macrophage chemokine secretion through toll-like receptor 4. J Immunol 167:2887–2894

Smirnova I, Mann N, Dols A, Derkx HH, Hibberd ML, Levin M, Beutler B (2003) Assay of locus-specific genetic load implicates rare Toll-like receptor 4 mutations in meningococcal susceptibility. Proc Natl Acad Sci USA 100:6075–6080

Stuart LM, Deng J, Silver JM, Takahashi K, Tseng AA, Hennessy EJ, Ezekowitz RA, Moore KJ (2005) Response to Staphylococcus aureus requires CD36-mediated phagocytosis triggered by the COOH-terminal cytoplasmic domain. J Cell Biol 170:477–485

Suzuki N, Suzuki S, Duncan GS, Millar DG, Wada T, Mirtsos C, Takada H, Wakeham A, Itie A, Li S, Penninger JM, Wesche H, Ohashi PS, Mak TW, Yeh WC (2002) Severe impairment of interleukin-1 and Toll-like receptor signalling in mice lacking IRAK-4. Nature 416:750–756

Suzuki N, Suzuki S, Millar DG, Unno M, Hara H, Calzascia T, Yamasaki S, Yokosuka T, Chen NJ, Elford AR, Suzuki J, Takeuchi A, Mirtsos C, Bouchard D, Ohashi PS, Yeh WC, Saito T (2006) A critical role for the innate immune signaling molecule IRAK-4 in T cell activation. Science 311:1927–1932

Swantek JL, Tsen MF, Cobb M, Thomas JA (2000) IL-1 receptor-associated kinase modulates host responsiveness to endotoxin. J Immunol 164:4301–4306

Takaesu G, Kishida S, Hiyama A, Yamaguchi K, Shibuya H, Irie K, Ninomiya-Tsuji J, Matsumoto K (2000) TAB2, a novel adaptor protein, mediates activation of TAK1 MAPKKK by linking TAK1 to TRAF6 in the IL-1 signal transduction pathway. Mol Cell 5:649–658.

Takaesu G, Ninomiya-Tsuji J, Kishida S, Li X, Stark G, Matsumoto M (2001) Interleukin-1 (IL-1) receptor-associated kinase leads to activation of tak 1 by inducing tab2 translocation in the IL-1 signaling pathway. Mol Cell Biol 21:2475–2484

Takaoka A, Yanai H, Kondo S, Duncan G, Negishi H, Mizutani T, Kano S, Honda K, Ohba Y, Mak TW, Taniguchi T (2005) Integral role of IRF-5 in the gene induction programme activated by Toll-like receptors. Nature 434:243–249

Takatsuna H, Kato H, Gohda J, Akiyama T, Moriya A, Okamoto Y, Yamagata Y, Otsuka M, Umezawa K, Semba K, Inoue J (2003) Identification of TIFA as an adapter protein that links tumor necrosis factor receptor-associated factor 6 (TRAF6) to interleukin-1 (IL-1) receptor-associated kinase-1 (IRAK-1) in IL-1 receptor signaling. J Biol Chem 278:12144–12150

Takeda K, Akira S (2005) Toll-like receptors in innate immunity. Int Immunol 17:1–14

Takeuchi O, Kaufmann A, Grote K, Kawai T, Hoshino K, Morr M, Muhlradt PF, Akira S (2000) Cutting edge: preferentially the R-stereoisomer of the mycoplasmal lipopeptide macrophage-activating lipopeptide-2 activates immune cells through a toll-like receptor 2- and MyD88-dependent signaling pathway. J Immunol 164:554–557

Takeuchi O, Sato S, Horiuchi T, Hoshino K, Takeda K, Dong Z, Modlin RL, Akira S (2002) Cutting edge: role of Toll-like receptor 1 in mediating immune response to microbial lipoproteins. J Immunol 169:10–14

Tal G, Mandelberg A, Dalal I, Cesar K, Somekh E, Tal A, Oron A, Itskovich S, Ballin A, Houri S, Beigelman A, Lider O, Rechavi G, Amariglio N (2004) Association between common Toll-like receptor 4 mutations and severe respiratory syncytial virus disease. J Infect Dis 189:2057–2063

Tapping RI, Tobias PS (2003) J Endotoxin Res 9:264–268

Termeer C, Benedix F, Sleeman J, Fieber C, Voith U, Ahrens T, Miyake K, Freudenberg M, Galanos C, Simon JC (2002) Oligosaccharides of hyaluronan activate dendritic cells via Toll-like receptor 4. J Exp Med 195:99–111

Thomas JA, Allen JL, Tsen M, Dubnicoff T, Danao J, Liao XC, Cao Z, Wasserman SA (1999) Impaired cytokine signaling in mice lacking the IL-1 receptor-associated kinase. J Immunol 163:978–985

Travassos LH, Girardin SE, Philpott DJ, Blanot D, Nahori MA, Werts C, Boneca IG (2004) Toll-like receptor 2-dependent bacterial sensing does not occur via peptidoglycan recognition. EMBO Rep 5:1000–1006

Tsan MF, Gao B (2004) Endogenous ligands of Toll-like receptors. J Leukoc Biol 76:514–519

Uehori J, Matsumoto M, Tsuji S, Akazawa T, Takeuchi O, Akira S, Kawata T, Azuma I, Toyoshima K, Seya T (2003) Simultaneous blocking of human Toll-like receptors 2 and 4 suppresses myeloid dendritic cell activation induced by *Mycobacterium bovis* bacillus Calmette–Guerin peptidoglycan. Infect Immun 71:4238–4249

Uematsu S, Sato S, Yamamoto M, Hirotani T, Kato H, Takeshita F, Matsuda M, Coban C, Ishii KJ, Kawai T, Takeuchi O, Akira S (2005) Interleukin-1 receptor-associated kinase-1 plays an essential role for Toll-like receptor (TLR)7- and TLR9-mediated interferon-alpha induction. J Exp Med 201:915–923

Underhill DM, Ozinski A (2002) Toll-like receptors:key mediators of microbe detection. Curr Opin Immunol 14:103–110

Vabulas RM, Ahmad-Nejad P, Costa C da, Miethke T, Kirschning CJ, Hacker H, Wagner H (2001) Endocytosed HSP60s use toll-like receptor 2 (TLR2) and TLR4 to activate the toll/interleukin-1 receptor signaling pathway in innate immune cells. J Biol Chem 276:31332–31339

Vabulas RM, Braedel S, Hilf N, Singh-Jasuja H, Herter S, Ahmas-Nejad P, Kirschning CJ, Da Costa C, Rammensee HG, Wagner H, Schild H (2002a) The endoplasmic reticulum-resident heat shock protein Gp96 activates dendritic cells via the Toll-like receptor 2/4 pathway. J Biol Chem 277:20847–20853

Vabulas RM, Ahmad-Nejad P, Ghose S, Kirschning CJ, Issels RD, Wagner H (2002b) HSP70 as endogenous stimulus of the Toll/interleukin-1 receptor signal pathway. J Biol Chem 277:15107–15112

Vasselon T, Detmers PA, Charron D, Haziot A (2004) TLR2 recognizes a bacterial lipopeptide through direct binding. J Immunol 173:7401–7405

Viriyakosol S, Tobias PS, Kitchens RL, Kirkland TN (2001) MD-2 binds to bacterial lipopolysaccharide. J Biol Chem 276:38044–38051

Viriyakosol S, Fierer J, Brown GD, Kirkland TN (2005) Innate immunity to the pathogenic fungus *Coccidioides posadasii* is dependent on the Toll-like receptor 2 and dectin-1. Infect Immun 73:1553–1560

Visintin A, Latz E, Monks BG, Espevik T, Golenbock DT (2003) Lysines 128 and 132 enable lipopolysaccharide binding to MD-2, leading to Toll-like receptor-4 aggregation and signal transduction. J Biol Chem 278:48313–48320

Vogel S, Hirschfeld MJ, Perera PY (2001) Signal integration in lipopolysaccharide (LPS)-stimulated murine macrophages. J Endotoxin Res 7:237–241

Vogel SN, Fitzgerald KA, Fenton MJ (2003) TLRs: differential adapter utilization by toll-like receptors mediates TLR-specific patterns of gene expression. Mol Interv 3:466–477

Vogel SN, Awomoyi AA, Rallabhandi P, Medvedev AE (2005) Mutations in TLR4 signaling that lead to increased susceptibility to infection in humans: an overview. J Endotoxin Res 11:333–339

Wang C, Deng L, Hong M, Akkaraju GR, Inoue J, Chen ZJ (2001) TAK1 is a ubiquitin-dependent kinase of MKK and IKK. Nature 412:346–351

Wang H, Bloom O, Zhang M, Vishnubhakat JM, Ombrellino M, Che J, Frazier A, Yang H, Ivanova S, Borovikova L, Manogue KR, Faist E, Abraham E, Andersson J, Andersson U, Molina PE, Abumrad NN, Sama A, Tracey KJ (1999) HMG-1 as a late mediator of endotoxin lethality in mice. Science 285:248–251

Wesche H, Henzel WJ, Shillinglaw W, Li S, Cao Z (1997) MyD88: an adapter that recruits IRAK to the IL-1 receptor complex. Immunity 7:837–847

Wesche H, Gao X, Li X, Kirschning CJ, Stark GR, Cao Z (1999) IRAK-M is a novel member of the pelle/interleukin-1 receptor associated kinase (IRAK) family. J Biol Chem 274:19403–19410

Wietek C, O'Neill L (2002) IRAK-4: a new drug target in inflammation, sepsis, and autoimmunity. Mol Interv 2:212–215

Wright SD, Ramos RA, Tobias PS, Ulevitch RJ, Mathison JC (1990) CD14, a receptor for complexes of lipopolysaccharide (LPS) and LPS binding protein. Science 249:1431–1433

Yamamoto M, Sato S, Hemmi H, Sanjo H, Uematsu S, Kaisho T, Hoshino K, Takeuchi O, Kobayashi M, Fujita T, Takeda K, Akira S (2002a) Essential role for TIRAP in activation of the signalling cascade shared by TLR2 and TLR4. Nature 420:324–329

Yamamoto M, Sato S, Mori K, Hoshino K, Takeuchi O, Takeda K, Akira S (2002b) Cutting edge: a novel Toll/IL-1 receptor domain-containing adapter that preferentially activates the IFN-beta promoter in the Toll-Like receptor signaling. J Immunol 169:6668–6672

Yamamoto M, Sato S, Hemmi H, Hoshino K, Kaisho T, Sanjo H, Takeuchi O, Sugiyama M, Okabe M, Takeda K, Akira S (2003a) Role of adaptor TRIF in the MyD88-independent toll-like receptor signaling pathway. Science 301:640–643

Yamamoto M, Sato S, Hemmi H, Uematsu S, Hoshino K, Kaisho T, Takeuchi O, Takeda K, Akira S (2003b) TRAM is specifically involved in the Toll-like receptor 4-mediated MyD88-independent signaling pathway. Nat Immunol 4:1144–1150

Yamamoto M, Takeda K, Akira S (2004) TIR domain-containing adaptors define the specificity of TLR signaling. Mol Immunol 40:861–868

Yamin T-T, Miller DK (1997) The interleukin-1 receptor-associated kinase is degraded by proteosomes following its phosphorylation. J Biol Chem 272:21540–21547

Yang H, Young DW, Gusovsky F, Chow JC (2000) Cellular events mediated by lipopolysaccharide-stimulated Toll-like receptor 4: MD-2 is required for activation of mitogen-activated protein kinases and Elk-1. J Biol Chem 275:20861–20866

Yang IA, Barton SJ, Rorke S, Cakebread JA, Keith TP, Clough JB, Holgate ST, Holloway JW (2004) Toll-like receptor 4 polymorphism and severity of atopy in asthmatics. Genes Immun 5:41–45

Yarovinsky F, Zhang D, Andersen JF, Bannenberg GL, Serhan CN, Hayden MS, Hieny S, Sutterwala FS, Flavell RA, Ghosh S, Sher A (2005) TLR11 activation of dendritic cells by a protozoan profilin-like protein. Science 308:1626–1629

Yoshida H, Jono H, Kai H, Li JD (2005) The tumor suppressor cylindromatosis (CYLD) acts as a negative regulator for toll-like receptor 2 signaling via negative cross-talk with TRAF6 AND TRAF7. J Biol Chem 280:41111–41121

Zarember KA, Godowski PJ (2002) Tissue expression of human Toll-like receptors and differential regulation of Toll-like receptor mRNAs in leukocytes in response to microbes, their products, cytokines. J Immunol 168:554–561

Zhang D, Zhang G, Hayden MS, Greenblatt MB, Bussey C, Flavell RA, Ghosh S (2004) A toll-like receptor that prevents infection by uropathogenic bacteria. Science 303:1522–1526

Chapter 8
NLRs: a Cytosolic Armory of Microbial Sensors Linked to Human Diseases

Mathias Chamaillard

1 Introduction 170
2 NLRs, a Conserved Cytosolic Arm of the Innate Immune System 171
3 Physiological Role of NLRs in Innate and Adaptive Immunity: NLRs Join TLRs.......... 174
   3.1 Host Sensing of Non-TLR PAMPs: Lessons from NOD1 and NOD2 Studies 174
   3.2 NLRs Promote Maturation of TLR-Induced Il-1β and IL-18 Release 176
4 What Can we Learn from NLRs Linked to Human Diseases? 177
   4.1 NOD1 and NOD2 Mutations Linked to Chronic Inflammatory Diseases 178
   4.2 Auto-Inflammatory Diseases 179
   4.3 Reproduction Diseases 179
5 Concluding Remarks: Towards the Development of "Magic" Bullets 179
References 180

Abstract In mammals, a tissue-specific set of Nod-like receptors (NLRs) enables collectively a swift and differential cytosolic detection of evolutionary distant microbial- and/or danger-associated molecular patterns from the extracellular and intracellular microenvironment. Repressing and de-repressing this surveillance machinery contribute to vital immune homeostasis and protective responses within specific tissues. Conversely, defective biology of NLR signaling pathways drives the development of recurrent infectious and/or inflammatory diseases by failing to mount barrier functions, to instruct the adaptive immune response and/or to ignore self and non-self antigens. Better decoding of microbial shedding and immune escape strategies will provide clues for the development of rational therapies striving to cure and prevent common and emerging immunopathologies in humans.

INSERM U795, Physiopathology of Inflammatory Bowel Disease, Swynghedauw Hospital, Rue A. Verhaeghe, 59037 Lille, France, *m-chamaillard@chru-lille.fr*

1 Introduction

Eradication of infectious agents represents an essential barrier function for the survival of all mammals. Infected host mammalian cells elicit multiple defense programs, which can be broadly separated into innate and adaptive immune response. During first days of infection, microbial sensing is conferred by extra- and intra-cellular detection of so-called pathogen-associated molecular patterns (PAMPs) by specialized innate immune sensors, also called pathogen-recognition molecules (PRMs; Medzhitov 2001). In mammals, the Toll-like receptor (TLR) family members are *bona fide* membrane-bound PRMs that contain leucine-rich repeats (LRRs) in their ectodomains for PAMP recognition outside of the cell, at the cell surface, or on the luminal side of intracellular vesicular compartments (Barton and Medzhitov 2003; Akira et al. 2006). In contrast to TLRs, the family of cytosolic nucleotide-binding oligomerization domain leucine-rich repeat-containing proteins (so-called NLRs) provides a surveillance mechanism inside the infected cells (Fig. 1; Inohara et al. 2005; Ting and Davis 2005; Meylan et al. 2006). Upon exposure to microbes and/or 'danger' signals, activation of the membrane-bound TLRs and NLRs signaling pathways confer mutually host

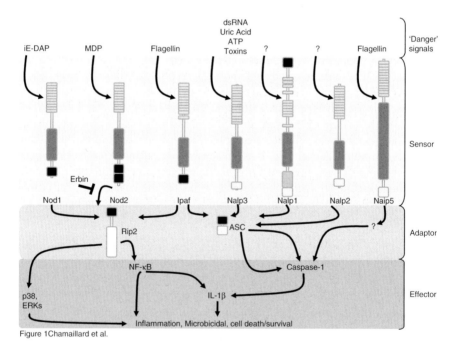

Fig. 1 Schematic of NLR signalling

surveillance by controlling antimicrobial processes, inflammatory cytokines and caspases. Similar sensing mechanisms have also been identified inside and outside plant cells, indicating common defense strategies in mammals and plants. Finally during weeks after infection, TLRs and NLRs promote the development of adaptive immunity mediated by T and B lymphocytes, a unique immune arm evolved by mammalian hosts for a specific elimination of infectious agents. Herein, we review recent findings on the biological roles of mammalian NLRs in health and disease.

2   NLRs, a Conserved Cytosolic Arm of the Innate Immune System

Initial *in silico* searches of genomic databases for proteins related to Apaf-1 revealed a family of NLRs proteins, which comprises 22 members in humans (Table 1). Mainly expressed in immunocytes, most of the NLRs possess a tripartite structure including a C-terminal leucine-rich repeat domain that is required for ligand interaction, a central nucleotide-binding oligomerisation domain [NOD (also called NACHT for neuronal apoptosis inhibitor protein), CIITA, HET-E, TP1] for oligomerization and a N-terminal specific domain to interact with downstream effectors (Table 1). NLRs are sub-classified accordingly to their effector domains: (a) the NOD proteins which contain a caspase-activating and recruitment domain (CARD), (b) the NALP proteins which contain a pyrin domain (PYD), (c) the NAIP proteins which contain a baculovirus inhibitor of apoptosis protein repeat (BIR) and (d) CIITA (Table 1; Inohara et al. 2005). The biological function of NLRs is mainly conferred by their effector domain, which allows homophilic or heterophilic interactions with specific downstream protein-interacting partners. Besides tripartite NLRs, several NLRs display a more complex structure (Table 1), such as CIITA that contains an amino-terminal transcriptional activation domain. CIITA is the first described NLR and controls histocompatibility complex class (MHC) genes, particularly class II (MHC-II). NOD9 do not present effector domain and NALP1 contains an amino-terminal PYD, a carboxyl-terminal domain homologous to a unique region (NC), a truncated PYD and a CARD (Table 1; Inohara et al. 2005). Interestingly, the modular structure of NLRs shares similarities with the disease resistance (*R*) genes found in plants. Both *R* gene products located at the membrane, like TLRs, or in the cytosol, like NLRs, contain typically carboxyl-terminal LRRs. As in mammalian cells, the *R* gene products represent detected pathogens and elicit the hypersensitivity response (Inohara et al. 2005). The structural similarities that membrane-bound and cytosolic *R* gene products share with TLRs and NLRs, respectively, suggest that NLRs play an important role in the regulation of multiple immune response aspects in mammals.

Table 1 The NLR family

HUGO symbols	Subfamily	Other names and aliases	Location	Domain organization	Effector domain	Downstream effector	Caspase-1 activation	NF-κB	Agonists
CARD4	NODs	NOD1; CLR7.1	7p14.3	CARD-NACHT-NAD-LRR	CARD	RICK/RIP2	No effect	Activates	γ-D-Glu-meso-diaminopimelic acid
CARD15		NOD2; CLR16.3	16q12.1	CARD2x-NACHT-NAD-LRR	CARD	RICK/RIP2	No effect	Activates	MurNAc-L-Ala-D-isoGln
None		NOD3; CLR16.2	16p13.3	CARD-NACHT-NAD-LRR	CARD				
None		NOD27; CLR16.1	16q13	CARD-NACHT-LRR	CARD				
CARD12		CLAN; CLR2.1; IPAF	2p22.3	CARD-NACHT-LRR	CARD	ASC/PYCARD/TMS	Activates		Flagellin
None	ND	NOD9; CLR11.3	11q23.3	X-NACHT-LRR	X				
NALP1	NALPs	DEFCAP; NAC; CARD7; CLR17.1	17p13.2	PYD-NACHT-NAD-LRR-FIIND-CARD	PYD	ASC/PYCARD/TMS	Activates		
NALP2		PYPAF2; NBS1; PAN1; CLR19.9	19q13.42	PYD-NACHT-NAD-LRR	PYD	ASC/PYCARD/TMS	Activates	Inhibits	
CIAS1		PYPAF1; Cryopyrin; CLR1.1; NALP3		PYD-NACHT-NAD-LRR	PYD	ASC/PYCARD/TMS	Activates		Bacterial/viral RNA, toxins, ATP, uric acid crystals
NALP4		PYPAF4; PAN2; RNH2; CLR19.5	19q13.43	PYD-NACHT-NAD-LRR	PYD				
NALP5		PYPAF8; MATER; PAN11; CLR19.8	19q13.43	PYD-NACHT-NAD-LRR	PYD				
NALP6		PYPAF5; PAN3; CLR11.4	11p15.5	PYD-NACHT-NAD-LRR	PYD	ASC/PYCARD/TMS	Activates	Activates	
NALP7		PYPAF3; NOD12; PAN7; CLR19.4	Xq26.2	PYD-NACHT-NAD-LRR	PYD		Inhibits		
NALP8		PAN4; NOD16; CLR19.2	19q13.43	PYD-NACHT-NAD-LRR	PYD				

NALP9		NOD6; PAN12; CLR19.1	19q13.42	PYD-NACHT-NAD-LRR	PYD
NALP10		PAN5; NOD8; Pynod; CLR11.1	11p15.4	PYD-NACHT-NAD	PYD
NALP11		PYPAF6; NOD17; PAN10; CLR19.6	19q13.43	PYD-NACHT-NAD-LRR	PYD
NALP12		PYPAF7; Monarch1; RNO2; PAN6; CLR19.3	19q13.42	PYD-NACHT-NAD-LRR	PYD	ASC/PYCARD/TMS	Activates
NALP13		NOD14; PAN13; CLR19.7	19q13.43	PYD-NACHT-NAD-LRR	PYD
NALP14		NOD5; PAN8; CLR11.2	11p15.4	PYD-NACHT-NAD-LRR	PYD
Birc1	NAIPs	Naip5; CLR5.1	5q13.2	BIR3x-NACHT-LRR	BIR3x	Activates	Flagellin
CIITA	CIITA	MHC2TA; C2TA	16p13.13	(CARD)-AD-NACHT-NAD-LRR

3 Physiological Role of NLRs in Innate and Adaptive Immunity: NLRs Join TLRs

3.1 Host Sensing of Non-TLR PAMPs: Lessons from NOD1 and NOD2 Studies

Independently of TLRs, the nucleotide-binding oligomerisation domain proteins 2 and 1 (NOD2 and NOD1; also referred as CARD15 and CARD4, respectively) are cytoplasmic proteins involved in bacterial peptidoglycan (PGN) sensing through the detection of the muropeptide N-acetylmuramic-L-Ala-D-isoGln (MDP; Girardin et al. 2003a; Inohara et al. 2003; Kobayashi et al. 2005) and dipeptide γ-D-glutamyl-meso-diaminopimelic acid (iE-DAP; Chamaillard et al. 2003a; Girardin et al. 2003b; Fig. 2). Whereas MDP is a conserved structure shared by almost all Gram-positive and Gram-negative bacteria, the dipeptide iE-DAP is primarily found in Gram-negative bacteria and certain Gram-positive bacteria, such as the food-borne pathogen *Listeria monocytogenes*. Hence unlike NOD2 that can be considered as a general sensor of bacterial microorganisms, NOD1 detect solely Gram-negative bacteria and a specific subset of Gram-positive bacteria that includes *L. monocytogenes*. Upon exposure to their respective bacterial agonists, NOD2 and NOD1 oligomerize and the downstream serine-threonine kinase RICK (also called RIP2/RIPK2/CARDIAK) is recruited through homophilic CARD-CARD interactions. Once activated the RIP2 signaling pathways leads to the activation of the NF-κB transcription factor through the ubiquitination of the IKKγ/NEMO

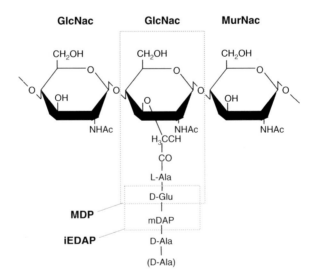

Fig. 2 NOD1 and NOD2 are the sensing bacterial peptidoglycans

subunit of the signalosome (Inohara et al. 2000; Ogura et al. 2001b; Kobayashi et al. 2002; Abbott et al. 2004). Recent reports identified also accessory molecules that regulate the Rip2 signaling pathway, such as Erbin, Tak1, Grim-19 and Centaurin-beta1 (Chen et al. 2004; Barnich et al. 2005; McDonald et al. 2005; Yamamoto-Furusho et al. 2006; Fig. 1). Accordingly, mice deficient for NOD1, NOD2 and RICK are highly susceptible to infection by the Gram-positive facultative intracellular pathogen, *L. monocytogenes* (Kobayashi et al. 2002, 2005). Interestingly whereas *Nod1*-deficient mice are only susceptible to systemic infection by *L. monocytogenes* (Chamaillard, unpublished data), *Nod2*-deficient mice showed an increased susceptibility to orogastric infection by this pathogenic bacterium via the regulation of certain Paneth cells-derived cationic antimicrobial peptides (Kobayashi et al. 2005; Fig. 3). Furthermore, *Nod2*-deficient mice displayed a severe deficiency in the production of antigen specific immunoglobulin (Kobayashi et al. 2005) and experienced an enhanced release of IL-12 and IL-18, leading to an increased proliferation and survival of $CD4^+$ T cells (Watanabe et al. 2006). Consistently, *Nod2*-deficient mice and mutant mice bearing the orthologue of the major CD-associated

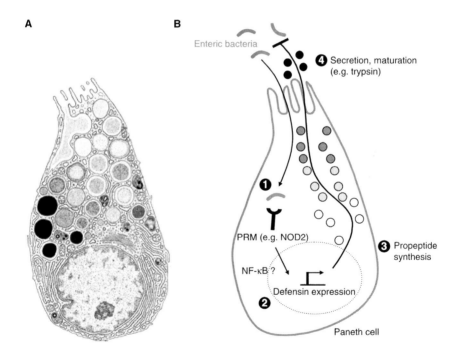

Fig. 3 Schematic of NOD2 signalling in Paneth cells. **A** An electron microscopy picture of Paneth cell is represented. **B** Following exposure to bacterial moieties (*1*), pathogen-recognition molecules (e.g. NOD2) expressed in Paneth cells regulate the expression (*2*) of immature α-defensins (*3*), which are processed (*4*) and secreted outside the cells to maintain sterility in the lumen of the crypt of Lyberkühn (*5*)

$NOD2^{3020ins}$ allele showed increased susceptibility to antigen-specific colitis (Watanabe et al. 2006) and to DSS-induced colitis (Maeda et al. 2005), respectively. More recently, professional immunocytes carrying NOD2 mutations or lacking NOD2, but not NOD1, showed impaired responsiveness to MDP (Inohara et al. 2003; Li et al. 2004; Netea et al. 2005; van Heel et al. 2005), but also to *Mycobacterium tuberculosis* (Ferwerda et al. 2005). However, NOD2 was dispensable for cytokine production in response to *L. monocytogenes*, as shown by the normal inflammatory cytokines production of *Nod2*-deficient macrophages compared with wild-type macrophages. These results indicate an essential role of NOD2 in innate and adaptive immunity at the intestinal mucosal interface. Further, recent reports unravelled a NOD1-dependent sensing of the non-invasive Gram-negative pathogen *Helicobacter pylori* (Viala et al. 2004), through the detection of muropeptides injected into host cells by a bacterial type IV secretion system. Similarly to the physiological role of NOD2, NOD1 is required for expression of certain β-defensins by gastric epithelial cells during *H. pylori* infection (Boughan et al. 2006). However the function of NOD1 in adaptive immunity still remains elusive. Finally, bacterial muropeptides are know to synergistically promote the TLR-dependent production of inflammatory cytokines and chemokines (Chamaillard et al. 2003a; Kobayashi et al. 2005; van Heel et al. 2005a, b), indicating that NOD1 and NOD2 might be crucial in initiating the immune response towards invasive bacteria.

3.2 NLRs Promote Maturation of TLR-Induced Il-1β and IL-18 Release

Caspase-1 activation is crucial to repel pathogenic infection and to promote the maturation of the adaptive immune system (Kuida et al. 1995; Gu et al. 1997). Upon microbial recognition, the expression of IL-1β and IL-18 is mainly induced in a TLR-dependent manner. However, the maturation of such cytokines requires a simultaneous TLR-independent activation of inflammatory caspases, such as caspase-1 and caspase-5 (and caspase-11 in mice). Several recent reports identified a caspases-activating molecular platform, referred as inflammasome (Martinon et al. 2002; Agostini et al. 2004; Mariathasan et al. 2004; Martinon and Tschopp 2004). The inflammasome assembly is conferred by homophilic interactions with the PYD and CARD. ASC can be linked to NALP1, NALP2 and/or NALP3 through a PYD–PYD interaction and to caspase-1 through a CARD–CARD interaction (Fig. 1). An additional molecular assembly has been reported for NALP1, but not NALP3, which can recruit the caspase-5 through homophilic CARD–CARD interactions. However, the cellular trafficking of the inflammasome and the additional signaling partners, such as the adaptor ASC2 (Damiano et al. 2004) and additional NLRs, namely Ipaf (Geddes et al. 2001; Poyet et al. 2001) and Naip5 (Zamboni et al. 2006), should be further described.

The ice protease-activating factor (Ipaf, also known as CARD12/CLAN) is required for cytosolic sensing of bacterial flagellin (Franchi et al. 2006; Miao et al. 2006; Fig. 1), independently of TLR5, which is essential for flagellin-induced NF-κB activation (Feuillet et al. 2006; Uematsu et al. 2006). Intracellular exposure to *Salmonella typhimurium*, *Legionella pneumophila* and flagellin triggered caspase 1 activation and cell death in an Ipaf-dependent manner (Mariathasan et al. 2004; Amer et al. 2006; Franchi et al. 2006). Ipaf signals through the apoptosis-associated speck-like (ASC) protein containing a Card and also known as PYCARD/TMS, an adaptor protein composed of an amino-terminal PYD and a carboxyl-terminal CARD (Srinivasula et al. 2002; Masumoto et al. 2003). Consistently, ASC-deficient macrophages failed to activate caspase-1 and to release IL-1β and IL18 in response to flagellated bacteria.

More recently, the activation of another NLR family member, cryopyrin (also named NALP3/PYPAF1), led to caspase-1 activation and cell death in an ASC-dependent manner (Dowds et al. 2003; Fig. 1, Table 1). Independently of TLR7, cryopyrin is required for intracellular sensing of purine-like structures, including bacterial and viral RNAs (Kanneganti et al. 2006a, b) and the endogenous 'danger' signals monosodium urate or calcium pyrophosphate dehydrate crystals (Martinon et al. 2006). Finally, independently of Ipaf, a recent report indicated that ASC- and caspase-1-deficient mice showed enhanced susceptibility to *Francisella tularensis* (Mariathasan et al. 2005), a highly infectious Gram-negative coccobacillus that causes the zoonosis tularemia.

In addition to Ipaf and Cryopyrin, several genetic determinants have been implicated in host responsiveness towards the intracellular replication of *L. pneumophila*, the infectious agent causing Legionnaires' disease, and subsequent cell death induced by this pathogenic bacterium. Notably, naturally occurring mutations in the gene encoding the NLR member Birc1e (baculovirus inhibitor of apoptosis repeat-containing 1), also known as Naip5 (Diez et al. 2003; Wright et al. 2003). However, further work should clarify whether Ipaf and Birc1e function together or independently in preventing cytosolic replication of *L. pneumophila* (Amer et al. 2006; Zamboni et al. 2006).

Finally, a subset of pathogenic bacteria has also evolved several counter-acting mechanisms by promoting cell death of immunocytes through the release of caspases-activating microbial and/or danger signals. Notably, natural mutations in the Nalp1b-encoding genes have been associated to increased toxin-induced cell death in macrophages (Boyden and Dietrich 2006). Taken together, the release of mature Il-1β and IL-18 represents a unique role for NLRs in cooperating with TLRs-mediated immunity.

4 What Can we Learn from NLRs Linked to Human Diseases?

Both loss- and gain-of-function mutations in NLRs proteins have been associated with the pathogenesis of human diseases.

4.1 NOD1 and NOD2 Mutations Linked to Chronic Inflammatory Diseases

Multiple genetic variations of the *NOD2* gene have been associated with susceptibility to several inflammatory diseases, including atopic eczema, Blau syndrome (BS), Crohn's disease (CD), early-onset sarcoidosis (EOS) and graft-versus-host disease (Hugot et al. 2001; Miceli-Richard et al. 2001; Ogura et al. 2001a; Holler et al. 2004; Kanazawa et al. 2005; Weidinger et al. 2005a). Three mutations in the NOD motif of *NOD2* (R334Q-W, L469F) have been associated with EOS and BS, rare autosomal dominant disorders characterized by associated arthritis, uveitis, skin rashes and granuloma (Miceli-Richard et al. 2001; Kanazawa et al. 2005; Fig. 4). An increased basal and MDP-induced NF-κB activity is associated in vitro with the EOS/BS-causing mutations, indicating that those variants might behave as hyper-responsive alleles (Chamaillard et al. 2003b; Tanabe et al. 2004). These results suggest that such mutations of conserved amino acid in the NOD motif can trigger a highly penetrant uncontrolled inflammatory signaling, resulting in early-onset dominant inflammatory diseases. Consistently with the recessive nature of CD susceptibility (Kuster et al. 1989; Monsen et al. 1990; Orholm et al. 1993), individuals with one of the three major independent disease-associated alleles (namely R702W, G908R, 1007fs) have a 2- to 4-fold increased risk of developing CD, whereas homozygous or compound heterozygous carriers have an up to 40-fold increase in genotype-relative risk (Hugot et al. 2001; Fig. 4). Taken as a whole, those results revealed opposite physiopathological NOD2-dependent mechanisms which might influence the development of distinct inflammatory diseases (Chamaillard et al. 2003b). More recently and similarly to CD-associated NOD2 mutations, recent reports identified a complex intronic polymorphism of the *NOD1* gene that is associated with the pathogenesis of IBD, asthma and atopic eczema (Hysi et al. 2005; McGovern et al. 2005; Weidinger et al. 2005b).

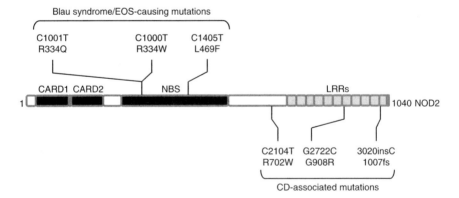

Fig. 4 Schematic of loss- and gain-of-function NOD2 mutations associated with distinct inflammatory disorders

4.2 Auto-Inflammatory Diseases

Mutations in the *CIAS1* gene, encoding for NALP3, are associated with the clinical expression of a broad phenotypic spectrum of dominantly inherited auto-inflammatory disorders, including Muckle–Wells syndrome (MWS), familial cold autoinflammatory syndrome (FCAS) and chronic neurologic cutaneous and articular syndrome [CINCA, also called NOMID (for neonatal onset multisystem inflammatory disease); Hoffman et al. 2001; Aganna et al. 2002; Aksentijevich et al. 2002; Dode et al. 2002; Feldmann et al. 2002]. All these disorders are characterized by recurrent inflammatory crises together with fever, rash and arthritis. At least 22 distinct missense mutations, including 20 de novo mutational events, of the *CIAS1* gene are associated with MWS, FCAS and CINCA (Aksentijevich et al. 2002). All of these mutations are within the NOD and NOD-associated domains. Notably, three mutations (V198M, D303N, R260W) predispose to the three syndromes, suggesting the existence of modifying genes and/or environmental factors modulating the phenotypic expression. Like NOD2 mutations causing EOS and BS, cryopyrin mutations are associated with an increased IL-1β release (Agostini et al. 2004; Dowds et al. 2004). Consistently, patients with auto-inflammatory syndromes experience clinical improvement after treatment by a specific antagonist of IL-1β signaling (Hawkins et al. 2004; Hoffman et al. 2004; Ramos et al. 2005; Boschan et al. 2006; Goldbach-Mansky et al. 2006; Matsubara et al. 2006), indicating a fundamental role of IL-1β in the pathogenesis of auto-inflammatory diseases.

4.3 Reproduction Diseases

Mutations in the gene encoding for NALP7 (also known as Pypaf3/Nod12/Pan7/Clr19.4) cause familial and recurrent hydatidiform moles (Bestor and Bourc'his 2006; Djuric et al. 2006; Murdoch et al. 2006), tumors that forms in the uterus as a mass of cysts resembling a bunch of grapes. Unlike NALP1-3, IPAF and NAIP5, NALP7 is a negative regulator of IL-1β signaling (Kinoshita et al. 2005) that promote tumorigenesis (Okada et al. 2004). Further work should now determine: (a) how the NALP7 signaling pathway might be activated, (b) whether NALP7 might interfere with the NALP1-, NALP3-, IPAF- and NAIP5-inflammasome and (c) whether, like other NLRs, NALP7 might sense microbial infections.

5 Concluding Remarks: Towards the Development of "Magic" Bullets

Taken as a whole, NLRs and TLRs evolved complementary functions to confer a potent innate and adaptive immune system towards microbial and 'danger' signals. NLRs represent a cytosolic armory of microbial sensors linked to several human diseases. Lastly, further work should now focus on better understanding the physiological roles of normal and mutant NLRs and on developing novel therapeutic strategies aiming to restore abnormal NLRs functions.

References

Abbott DW, Wilkins A, Asara JM, Cantley LC (2004) The Crohn's disease protein, NOD2, requires RIP2 in order to induce ubiquitinylation of a novel site on NEMO. Curr Biol 14:2217–2227

Aganna E, Martinon F, Hawkins PN, Ross JB, Swan DC, Booth DR, Lachmann HJ, Bybee A, Gaudet R, Woo P, Feighery C, Cotter FE, Thome M, Hitman GA, Tschopp J, McDermott MF (2002) Association of mutations in the NALP3/CIAS1/PYPAF1 gene with a broad phenotype including recurrent fever, cold sensitivity, sensorineural deafness, and AA amyloidosis. Arthritis Rheum 46:2445–2452

Agostini L, Martinon F, Burns K, McDermott MF, Hawkins PN, Tschopp J (2004) NALP3 forms an IL-1beta-processing inflammasome with increased activity in Muckle–Wells autoinflammatory disorder. Immunity 20:319–325

Akira S, Uematsu S, Takeuchi O (2006) Pathogen recognition and innate immunity. Cell 124:783–801

Aksentijevich I, Nowak M, Mallah M, Chae JJ, Watford WT, Hofmann SR, Stein L, Russo R, Goldsmith D, Dent P, Rosenberg HF, Austin F, Remmers EF, Balow JE Jr, Rosenzweig S, Komarow H, Shoham NG, Wood G, Jones J, Mangra N, Carrero H, Adams BS, Moore TL, Schikler K, Hoffman H, Lovell DJ, Lipnick R, Barron K, O'Shea JJ, Kastner DL, Goldbach-Mansky R (2002) De novo CIAS1 mutations, cytokine activation, and evidence for genetic heterogeneity in patients with neonatal-onset multisystem inflammatory disease (NOMID): a new member of the expanding family of pyrin-associated autoinflammatory diseases. Arthritis Rheum 46:3340–3348

Amer A, Franchi L, Kanneganti TD, Body-Malapel M, Ozoren N, Brady G, Meshinchi S, Jagirdar R, Gewirtz A, Akira S, Nunez G (2006) Regulation of Legionella phagosome maturation and infection through flagellin and host IPAF. J Biol Chem 281:35217–35223

Barnich N, Hisamatsu T, Aguirre JE, Xavier R, Reinecker HC, Podolsky DK (2005) GRIM-19 interacts with nucleotide oligomerization domain 2 and serves as downstream effector of antibacterial function in intestinal epithelial cells. J Biol Chem 280:19021–19026

Barton GM, Medzhitov R (2003) Toll-like receptor signaling pathways. Science 300:1524–1525

Bestor TH, Bourc'his D (2006) Genetics and epigenetics of hydatidiform moles. Nat Genet 38:274–276

Boschan C, Witt O, Lohse P, Foeldvari I, Zappel H, Schweigerer L (2006) Neonatal-onset multisystem inflammatory disease (NOMID) due to a novel S331R mutation of the CIAS1 gene and response to interleukin-1 receptor antagonist treatment. Am J Med Genet A 140:883–886

Boughan PK, Argent RH, Body-Malapel M, Park JH, Ewings KE, Bowie AG, Ong SJ, Cook SJ, Sorensen OE, Manzo BA, Inohara N, Klein NJ, Nunez G, Atherton JC, Bajaj-Elliott M (2006) Nucleotide-binding oligomerization domain-1 and epidermal growth factor receptor: critical regulators of beta-defensins during *Helicobacter pylori* infection. J Biol Chem 281:11637–11648

Boyden ED, Dietrich WF (2006) Nalp1b controls mouse macrophage susceptibility to anthrax lethal toxin. Nat Genet 38:240–244

Chamaillard M, Hashimoto M, Horie Y, Masumoto J, Qiu S, Saab L, Ogura Y, Kawasaki A, Fukase K, Kusumoto S, Valvano MA, Foster SJ, Mak TW, Nunez G, Inohara N (2003a) An essential role for NOD1 in host recognition of bacterial peptidoglycan containing diaminopimelic acid. Nat Immunol 4:702–707

Chamaillard M, Philpott D, Girardin SE, Zouali H, Lesage S, Chareyre F, Bui TH, Giovannini M, Zaehringer U, Penard-Lacronique V, Sansonetti PJ, Hugot JP, Thomas G (2003b) Gene–environment interaction modulated by allelic heterogeneity in inflammatory diseases. Proc Natl Acad Sci USA 100:3455–3460

Chen CM, Gong Y, Zhang M, Chen JJ (2004) Reciprocal cross-talk between Nod2 and TAK1 signaling pathways. J Biol Chem 279:25876–25882

Damiano JS, Oliveira V, Welsh K, Reed JC (2004) Heterotypic interactions among NACHT domains: implications for regulation of innate immune responses. Biochem J 381:213–219

Diez E, Lee SH, Gauthier S, Yaraghi Z, Tremblay M, Vidal S, Gros P (2003) Birc1e is the gene within the Lgn1 locus associated with resistance to *Legionella pneumophila*. Nat Genet 33:55–60

Djuric U, El-Maarri O, Lamb B, Kuick R, Seoud M, Coullin P, Oldenburg J, Hanash S, Slim R (2006) Familial molar tissues due to mutations in the inflammatory gene, NALP7, have normal postzygotic DNA methylation. Hum Genet 120:390–395

Dode C, Le Du N, Cuisset L, Letourneur F, Berthelot JM, Vaudour G, Meyrier A, Watts RA, Scott DG, Nicholls A, Granel B, Frances C, Garcier F, Edery P, Boulinguez S, Domergues JP, Delpech M, Grateau G (2002) New mutations of CIAS1 that are responsible for Muckle-Wells syndrome and familial cold urticaria: a novel mutation underlies both syndromes. Am J Hum Genet 70:1498–1506

Dowds TA, Masumoto J, Chen FF, Ogura Y, Inohara N, Nunez G (2003) Regulation of cryopyrin/Pypaf1 signaling by pyrin, the familial Mediterranean fever gene product. Biochem Biophys Res Commun 302:575–580

Dowds TA, Masumoto J, Zhu L, Inohara N, Nunez G (2004) Cryopyrin-induced interleukin 1beta secretion in monocytic cells: enhanced activity of disease-associated mutants and requirement for ASC. J Biol Chem 279:21924–21928

Feldmann J, Prieur AM, Quartier P, Berquin P, Certain S, Cortis E, Teillac-Hamel D, Fischer A, Saint Basile G de (2002) Chronic infantile neurological cutaneous and articular syndrome is caused by mutations in CIAS1, a gene highly expressed in polymorphonuclear cells and chondrocytes. Am J Hum Genet 71:198–203

Ferwerda G, Girardin SE, Kullberg BJ, Le Bourhis L, Jong DJ de, Langenberg DM, Crevel R van, Adema GJ, Ottenhoff TH, Van der Meer JW, Netea MG (2005) NOD2 and toll-like receptors are nonredundant recognition systems of *Mycobacterium tuberculosis*. PLoS Pathog 1:279–285

Feuillet V, Medjane S, Mondor I, Demaria O, Pagni PP, Galan JE, Flavell RA, Alexopoulou L (2006) Involvement of Toll-like receptor 5 in the recognition of flagellated bacteria. Proc Natl Acad Sci USA 103:12487–12492

Franchi L, Amer A, Body-Malapel M, Kanneganti TD, Ozoren N, Jagirdar R, Inohara N, Vandenabeele P, Bertin J, Coyle A, Grant EP, Nunez G (2006) Cytosolic flagellin requires Ipaf for activation of caspase-1 and interleukin 1beta in salmonella-infected macrophages. Nat Immunol 7:576–582

Geddes BJ, Wang L, Huang WJ, Lavellee M, Manji GA, Brown M, Jurman M, Cao J, Morgenstern J, Merriam S, Glucksmann MA, DiStefano PS, Bertin J (2001) Human CARD12 is a novel CED4/Apaf-1 family member that induces apoptosis. Biochem Biophys Res Commun 284:77–82

Girardin SE, Boneca IG, Viala J, Chamaillard M, Labigne A, Thomas G, Philpott DJ, Sansonetti PJ (2003a) Nod2 is a general sensor of peptidoglycan through muramyl dipeptide (MDP) detection. J Biol Chem 278:8869–8872

Girardin SE, Boneca IG, Carneiro LA, Antignac A, Jehanno M, Viala J, Tedin K, Taha MK, Labigne A, Zahringer U, Coyle AJ, DiStefano PS, Bertin J, Sansonetti PJ, Philpott DJ (2003b) Nod1 detects a unique muropeptide from gram-negative bacterial peptidoglycan. Science 300:1584–1587

Goldbach-Mansky R, Dailey NJ, Canna SW, Gelabert A, Jones J, Rubin BI, Kim HJ, Brewer C, Zalewski C, Wiggs E, Hill S, Turner ML, Karp BI, Aksentijevich I, Pucino F, Penzak SR,

Haverkamp MH, Stein L, Adams BS, Moore TL, Fuhlbrigge RC, Shaham B, Jarvis JN, O'Neil K, Vehe RK, Beitz LO, Gardner G, Hannan WP, Warren RW, Horn W, Cole JL, Paul SM, Hawkins PN, Pham TH, Snyder C, Wesley RA, Hoffmann SC, Holland SM, Butman JA, Kastner DL (2006) Neonatal-onset multisystem inflammatory disease responsive to interleukin-1beta inhibition. N Engl J Med 355:581–592

Gu Y, Kuida K, Tsutsui H, Ku G, Hsiao K, Fleming MA, Hayashi N, Higashino K, Okamura H, Nakanishi K, Kurimoto M, Tanimoto T, Flavell RA, Sato V, Harding MW, Livingston DJ, Su MS (1997) Activation of interferon-gamma inducing factor mediated by interleukin-1beta converting enzyme. Science 275:206–209

Hawkins PN, Lachmann HJ, Aganna E, McDermott MF (2004) Spectrum of clinical features in Muckle–Wells syndrome and response to anakinra. Arthritis Rheum 50:607–612

Heel DA van, Ghosh S, Butler M, Hunt KA, Lundberg AM, Ahmad T, McGovern DP, Onnie C, Negoro K, Goldthorpe S, Foxwell BM, Mathew CG, Forbes A, Jewell DP, Playford RJ (2005a) Muramyl dipeptide and toll-like receptor sensitivity in NOD2-associated Crohn's disease. Lancet 365:1794–1796

Heel DA van, Ghosh S, Hunt K, Mathew C, Forbes A, Jewell D, Playford R (2005b) Synergy between TLR9 and NOD2 innate immune responses is lost in genetic Crohn's disease. Gut 54:1553–1557

Hoffman HM, Mueller JL, Broide DH, Wanderer AA, Kolodner RD (2001) Mutation of a new gene encoding a putative pyrin-like protein causes familial cold autoinflammatory syndrome and Muckle–Wells syndrome. Nat Genet 29:301–305

Hoffman HM, Rosengren S, Boyle DL, Cho JY, Nayar J, Mueller JL, Anderson JP, Wanderer AA, Firestein GS (2004) Prevention of cold-associated acute inflammation in familial cold autoinflammatory syndrome by interleukin-1 receptor antagonist. Lancet 364:1779–1785

Holler E, Rogler G, Herfarth H, Brenmoehl J, Wild PJ, Hahn J, Eissner G, Scholmerich J, Andreesen R (2004) Both donor and recipient NOD2/CARD15 mutations associate with transplant-related mortality and GvHD following allogeneic stem cell transplantation. Blood 104:889–894

Hugot JP, Chamaillard M, Zouali H, Lesage S, Cezard JP, Belaiche J, Almer S, Tysk C, O'Morain CA, Gassull M, Binder V, Finkel Y, Cortot A, Modigliani R, Laurent-Puig P, Gower-Rousseau C, Macry J, Colombel JF, Sahbatou M, Thomas G (2001) Association of NOD2 leucine-rich repeat variants with susceptibility to Crohn's disease. Nature 411:599–603

Hysi P, Kabesch M, Moffatt MF, Schedel M, Carr D, Zhang Y, Boardman B, Mutius E von, Weiland SK, Leupold W, Fritzsch C, Klopp N, Musk AW, James A, Nunez G, Inohara N, Cookson WO (2005) NOD1 variation, immunoglobulin E and asthma. Hum Mol Genet 14:935–941

Inohara N, Koseki T, Lin J, Peso L del, Lucas PC, Chen FF, Ogura Y, Nunez G (2000) An induced proximity model for NF-kappa B activation in the Nod1/RICK and RIP signaling pathways. J Biol Chem 275:27823–27831

Inohara N, Ogura Y, Fontalba A, Gutierrez O, Pons F, Crespo J, Fukase K, Inamura S, Kusumoto S, Hashimoto M, Foster SJ, Moran AP, Fernandez-Luna JL, Nunez G (2003) Host recognition of bacterial muramyl dipeptide mediated through NOD2. Implications for Crohn's disease. J Biol Chem 278:5509–5512

Inohara N, Chamaillard M, McDonald C, Nunez G (2005) NOD-LRR proteins: role in host–microbial interactions and inflammatory disease. Annu Rev Biochem 74:355–383

Kanazawa N, Okafuji I, Kambe N, Nishikomori R, Nakata-Hizume M, Nagai S, Fuji A, Yuasa T, Manki A, Sakurai Y, Nakajima M, Kobayashi H, Fujiwara I, Tsutsumi H, Utani A, Nishigori C, Heike T, Nakahata T, Miyachi Y (2005) Early-onset sarcoidosis and CARD15 mutations with constitutive nuclear factor-kappaB activation: common genetic etiology with Blau syndrome. Blood 105:1195–1197

Kanneganti TD, Ozoren N, Body-Malapel M, Amer A, Park JH, Franchi L, Whitfield J, Barchet W, Colonna M, Vandenabeele P, Bertin J, Coyle A, Grant EP, Akira S, Nunez G (2006a) Bacterial RNA and small antiviral compounds activate caspase-1 through cryopyrin/Nalp3. Nature 440:233–236

Kanneganti TD, Body-Malapel M, Amer A, Park JH, Whitfield J, Taraporewala ZF, Miller D, Patton JT, Inohara N, Nunez G (2006b) Critical role for cryopyrin/Nalp3 in activation of caspase-1 in response to viral infection and double-stranded RNA. J Biol Chem 281:36560–36568

Kinoshita T, Wang Y, Hasegawa M, Imamura R, Suda T (2005) PYPAF3, a PYRIN-containing APAF-1-like protein, is a feedback regulator of caspase-1-dependent interleukin-1beta secretion. J Biol Chem 280:21720–21725

Kobayashi K, Inohara N, Hernandez LD, Galan JE, Nunez G, Janeway CA, Medzhitov R, Flavell RA (2002) RICK/Rip2/CARDIAK mediates signalling for receptors of the innate and adaptive immune systems. Nature 416:194–199

Kobayashi KS, Chamaillard M, Ogura Y, Henegariu O, Inohara N, Nunez G, Flavell RA (2005) Nod2-dependent regulation of innate and adaptive immunity in the intestinal tract. Science307:731–734

Kuida K, Lippke JA, Ku G, Harding MW, Livingston DJ, Su MS, Flavell RA (1995) Altered cytokine export and apoptosis in mice deficient in interleukin-1 beta converting enzyme. Science 267:2000–2003

Kuster W, Pascoe L, Purrmann J, Funk S, Majewski F (1989) The genetics of Crohn disease: complex segregation analysis of a family study with 265 patients with Crohn disease and 5,387 relatives. Am J Med Genet 32:105–108

Li J, Moran T, Swanson E, Julian C, Harris J, Bonen DK, Hedl M, Nicolae DL, Abraham C, Cho JH (2004) Regulation of IL-8 and IL-1beta expression in Crohn's disease associated NOD2/CARD15 mutations. Hum Mol Genet 13:1715–1725

Maeda S, Hsu LC, Liu H, Bankston LA, Iimura M, Kagnoff MF, Eckmann L, Karin M (2005) Nod2 mutation in Crohn's disease potentiates NF-kappaB activity and IL-1beta processing. Science 307:734–738

Mariathasan S, Newton K, Monack DM, Vucic D, French DM, Lee WP, Roose-Girma M, Erickson S, Dixit VM (2004) Differential activation of the inflammasome by caspase-1 adaptors ASC and Ipaf. Nature430:213–218

Mariathasan S, Weiss DS, Dixit VM, Monack DM (2005) Innate immunity against *Francisella tularensis* is dependent on the ASC/caspase-1 axis. J Exp Med 202:1043–1049

Martinon F, Tschopp J (2004) Inflammatory caspases: linking an intracellular innate immune system to autoinflammatory diseases. Cell117:561–574

Martinon F, Burns K, Tschopp J (2002) The inflammasome: a molecular platform triggering activation of inflammatory caspases and processing of proIL-beta. Mol Cell 10:417–426

Martinon F, Petrilli V, Mayor A, Tardivel A, Tschopp J (2006) Gout-associated uric acid crystals activate the NALP3 inflammasome. Nature 440:237–241

Masumoto J, Dowds TA, Schaner P, Chen FF, Ogura Y, Li M, Zhu L, Katsuyama T, Sagara J, Taniguchi S, Gumucio DL, Nunez G, Inohara N (2003) ASC is an activating adaptor for NF-kappa B and caspase-8-dependent apoptosis. Biochem Biophys Res Commun 303:69–73

Matsubara T, Hasegawa M, Shiraishi M, Hoffman HM, Ichiyama T, Tanaka T, Ueda H, Ishihara T, Furukawa S (2006) A severe case of chronic infantile neurologic, cutaneous, articular syndrome treated with biologic agents. Arthritis Rheum 54:2314–2320

McDonald C, Chen FF, Ollendorff V, Ogura Y, Marchetto S, Lecine P, Borg JP, Nunez G (2005) A role for Erbin in the regulation of Nod2-dependent NF-kappaB signaling. J Biol Chem280:40301–40309

McGovern DP, Hysi P, Ahmad T, van Heel DA, Moffatt MF, Carey A, Cookson WO, Jewell DP (2005) Association between a complex insertion/deletion polymorphism in NOD1 (CARD4) and susceptibility to inflammatory bowel disease. Hum Mol Genet14:1245–1250

Medzhitov R (2001) Toll-like receptors and innate immunity. Nat Rev Immunol1:135–145

Meylan E, Tschopp J, Karin M (2006) Intracellular pattern recognition receptors in the host response. Nature 442:39–44

Miao EA, Alpuche-Aranda CM, Dors M, Clark AE, Bader MW, Miller SI, Aderem A (2006) Cytoplasmic flagellin activates caspase-1 and secretion of interleukin 1beta via Ipaf. Nat Immunol 7:569–575

Miceli-Richard C, Lesage S, Rybojad M, Prieur AM, Manouvrier-Hanu S, Hafner R, Chamaillard M, Zouali H, Thomas G, Hugot JP (2001) CARD15 mutations in Blau syndrome. Nat Genet 29:19–20

Monsen U (1990) Inflammatory bowel disease. An epidemiological and genetic study. Acta Chir Scand Suppl 559:1–42

Murdoch S, Djuric U, Mazhar B, Seoud M, Khan R, Kuick R, Bagga R, Kircheisen R, Ao A, Ratti B, Hanash S, Rouleau GA, Slim R (2006) Mutations in NALP7 cause recurrent hydatidiform moles and reproductive wastage in humans. Nat Genet 38:300–302

Netea MG, Ferwerda G, Jong DJ de, Jansen T, Jacobs L, Kramer M, Naber TH, Drenth JP, Girardin SE, Kullberg BJ, Adema GJ, Van der Meer JW (2005) Nucleotide-binding oligomerization domain-2 modulates specific TLR pathways for the induction of cytokine release. J Immunol 174:6518–6523

Ogura Y, Bonen DK, Inohara N, Nicolae DL, Chen FF, Ramos R, Britton H, Moran T, Karaliuskas R, Duerr RH, Achkar JP, Brant SR, Bayless TM, Kirschner BS, Hanauer SB, Nunez G, Cho JH (2001a) A frameshift mutation in NOD2 associated with susceptibility to Crohn's disease. Nature 411:603–606

Ogura Y, Inohara N, Benito A, Chen FF, Yamaoka S, Nunez G (2001b) Nod2, a Nod1/Apaf-1 family member that is restricted to monocytes and activates NF-kappaB. J Biol Chem 276:4812–4818

Okada K, Hirota E, Mizutani Y, Fujioka T, Shuin T, Miki T, Nakamura Y, Katagiri T (2004) Oncogenic role of NALP7 in testicular seminomas. Cancer Sci 95:949–954

Orholm M, Iselius L, Sorensen TI, Munkholm P, Langholz E, Binder V (1993) Investigation of inheritance of chronic inflammatory bowel diseases by complex segregation analysis. BMJ 306:20–24

Poyet JL, Srinivasula SM, Tnani M, Razmara M, Fernandes-Alnemri T, Alnemri ES (2001) Identification of Ipaf, a human caspase-1-activating protein related to Apaf-1. J Biol Chem276:28309–28313

Ramos E, Arostegui JI, Campuzano S, Rius J, Bousono C, Yague J (2005) Positive clinical and biochemical responses to anakinra in a 3-yr-old patient with cryopyrin-associated periodic syndrome (CAPS). Rheumatology 44:1072–1073

Srinivasula SM, Poyet JL, Razmara M, Datta P, Zhang Z, Alnemri ES (2002) The PYRIN-CARD protein ASC is an activating adaptor for caspase-1. J Biol Chem 277:21119–21122

Tanabe T, Chamaillard M, Ogura Y, Zhu L, Qiu S, Masumoto J, Ghosh P, Moran A, Predergast MM, Tromp G, Williams CJ, Inohara N, Nunez G (2004) Regulatory regions and critical residues of NOD2 involved in muramyl dipeptide recognition. EMBO J23:1587–1597

Ting JP, Davis BK (2005) CATERPILLER: a novel gene family important in immunity, cell death, and diseases. Annu Rev Immunol23:387–414

Uematsu S, Jang MH, Chevrier N, Guo Z, Kumagai Y, Yamamoto M, Kato H, Sougawa N, Matsui H, Kuwata H, Hemmi H, Coban C, Kawai T, Ishii KJ, Takeuchi O, Miyasaka M, Takeda K, Akira S (2006) Detection of pathogenic intestinal bacteria by Toll-like receptor 5 on intestinal CD11c+ lamina propria cells. Nat Immunol 7:868–874

Viala J, Chaput C, Boneca IG, Cardona A, Girardin SE, Moran AP, Athman R, Memet S, Huerre MR, Coyle AJ, DiStefano PS, Sansonetti PJ, Labigne A, Bertin J, Philpott DJ, Ferrero RL (2004) Nod1 responds to peptidoglycan delivered by the *Helicobacter pylori* cag pathogenicity island. Nat Immunol 5:1166–1174

Watanabe T, Kitani A, Murray PJ, Wakatsuki Y, Fuss IJ, Strober W (2006) Nucleotide binding oligomerization domain 2 deficiency leads to dysregulated TLR2 signaling and induction of antigen-specific colitis. Immunity 25:473–485

Weidinger S, Klopp N, Rummler L, Wagenpfeil S, Baurecht HJ, Gauger A, Darsow U, Jakob T, Novak N, Schafer T, Heinrich J, Behrendt H, Wichmann HE, Ring J, Illig T (2005a) Association of CARD15 polymorphisms with atopy-related traits in a population-based cohort of Caucasian adults. Clin Exp Allergy 35:866–872

Weidinger S, Klopp N, Rummler L, Wagenpfeil S, Novak N, Baurecht HJ, Groer W, Darsow U, Heinrich J, Gauger A, Schafer T, Jakob T, Behrendt H, Wichmann HE, Ring J, Illig T (2005b)

Association of NOD1 polymorphisms with atopic eczema and related phenotypes. J Allergy Clin Immunol 116:177–184

Wright EK, Goodart SA, Growney JD, Hadinoto V, Endrizzi MG, Long EM, Sadigh K, Abney AL, Bernstein-Hanley I, Dietrich WF (2003) Naip5 affects host susceptibility to the intracellular pathogen Legionella pneumophila. Curr Biol 13:27–36

Yamamoto-Furusho JK, Barnich N, Xavier R, Hisamatsu T, Podolsky DK (2006) Centaurin Beta 1 down-regulates NOD1 and NOD2-dependent NF-kB activation. J Biol Chem 281:36060–36070

Zamboni DS, Kobayashi KS, Kohlsdorf T, Ogura Y, Long EM, Vance RE, Kuida K, Mariathasan S, Dixit VM, Flavell RA, Dietrich WF, Roy CR (2006) The Birc1e cytosolic pattern-recognition receptor contributes to the detection and control of Legionella pneumophila infection. Nat Immunol 7:318–325

Chapter 9
Antimicrobial Peptides as First-Line Effector Molecules of the Human Innate Immune System

Regine Gläser(✉), Jürgen Harder, and Jens-Michael Schröder

1	Introduction	188
2	Epithelial Antimicrobial Peptides and Proteins	189
	2.1 Lysozyme	189
	2.2 Human Beta Defensins	189
	2.3 Human Alpha Defensins	194
	2.4 RNases	195
	2.5 S100 Proteins: S100 A7 (Psoriasin)	196
	2.6 Others	198
3	Phagocyte Antimicrobial Peptides	200
	3.1 Human Alpha Defensins	200
	3.2 Cathelicidins	200
	3.3 S100 Proteins: S100 A8/9 (Calprotectin) and S100A12 (Calgranulin C)	201
	3.4 Others	202
4	Putative Action of Antimicrobial Peptides in the Healthy Human	202
5	Antimicrobial Peptides and Diseases	205
	5.1 Skin Diseases	205
	5.2 Wound Healing	206
	5.3 Diseases of the Airway Epithelia: Cystic Fibrosis	207
	5.4 Gastrointestinal Diseases: Inflammatory Bowel Diseases	208
	5.5 Diseases Associated with Phagocyte Dysfunction	209
6	General Conclusion and Future Aspects	210
References		210

Abstract Findings of the past two decades clearly document that epithelial cells have the capacity to mount a "chemical barrier" apart from the physical defense shield against invading microorganisms. This "chemical barrier" includes preformed antimicrobial proteins present at the uppermost layers of the epithelium as well as newly synthesized compounds that are produced upon stimulation after contact with pathogenic bacteria or bacterial products, endogenous proinflammatory cytokines and/or the disruption of the physical barrier by wounding with subsequently

---

Department of Dermatology, Venerology and Allergology, Clinical Research Unit, University Hospital Schleswig-Holstein, Campus Kiel, Schittenhelmstrasse 7, 24105 Kiel, Germany, *rglaeser@dermatology.uni-kiel.de*

released growth factors. This chapter introduces the reader into the field by giving an overview of the most important human epithelial and phagocyte derived antimicrobial peptides. Furthermore, strategies for the putative action of antimicrobial peptides in the healthy human are presented. The third part of the review gives an overview of several diseases which are in connection with a decreased or impaired antimicrobial peptide expression: skin diseases and wound healing, diseases of the airway epithelia and the gastrointestinal tract as well as diseases associated with phagocyte dysfunction. Exogenous application of antimicrobial peptides could be a promising therapeutic option in the near future for the treatment of patients with epithelial infections and chronic wounds but a much more promising option would be the promotion of the endogenous expression of antimicrobial peptides.

1 Introduction

Barrier organs such as the skin, the airways and the gastrointestinal tract, are always in contact with the environment and are covered with a characteristic microflora (Noble 1992). Depending on localization, the composition of this flora varyies qualitatively and quantitatively, between 10^2 and 10^7 microorganisms/cm^2 on the skin and between 10^{13} and 10^{14} microbes/cm^3 faeces in the gut.

With respect to the capability of bacteria for doubling their number within 20 min under optimal conditions, it might be an enigma that healthy body surfaces usually do not show excessive microbial growth and signs of infection. This unexpected phenomenon might be explained by considering body surfaces as defense organs, in which particular strategies have evolved to protect them from infection. One of the most important parts of this strategy is the existence of an intact physical barrier consisting of the stratum corneum in the skin and the mucus in non-cornified epithelia (Elias 2005). Both desquamation of corneocytes and secretion of mucus lead to a permanent regeneration of these body surfaces with simultaneous elimination of microorganisms adhering to these epithelial layers. Infiltration of microorganisms into the living epithelial cells is thus inhibited by this phenomenon.

Although for long time the physical barrier was believed to represent the sole component protecting body surfaces from infection, today a number of hints point out that this is not the case. For example, bacteria can produce a number of enzymes that can degrade essential elements of the epithelial cells like lipids, proteins and glycoconjugates, making it possible to overcome the physical barrier. These observations led to the hypothesis that, apart from the physical barrier, a "chemical barrier" of epithelia should also exist, consisting of molecules produced in the strategically optimal located uppermost parts of the barrier organs where they control growth of bacteria and thus inhibit infection of body surfaces. Because healthy epithelia do not contain any blood-derived leukocytes such as neutrophils, which contain a number of bactericidal compounds (Ganz 2004), the actual epithelial cells might be the source of these "chemical barrier" compounds.

Today it is widely accepted that the epithelial production of antimicrobial peptides and proteins represents a general phenomenon of the epithelial defense system. Although originally discovered in plants and invertebrates, it was documented in 1987 that vertebrate skin might also use antimicrobial peptides as part of a "chemical defense system" to protect the epithelium from infection. Zasloff et al. wondered why frogs, living in a laboratory pond under non-sterile conditions, had no signs of skin infection after surgical treatment – despite the huge number of microbes present. Based on this unexpected finding Zasloff was able to isolate a broad-spectrum antimicrobial peptide from the skin of the African clawed frog *Xenopus laevis* which he termed "Magainin" (Zasloff 1987). During recent years a number of antimicrobial peptides have been characterized from human skin and other epithelia, to which the reader is introduced in this chapter. An overview of the most important human epithelial antimicrobial peptides is provided in Table 1.

2 Epithelial Antimicrobial Peptides and Proteins

2.1 Lysozyme

Lysozyme represents the first antimicrobial protein which was identified in human skin (Klenha and Krs 1967; Ogawa et al. 1971), nearly 50 years after its initial discovery as "bacteriolytic activity" in nasal secretions by Alexander Fleming (1922). In the skin lysozyme is mainly located in the cytoplasm of epidermal cells in the granular layers and in the stratum corneum (Ogawa et al. 1971; Papini et al. 1982). Also, pilosebaceous follicle cells and hair bulb cells as well as parts of the eccrine and apocrine sweat glands were shown to be lysozyme-positive (Ezoe and Katsumata 1990).

Although lysozyme is mainly directed against Gram-positive bacteria (e.g. *Staphylococcus aureus*; Kern et al. 1951) it is also active against Gram-negative bacteria, e.g. *Echerichia coli* (Ellison and Giehl 1991) and *Pseudomonas aeruginosa* (Cole et al. 2002), suggesting that it might contribute to control the growth of bacteria in healthy skin. However, the significance of lysozyme in the cutaneous defense system is still unclear because lysozyme was shown to be expressed exclusively in the cytoplasm and could not be identified within the stratum corneum (Ogawa et al. 1971) or in skin-derived washing fluids (Gläser et al. 2005).

2.2 Human Beta Defensins

2.2.1 hBD-1

The first discovered human β-defensin was originally isolated from human blood filtrate as a peptide with significant sequence homology to bovine β-defensins (Bensch et al. 1995). HBD-1 is a cationic peptide of 36 amino acid residues,

Table 1 Human antimicrobial proteins mentioned in this chapter

	Cellular source	Inducible expression	Antimicrobial activity Bacteria Gram⁺	Gram⁻	Fungi
α-Defensins					
HNP1-4	Neutrophils		++	++	++
HD-5, -6	Intestinal tract		+++	+++	+++
β-Defensins					
hBD-1	Keratinocytes, airway epithelia, urogenital tract		+	++	?
hBD-2	Keratinocytes, airway epithelia, intestinal tract	x	(+)[a]	+++	++
hBD-3	Keratinocytes, airway epithelia	x	+++	+++	+++
hBD-4	Airway epithelia, keratinocytes (mRNA)	x	++	++	+
Cathelicidins					
LL-37	Neutrophils, keratinocytes, airway epithelia, urogenital tract	x	++	++	++
Ribonucleases					
RNase 7	Keratinocytes, airway epithelia	x	+++	+++	+++
RNase 8	?		+++	+++	+++
S-100 Proteins					
Calprotectin (S100A8/A9)	Leukocytes, keratinocytes	x	+	+	++
Psoriasin (S100A7)	Keratinocytes, sebocytes, airway epithelia, urogenital tract	x	(+)[a]	++[b]	(+)[a]
Others					
Antileukoprotease (ALP)	Keratinocytes, airway epithelia		++	++	++
Adrenomedullin	Keratinocytes, sebocytes, sweat glands, intestinal tract		+	++	?
Dermcidin (DCD-1)	Sweat glands		+++	+++	++
Elafin	Keratinocytes, airway epithelia	x	+	+	?
Histatins	Oral epithelia		++	++	+++
Lactoferrin	Neutrophils, body fluids		++	++	++
Lysozyme	Skin, airway epithelia, body fluids		++	++	+

[a] in high concentrations.
[b] *Escherichia coli*, others in high concentrations.

containing six cysteines and forming three characteristic intramolecular disulfide bonds. In blood plasma and urine, several forms of hBD-1 have been isolated (Valore et al. 1998). Until now, natural hBD-1 protein has not yet been isolated from human skin. Fulton et al. detected hBD-1 mRNA expression in the suprabasal keratinocytes and sweat ducts of human skin using in situ hybridization (Fulton et al. 1997). A more detailed investigation confirmed the expression of hBD-1 in human keratinocytes by immunohistochemistry where hBD-1 was found to be consistently expressed in suprabasal keratinocytes of interfollicular skin samples derived from various body sites (Ali et al. 2001). *In vitro* induction of keratinocyte differentiation by calcium treatment led to the upregulation of hBD-1 gene expression (Frye et al. 2001; Abiko et al. 2003; Harder et al. 2004), an observation which may explain why hBD-1 peptide shows strongest expression in the more differentiated terminal layers of human skin (Ali et al. 2001). Overexpression of hBD-1 in keratinocytes resulted in increased expression of differentiation markers, suggesting that hBD-1 promotes differentiation of keratinocytes (Frye et al. 2001) but, in contrast to hBD-2 and hBD-3, gene expression of hBD-1 in keratinocytes was not markedly inducible by proinflammatory cytokines or by bacteria (Harder et al. 2004).

Only a few studies investigated the antimicrobial spectrum of hBD-1. Recombinant and natural hBD-1 forms exhibit salt-sensitive antimicrobial activity against various strains of *E. coli* at micromolar concentrations, even in normal urine (Valore et al. 1998). Singh et al. (1998) reported antimicrobial activity of a recombinant baculovirus-derived hBD-1-preparation against *P. aeruginosa*. In other studies, only a minor antimicrobial activity of native hBD-1 was detected (Zucht et al. 1998) and no activity of hBD-1 against pathogenic Gram-positive bacteria such as *S. aureus* was reported.

HBD-1 exhibits chemotactic activity for cells stably transfected with the chemokine receptor CCR 6 as shown for hBD-2, suggesting that hBD-1 may recruit immature dendritic cells and memory T cells to sites of microbial invasion (Yang et al. 1999) by linking the innate with the adaptive immune system.

2.2.2 hBD-2

HBD-2 was originally isolated from lesional psoriatic scale extracts using an *E. coli* affinity column (Harder et al. 1997) and the high expression of hBD-2 in psoriasis was confirmed by in situ hybridization and immunohistochemistry in lesional psoriatic keratinocytes (Liu et al. 1998). HBD-2 was found to be the first inducible member of the human defensin family. Endogenous proinflammatory cytokines like IL-1α, IL-1β, TNF-α, IL-17 and exogenous stimuli like bacteria such as *P. aeruginosa* have proven to be the most effective inducers for hBD-2 expression (Harder et al. 1997; Huh et al. 2002; Liu et al. 2002; Sorensen et al. 2003; Kao et al. 2004). The bacterial factors as well as their appropriate receptors on keratinocytes which are involved in the bacteria-mediated hBD-2 induction have not yet been identified. Recently, it has been shown that human keratinocytes express various Toll-like receptors (TLRs) from which TLR-2 mediates nuclear factor kappa B (NF-κB)-dependent gene expression in keratinocytes stimulated with *S. aureus* and its cell wall components (Mempel et al. 2003). These investigations indicate that TLRs may play a role for the bacteria-mediated induction of antimicrobial proteins in keratinocytes. Interestingly, keratinocytes are able to discriminate between commensal and pathogenic bacteria: induction of hBD-2 in primary keratinocytes treated with the skin commensal *S. epidermidis* was suppressed by inhibitors of the c-Jun N-terminal kinase (JNK) and p38 pathways, whereas induction by the skin pathogen *Streptococcus pyogenes* was blocked by inhibitors of NF-κB (Chung and Dale 2004).

HBD-2 is also known to be present in the stratified epithelia of the oral cavity as well as in cultured gingival epithelial cells (Weinberg et al. 1998). Like skin keratinocytes, it is predominately localized in the upper epithelial layers, consistent with the formation of the stratified epithelial barrier (Dale and Krisanaprakornkit 2001).

The hBD-2 promoter contains several putative transcription factor binding sites, including NF-κB, activator protein (AP)-1 and -2, as well as NF-IL-6, which are all known to be involved in the induction and regulation of inflammatory responses (Liu et al. 1998; Harder et al. 2000).

In contrast to psoriatic skin, healthy skin extracts harbor only low amounts of hBD-2 peptide (Schröder and Harder 1999). In normal skin, hBD-2 immunoreactivity is localized to the uppermost layers of the epidermis and/or stratum corneum (Ali et al. 2001), confirming recent studies showing an upregulation of hBD-2 in cultured primary keratinocytes brought to differentiation by high calcium concentrations (Pernet et al. 2003; Harder et al. 2004). HBD-2 expression was shown to be variable in skin with interindividual and site-specific differences in the intensity of immunostaining (Ali et al. 2001). Ultrastructural analyses detected hBD-2 in lamellar bodies and intercellular spaces of IL-1a stimulated cultured primary keratinocytes, indicating that hBD-2 is released together with the lipid compounds stored preformed in lamellar bodies. Release of these lamellar bodies contents leads to accumulation of hBD-2 in the intercellular spaces of the uppermost epidermis, where a high local concentration of this defensin is achieved to effectively control microbial growth (Oren et al. 2003).

Natural hBD-2 has been reported to exhibit primarily antimicrobial activity against Gram-negative bacteria such as *E. coli* and *P. aeruginosa*. This activity depends on ion composition and it was documented that increasing concentrations of NaCl diminished hBD-2 activity (Bals et al. 1998; Singh et al. 1998; Harder et al. 2000; Tomita et al. 2000). The capacity of hBD-2 to kill bacteria *in vivo* was demonstrated in a mouse gene therapy study with bacterial infection of hBD-2-transfected tumor cells (Huang et al. 2002).

In addition to its antimicrobial function, hBD-2 shows selective chemotactic activity for cells stably transfected with human CCR6, a G protein-coupled chemokine receptor which is preferentially expressed by immature dendritic cells and memory T cells (Yang et al. 1999). HBD-2 mimics chemokine activity similar to the only yet known CCR6-ligand CCL-20 (MIP-3α). Therefore, hBD-2 may promote a secondary adaptive immune responses by recruiting dendritic and T cells to the site of microbial invasion through interaction with CCR6. In addition hBD-2 was found to be a specific chemoattractant for TNF-α-treated human neutrophils and to act also as a chemotaxin for mast cells, stimulating their histamine release and prostaglandin D synthesis (Niyonsaba et al. 2001).

2.2.3 hBD-3

HBD-3 was isolated as the third member of the human beta-defensin family from lesional psoriatic scales using a biochemical approach with the intention to identify an antimicrobial peptide directed against the Gram-positive germ *S. aureus*. The complete cDNA of hBD-3 was subsequently cloned from keratinocytes and lung epithelial cells (Harder et al. 2001). Using a bioinformatics approach and by functional genomic analysis, separate groups (Garcia et al. 2001; Jia et al. 2001) identified the hBD-3 gene. The 67 amino acid peptide precursor, containing a 22 amino acid signal peptide, is approximately 43% identical to hBD-2 (Garcia et al. 2001; Harder et al. 2001; Jia et al. 2001). Besides skin, gingival keratinocytes, tonsils, esophagus, trachea, adult heart, skeletal muscle, placenta and fetal thymus comprise the major hBD-3 mRNA-expressing tissues (Garcia et al. 2001; Harder et al. 2001; Jia et al. 2001).

In contrast to hBD2, interferon (IFN)-γ represents a powerful hBD-3-inducing cytokine in keratinocytes, whereas TNF-α induces hBD-3-mRNA expression only at a low level (Harder et al. 2001; Nomura et al. 2003). Expression of hBD-3 is also inducible in keratinocytes or tracheal epithelial cells by contact with bacteria (Garcia et al. 2001; Harder et al. 2001).

In contrast to hBD-2, hBD-3 exhibits a broad spectrum of potent antimicrobial activity against many potentially pathogenic Gram-negative and Gram-positive bacteria and fungi, including multiresistant *S. aureus* (MRSA) and vancomycin-resistant *Enterococcus faecium* (VRE) (Garcia et al. 2001; Harder et al. 2001; Hoover et al. 2003; Maisetta et al. 2003; Sahly et al. 2003). Recently it was shown in *S. aureus* that inactivation of the fmtC gene, which is associated with methicillin resistance, results in increased susceptibility of MRSA to hBD-3 (Midorikawa et al. 2003).

The mechanism of hBD-3 microbicidal activity is still unknown. Ultrastructural investigation of hBD-3-treated *S. aureus* revealed signs of perforation of the peripheral cell wall (Fig. 1) resembling those morphological effects seen when the germ is treated with penicillin (Harder et al. 2001).

HBD-3 acts also as a chemokine (Wu et al. 2003). In contrast to its antimicrobial activity, hBD-3-dependent chemotaxis of monocytes and CCR6-transfected HEK 293 cells strongly depends on the topology of disulfide connectivities in hBD3, suggesting that a defined 3D structure present in disulfide connectivities of the natural hBD-3 is required for productive binding and activation of the CCR6 receptor.

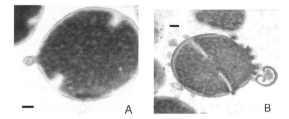

Fig. 1 Morphology of hBD-3 treated *Staphylococcus aureus*. Transmission electron microscopy of *S. aureus* treated with hBD-3 for 30 min (**A**) or 2 h (**B**). *Bars* 0.1 µm. Reproduced from Harder et al. (2001) with permission

2.2.4 hBD-4

The fourth member of the human β-defensin family was identified by screening the human genome database (Garcia et al. 2001). Very recently gene expression of hBD-4 was detected in primary keratinocytes (Harder et al. 2004). Synthetic hBD-4 revealed antimicrobial activity at micromolar concentrations against *P. aeruginosa* and *Staphylococcus carnosus* and hBD-4 gene expression was shown to be upregulated by bacteria in respiratory epithelial cells (Garcia et al. 2001). However, until now nothing is known about the expression of hBD-4 peptide in human skin and all attempts have failed to isolate hBD-4 peptide from psoriatic scale extracts as well as from healthy human skin-derived stratum corneum (Gläser, Harder and Schröder, unpublished observations). Therefore, further investigations need to be performed to elucidate the role of hBD-4 in the chemical skin defense system.

2.3 Human Alpha Defensins

2.3.1 Human Defensin-5 and -6

The human intestinal tract is constantly exposed to an enormous indigenous bacterial flora. It was recently recognized that antimicrobial peptides of the defensin family likely play a role in protection against microbial invasion at a variety of mucosal epithelial surfaces, including that of the intestinal tract. In contrast to the human neutrophil peptides (HNPs), human defensin (HD)-5 and -6 are mainly expressed in the intestinal Paneth cells, specialized secretory epithelial cells, located at the base of the small intestinal crypts (Bevins 2006). Apart from in Paneth cells, HD-5 is also expressed in some villous epithelial cells in duodenum, jejunum and ileum, whereas the protein is not expressed in the stomach or colon. The protein is stored only in its precursor form in ileal Paneth cells, and processing of the peptide to a mature form occurs during and/or after secretion through trypsin

(Ghosh et al. 2002; Cunliffe 2003). HD-5 acts as a potent antimicrobial agent against certain microorganisms by disrupting the target cell membrane (Ouellette 1997; Ericksen et al. 2005). The most compelling evidence for a role of HD-5 *in vivo* is evident from studies of mice transgenic for HD-5 which are completely immune to infection and systemic disease from orally administered *Salmonella typhimurium* (Salzman et al. 2003).

2.3.2 Dermcidin

Dermcidin (DCD-1) is a novel anionic antimicrobial peptide produced and secreted exclusively by human eccrine sweat glands (Schittek et al. 2001). Dermcidin is proteolytically processed to DCD-1, a 47 amino acids containing peptide exhibiting antimicrobial activity against different bacteria (i.e. *E. coli*, *E. faecalis*, *Staphylococcus aureus*) and the yeast *Candida albicans*. DCD-1 is found in human sweat in antimicrobial active concentrations, where antimicrobial activity is not affected by the low pH value and high salt concentrations present. These findings indicate that sweat glands also contribute to the innate immune responses of the skin by secreting antimicrobial proteins.

2.4 RNases

2.4.1 RNase 7

RNase 7, a member of the RNase A superfamily, was originally isolated from heel stratum corneum extracts as one of the principal cationic proteins of healthy human skin (Harder and Schröder 2002). Apart from skin, RNase 7 mRNA was shown to be expressed in various cells of epithelial origin including trachea, tonsils, pharynx, tongue and salivary glands, as well as in renal cells and the thymus.

RNase 7 exhibits a broad-spectrum antimicrobial activity at low micromolar concentrations against Gram-negative bacteria (*P. aeruginosa*, *E. coli*), Gram-positive bacteria (*S. aureus*, *Propionibacterium acnes*) and the yeast *C. albicans*. Of particular interest, a high efficacy against a Vancomycin-resistant strain of *E. faecium* was demonstrated at a concentration of only 20 nM (Harder and Schröder 2002). Therefore, RNase 7 represents one of the most potent and efficacious human antimicrobial proteins known so far. Blocking the ribonuclease activity of RNase 7 did not reduce the antibacterial activity against *E. coli* (Gläser, Harder and Schröder, unpublished data), indicating a different mechanism of action.

The proinflammatory cytokines IL-1β, IFN-γ and, to a lesser degree, also TNF-α have been shown to induce RNase 7 mRNA expression and it was shown that bacteria like *P. aeruginosa* or *S. aureus* can upregulate the expression of RNase 7 (Harder and Schröder 2002).

2.4.2 RNase 8

Recently, RNase 8 was identified as another novel antimicrobial protein of the human RNase A superfamily (Rudolph et al. 2006). Expression of RNase 8 has been detected in placenta (Zhang et al. 2002). Although initially thought to be inactive (Zhang et al. 2002), recombinant RNase 8 exhibits a broad-spectrum microbicidal activity against potential pathogenic microorganisms, including multidrug-resistant strains at micro- to nanomolar concentrations (Rudolph et al. 2006), indicating that RNase 8 may also contribute to the innate host defense.

2.5 S100 Proteins: S100 A7 (Psoriasin)

S100 proteins are believed to mediate a variety of functions in eukaryotic cells including differentiation, cell cycle progression, intracellular Ca^{2+} signalling and cytoskeletal membrane interactions as well as playing a role in leukocyte chemotaxis (Heizmann et al. 2002; Eckert et al. 2004). Recently a few studies also indicated that S100 proteins may play a putative role in the innate host defense (Murthy et al. 1993; Gottsch et al. 1999; Cole et al. 2001).

Interestingly, gut bacteria like *E. coli* rarely colonize on human skin and die rapidly on the skin surface, whereas other species do not (Casewell and Desai 1983). This distinction can be readily demonstrated by exposure of the fingertips to either *E. coli* or the common skin pathogen *S. aureus*, resulting in effective killing of *E. coli* but not of *S. aureus* (Fig. 2). This finding lead to the identification of the S100-protein psoriasin (S100-A7) as an *E. coli*-killing defense chemical of healthy human skin (Gläser et al. 2005).

Psoriasin shows antimicrobial activity *in vitro* preferentially against *E. coli* at low micromolar concentrations. This activity is inhibited by pretreatment of psoriasin with Zn^{2+}, which suggests that its antimicrobial activity is mediated by deprivation of the essential trace element zinc. In healthy volunteers a neutralizing psoriasin antibody increases the growth of *E. coli* on the skin, documenting that psoriasin acts as a principal bactericidal component of human skin.

Psoriasin mRNA and protein expression was shown to be upregulated in primary keratinocytes by proinflammatory cytokines and after contact with bacterial culture filtrates – an observation that could also be confirmed *in vivo* by treatment of healthy donor's skin with bacterial culture filtrates These findings are in agreement with immunohistochemical analyses (Fig. 3) as well as analyses of skin-washing fluids showing that psoriasin is focally expressed, particular in areas where a high bacterial colonization is well documented (Fig. 4). Apart from in keratinocytes, psoriasin is also expressed by sebaceous glands, suggesting that the protein is possibly secreted together with lipids (Gläser et al. 2005).

Studies using chemically synthesized N-acetylated S100A7 confirmed the *E. coli*-cidal action of natural N-acetylated psoriasin (Li et al. 2005) and it was shown that a recombinant His-tag-psoriasin fusion protein adheres to and reduces

Fig. 2 Healthy skin is resistant against *Escherichia coli* infection. Washed fingertips of a healthy volunteer were artificially inoculated with either *S. aureus* (left) or *E. coli* (right) for 30 min. The fingertips were pressed onto a nutrient agar plate to determine the number of colonies after overnight incubation. Reproduced from Gläser et al. (2005) with permission

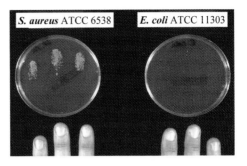

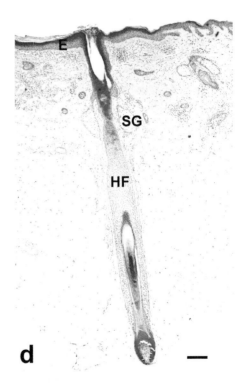

Fig. 3 Psoriasin is focally expressed in human skin and adnexal structures. Strong immunoreactivity with a monoclonal psoriasin antibody (*red staining*) is visible in the suprabasal keratinocytes of cheek epidermis (*E*), sebaceous glands (*SG*), and the hair follicle (*HF*). *Bar* 200 μm. Reproduced from Gläser et al. (2005) with permission

E. coli survival (Lee and Eckert 2007). Mutation of the conserved carboxyl-terminal EF-hand calcium-binding motif or heat denaturation only slightly reduced this His-tag-S100A7 antibacterial activity; and the activity could be destroyed by protease treatment. Interestingly, the central region of S100A7, including only amino acids 35–80, was shown to be sufficient for full antibacterial activity (Lee and Eckert 2007).

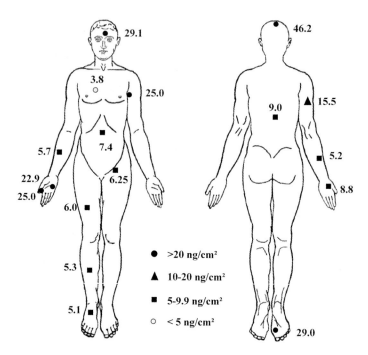

Fig. 4 Psoriasin secretion is depending from the body location. Standardized areas of eight healthy volunteers were rinsed with 10mM sodium phosphate buffer (pH 7.4) to determine (by ELISA) the median local concentration of psoriasin present at the skin surface. Reproduced from Gläser et al. (2005) with permission

2.6 Others

2.6.1 Adrenomedullin

The 52-amino-acid peptide adrenomedullin is involved in numerous physiological functions, including vasodilation, renal homeostasis, hormone regulation, neurotransmission and growth modulation (Zudaire et al. 2003). Adrenomedullin is expressed by many epithelia, including normal and neoplastic skin, in cells of the eccrine and apocrine sweat glands as well as in the sebaceous glands (Martinez et al. 1997).

Adrenomedullin exhibits high antimicrobial activity against *E. coli* and moderate activity against *S. aureus* (Allaker et al. 1999). Interestingly, adrenomedullin seems to be very effective in killing *P. acnes*, which plays a major role in the development of the common skin disease acne vulgaris. It was shown that adrenomedullin is secreted by keratinocytes *in vitro* but it is not clear whether concentrations within the antimicrobial range are reached *in vivo* (Allaker et al. 1999).

In oral epithelial cells as well as in gastric epithelial cells adrenomedullin expression is induced through bacterial challenge (Kapas et al. 2001; Allaker and

Kapas 2003) providing further evidence that adrenomedullin contributes to the defense mechanisms of different epithelial cells.

2.6.2 Antileukoprotease

The human serine protease inhibitor antileukoprotease (ALP) is known to be an antiproteolytic compound of human body fluids and mucous secretions (Fritz 1988). Expression of ALP in human skin was demonstrated by the isolation of ALP from healthy human stratum corneum and by detection in supernatants of cultured human primary keratinocytes (Wiedow et al. 1998). In addition to its antiprotease activity, ALP exhibits antimicrobial activity against a broad range of microorganisms like Gram-negative and Gram-positive bacteria as well as *C. albicans* (Wiedow et al. 1998; Wingens et al. 1998). These findings suggest that ALP not only protects skin against proteolysis but also controls cutaneous microbial growth.

2.6.3 Elafin

Another skin-derived serine protease inhibitor inducibly expressed in keratinocytes and termed elafin (Alkemade et al. 1994; Pfundt et al. 2000; Meyer-Hoffert et al. 2003; Sorensen et al. 2003), was reported to exhibit antimicrobial activity against *P. aeruginosa* and *S. aureus* (Simpson et al. 1999). The most active full-length form of synthetic Elafin (95 amino acids) exhibited killing activity against *P. aeruginosa* but only weak activity against *S. aureus*. Another report demonstrated only growth-inhibiting properties against three different *P. aeruginosa* strains and no bacteriostatic activity against *E. coli* using 57 amino-acids containing C-terminal recombinant elafin (Meyer-Hoffert et al. 2003). This truncated elafin form was originally purified from lesional psoriatic scale extracts (Wiedow et al. 1990). Due to its moderate antimicrobial activity found *in vitro*, it is not clear how effective elafin functions as an antibacterial agent *in vivo*. However, the antibacterial potential of elafin was shown in a mouse model, where adenoviral augmentation of elafin protected lungs against injury and infection mediated by *P. aeruginosa* (Simpson et al. 2001).

2.6.4 Histatins

Histatins are salivary histidine-rich cationic peptides, ranging from 7 to 38 amino acid residues in length, that exert a potent killing effect *in vitro* on *C. albicans* (Helmerhorst et al. 1997). *Candida* species are common commensal inhabitants of the oral cavity and it was shown that the oral yeast status is related to salivary histatin levels (Jainkittivong et al. 1998).

Histatins bind to a receptor on the fungal cell membrane and enter the cytoplasm where they target the mitochondrion. There they induce the non-lytic loss of ATP from actively respiring cells, leading to cell death. In addition, histatins have been shown to disrupt the cell cycle and lead to the generation of reactive oxygen species (Kavanagh and Dowd 2004).

3 Phagocyte Antimicrobial Peptides

3.1 *Human Alpha Defensins*

3.1.1 Human Neutrophil Peptides-1 to -3

Human neutrophil peptides (HNP)-1 to -3 constitute about 30% of the protein content of azurophil granules and exhibit a wide spectrum of antimicrobial activity against various bacteria, fungi and viruses (Martin et al. 1995). The antimicrobial features of HNP-1 to -3 together with their high abundance in neutrophilic granulocytes indicate that these antimicrobial peptides participate in the non-oxidative killing of phagocytosed bacteria in neutrophils. To date it is believed that HNP-1 to -3 are mainly produced in neutrophils and not in other cells and that the detection of these proteins in other cell preparations results from neutrophil contamination (Zhang et al. 2002; Mackewicz et al. 2003).

3.1.2 Human Neutrophil Peptide-4

HNP-4 was isolated and characterized as a novel antimicrobial peptide from the azurophilic granule fraction of discontinuous Percoll gradients (Wilde et al. 1989). *In vitro*, purified HNP-4 was shown to kill *E. coli*, *Streptococcus faecalis*, and *C. albicans* with a higher potency when compared with a mixture of the other human alpha defensins.

3.2 *Cathelicidins*

hCAP-18/LL-37 belongs to the cathelicidin family, a group of antimicrobial proteins originally isolated from porcine neutrophils, which share a highly conserved N-terminus, termed cathelin (Ritonja et al. 1989). hCAP-18/LL-37 is the only member of this family present in the human genome (Gudmundsson et al. 1996) and was first identified in a human bone marrow cDNA (Agerberth et al. 1995). The gene encodes a preproprotein of 18 kDa (hCAP-18) which is proteolytically processed by the serine protease proteinase 3, yielding the C-terminal 37 amino acids containing the antimicrobial LL-37 peptide (Sorensen et al. 2001).

LL-37 exhibits broad spectrum antimicrobial activity in the micromolar range against various Gram-negative and Gram-positive bacteria as well as fungi (Turner et al. 1998) but LL-37 is also cytotoxic in the micromolar range to eukaryotic cells under physiological salt conditions (Johansson et al. 1998). Both cytotoxic and antibacterial activity were shown to be inhibited by human Apolipoprotein A-I (Wang et al. 1998).

A recent report showed that the N-terminal cathelin-like prosequence of hCAP-18 exhibits antiprotease as well as antimicrobial activity (Zaiou et al. 2003), resulting in the innate host defense through the inhibition of bacterial growth and limitation of cysteine-proteinase-mediated tissue damage. Furthermore it was shown that LL-37 stimulates chemotaxis for neutrophils, monocytes and T cells via the formyl peptide-like receptor-1 (De et al. 2000) and that LL-37 induces mast cell chemotaxis (Niyonsaba et al. 2002). By recruiting effector cells to foci of inflammation and infection, these reports indicate an additional function of LL-37 besides its antimicrobial activity.

In contrast to healthy skin, LL-37 gene expression could be identified in keratinocytes of inflamed skin (Frohm et al. 1997), where it is stored in skin lamellar granules (Braff et al. 2005). Induction of hCAP18/LL-37 in keratinocytes occurs by insulin-like growth factor I and transforming growth factor (TGF)-alpha or 1,25-dihydroxyvitamin D3 (Gombart et al. 2005). LL-37 was also shown to be localized in the eccrine gland and sweat ductal epithelial cells, where antimicrobial activity against various bacteria in the sweat ionic environment indicates that this peptide may also contribute to the antibacterial activity of human sweat (Murakami et al. 2002). Very recently it was demonstrated that LL-37 is processed in sweat by a serine protease-dependent mechanism into multiple novel smaller antimicrobial peptides which show increased bactericidal and fungicidal activity, acting in a synergistic fashion (Murakami et al. 2004).

The relevance of cathelicidins in cutaneous host defence has been demonstrated in a mouse model. Mice deficient in the expression of cathelicidin antimicrobial peptide (CRAMP, the mouse homolog to human hCAP-18/LL-37) were more susceptible to skin infections caused by group A *Streptococcus* (GAS); and GAS mutants resistant to CRAMP produced more severe skin infections in wild-type mice (Nizet et al. 2001).

3.3 S100 Proteins: S100 A8/9 (Calprotectin) and S100A12 (Calgranulin C)

The heterodimeric complex of the two Ca^{2+}-binding S100 proteins S100A8 and A9, also known as calgranulin A and B or calprotectin, exhibits selective biostatic activity at high concentrations against *C. albicans* (Murthy et al. 1993). Zinc chelation was proposed as a potentially important host defense function of calprotectin (Clohessy and Golden 1995) and it was shown that intact calprotectin, consisting of both subunits, is necessary to form a zinc-binding site capable of inhibiting microbial growth (Sohnle et al. 2000).

Calgranulin C (S100 A12), representing a minor calgranulin in neutrophils, demonstrated filariacidal and filariastatic activity (Gottsch et al. 1999) and a short C-terminal peptide fragment of calgranulin c, named "calcitermin", exhibited bactericidal activity against Gram-negative bacteria, which interestingly was potentiated by adding Zn^{2+} (Cole et al. 2001).

3.4 Others

3.4.1 Neutrophil Gelatinase-Associated Lipocalin

The 25-kDa protein neutrophil gelatinase-associated lipocalin (NGAL) was initially isolated from the specific granules of human neutrophils (Kjeldsen et al. 1993; Kjeldsen et al. 1994). Later it was shown that NGAL exhibits bacteriostatic activity through its ability to bind bacterial ferric siderophores inhibiting the siderophore-mediated iron uptake by bacteria (Goetz et al. 2002). Since iron is essential for bacterial growth, its deprivation causes bacteriostatic effects. Expression of NGAL in human keratinocytes was shown to be upregulated by IL-1β, insulin-like growth factor (IGF)-I and transforming growth factor (TGF)-α (Sorensen et al. 2003).

3.4.2 Eosinophil-Derived RNases

Three proteins abundant in the cytoplasmic granules of human eosinophils, major basic protein (MBP), eosinophil cationic protein (ECP) and eosinophil-derived neurotoxin (EDN), were shown to exhibit antimicrobial activity. MBP exhibited antibacterial activity against *S. aureus* and *E. coli* (Lehrer et al. 1989), whereas ECP showed antiviral activity against respiratory syncytial virus (RSV) (Domachowske et al. 1998) in addition to its antibacterial activity against *S. aureus* (Lehrer et al. 1989). The antibacterial activity of MBP and ECP was shown to be modulated by incubation time, protein concentration, temperature and pH, causing outer and inner membrane permeabilization (Lehrer et al. 1989). EDN demonstrated potent antiviral activity against RSV (Domachowske et al. 1998).

3.4.3 Lactoferrin

Antimicrobial lactoferrin is present in different human body fluids where bacterial growth inhibition is caused by its ability to sequester iron (Weinberg 2001). While NGAL specifically binds to ferric siderophores, lactoferrin simply binds free iron, inhibiting the uptake of this essential trace element by bacteria.

4 Putative Action of Antimicrobial Peptides in the Healthy Human

The participation of at least some of the above-mentioned antimicrobial peptides as components of a "chemical barrier" would easily explain an unexpectedly constant number of microbes at body surfaces and the low infection rate:

Strategically it would be most important for the host to start defense reactions against infection as early as possible. Therefore one would postulate that a first-line defense takes place within the uppermost parts of the stratum corneum, at the "physical barrier", where the microorganisms try to adhere and start colonization (Fig. 5A). The epithelial expression of adhesins as well as the release of bacterial proteases could promote this adhesion. However, under physiological conditions, bacteria are unable to invade the epidermis. One important factor is clearly the process of desquamation and the additional presence of components of the "chemical barrier" in the stratum corneum itself.

Another strategy would be to inhibit bacterial growth at the surface by limiting nutrients and essential trace elements, such as iron and zinc. Keratinocytes are able to release NGAL and therefore to limit the bacterial iron availability indirectly via binding of bacterial siderophores (Goetz et al. 2002). In addition, the availability of zinc ions could be limited by zinc-binding antimicrobial proteins such as psoriasin (Gläser et al. 2005) or calprotectin (Clohessy and Golden 1995). Indeed, psoriasin is present at the skin surface, thus acting prior to bacterial invasion into the stratum corneum and subsequently into living skin areas. In addition to the depletion of essential nutrition elements, keratinocytes may produce factors limiting colonization and the formation of bacterial biofilms, which are usually absent at intact skin surfaces. Furthermore, keratinocytes may produce inhibitors of microbial proteases, which are essential for microbial invasion. ALP, which is present in healthy skin stratum corneum, represents a typical example for such antimicrobial protease inhibitors.

Once the stratum corneum layer is defective or missing, as is the case in micro-wounds (Fig. 5B), bacteria and bacterial products come into direct contact with living keratinocytes, which now enhances the production and release of inducible keratinocyte-derived antimicrobial peptides and proteins, like hBD-2, -3, RNase 7 and psoriasin (Harder et al. 1997, 2001; Harder and Schröder 2002; Gläser et al. 2005). This induction is mediated by not yet characterized bacterial "pathogen-associated molecular patterns" (PAMPs), which may induce only antimicrobial peptides and proteins, but not proinflammatory cytokines. When this scenario occurs at the skin surface, it will not take attention, because it is clinically invisible.

When there is a massive disturbance of living skin areas (Fig. 5C), e.g. by bigger wounds or invasive virulent microbes, PAMPs which are known to act via Toll-like receptors (TLRs) and/or putative other PRRs induce proinflammatory cytokines in keratinocytes and immune cells, leading to inflammation with the recruitment of leukocytes and activation of the adaptive immune system.

Therefore one would postulate that skin and other barrier organs actively try first to inhibit microbial growth, adherence and invasion in the uppermost parts of the epithelia. Once microbes overcome this first defense line, they are confronted with a second defense line consisting of antimicrobial peptides located in the stratum corneum and the uppermost layers of living epidermis. As soon as they overcome this second line of the "chemical defense system" the microbes activate responses causing visible inflammation.

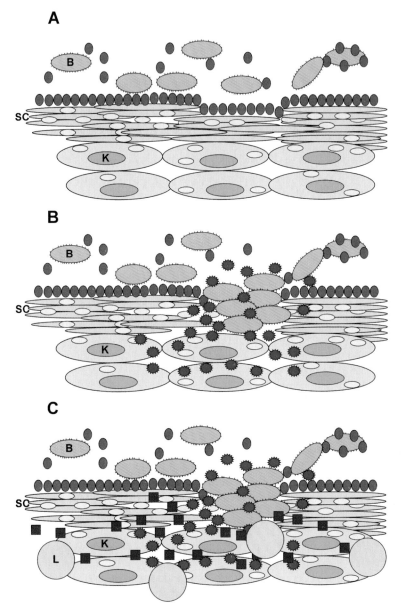

Fig. 5 Putative mechanism of the skin's "chemical barrier". **A** At healthy skin surfaces the adherence, colonization and growth of microbes is prevented by keratinocyte-derived antimicrobial proteins (AMP) and protease inhibitors. Bacterial growth inhibition is mediated by sequestration of essential trace metals (e.g. zinc, iron) and the release of preformed cationic AMPs. In addition skin-derived protease inhibitors block microbial proteases which are essential for tissue invasion. **B** After loss of the stratum corneum (microwounds) or by overcoming the first line of the "chemical barrier" (**A**), microbes have access to living keratinocytes. The contact of keratinocytes with pathogen-associated molecules (PAMs) results in the rapid release of inducible AMPs without any signs of inflammation. **C** Overcoming the second line of the "chemical barrier" (**B**) leads to the secretion of proinflammatory cytokines (e.g. interleukin-1, tumor necrosis factor-α), now with visible signs of inflammation. (SC-Stratum Corneum; K-Keratinocytes; L-Leucocytes; ●-Ion-DeprivingAMPs, Protease-Inhibitors; ⬭-PreformedCationicAMPs; ✹-InducibleAMPs; ■-Proinflammatorycytokines

5 Antimicrobial Peptides and Diseases

5.1 Skin Diseases

5.1.1 Psoriasis

The identification of various antimicrobial peptides from lesional skin of patients with the common chronic inflammatory skin disease psoriasis might explain the phenomenon that these patients suffer from significantly fewer skin infections than expected (Henseler and Christophers 1995). Psoriatic scale extracts are known to contain a broad spectrum of antimicrobial peptides when compared with healthy stratum corneum proteins (Harder and Schröder 2005) and several of these proteins, like hBD-2 (Harder et al. 1997), -3 (Harder et al. 2001), psoriasin (Madsen et al. 1991; Gläser et al. 2005), and calprotectin (Clohessy and Golden 1995), were originally characterized in psoriasis patients.

An enhanced expression of lysozyme in lesional skin of psoriasis patients was documented by immunohistochemical analysis (Gasior-Chrzan et al. 1994). Upregulation of the human cathelicidin gene was shown in inflammatory skin disorders like psoriasis, whereas in normal skin no induction was found (Frohm et al. 1997). By in situ hybridization and immunohistochemistry, the transcript and the peptide was located in keratinocytes throughout the epidermis of the inflammatory regions and LL-37 was detected in partially pure fractions derived from psoriatic scales by immunoblotting (Frohm et al. 1997). A strong cytoplasmic staining of antimicrobial ALP was reported in the suprabasal keratinocytes in lesional psoriatic epidermis whereas only weak expression was found in the stratum granulosum of healthy skin (Wingens et al. 1998). Strong induction of NGAL in the epidermis of psoriasis patients was also identified and the protein was confined to spatially distinct subpopulations of keratinocytes underlying areas of parakeratosis (Mallbris et al. 2002). High expression of hBD-2 in psoriasis was confirmed by in situ hybridization and immunohistochemistry in lesional psoriatic keratinocytes (Liu et al. 1998; Huh et al. 2002). RT-PCR, immunoblotting and immunohistochemical analysis revealed enhanced mRNA- and protein expression of hBD-2, -3 and LL-37 in lesional psoriatic skin (Ong et al. 2002; Nomura et al. 2003).

As yet, it is not clear which factors are responsible for the induction of these antimicrobial peptides in psoriasis. A likely endogenous inducer might be proinflammatory cytokines like TNF-α and/or IL-1β which are known to be elevated in psoriasis lesions (Gearing et al. 1990). Another candidate might be IL-22 (Wolk et al. 2004) which is critically involved in the IL-23-dependent activation of IL-17 (Zheng et al. 2007). IL-22-producing TH17-lymphocytes are believed to represent important immune cells in psoriasis (Liang et al. 2006). As yet it is not clear whether bacterial PAMs are also important trigger factors for the induction of antimicrobial peptides in psoriasis.

5.1.2 Atopic Dermatitis

Atopic dermatitis (AD) represents another common inflammatory skin disease often associated with an increased infection rate with bacteria, especially *S. aureus*. In contrast to psoriasis, hBD-2, -3 and LL-37 expression was shown not to be increased in acute and chronic lesions from patients suffering from this disease (Ong et al. 2002; Nomura et al. 2003) when compared with psoriasis. Elevated amounts of Th2 cytokines present in atopic skin are suggested to inhibit the expected induced expression of antimicrobial peptides in the inflamed skin; and it was shown that IL-4 and IL-13 are able to suppress the cytokine-mediated induction of hBD-2 and -3 (Ong et al. 2002; Nomura et al. 2003). Using primary keratinocytes from atopic dermatitis patients, it was recently shown that the deficiency in hBD-3 expression is an acquired defect and that neutralizing the Th2 cytokine milieu in skin explants from this patients resulted in augmentation of the innate immune response(Howell et al. 2006). These data suggest that the low expression of antimicrobial peptides may contribute to the increased susceptibility of skin infection in patients with AD. Representing the first skin disease in which a diminished production of antimicrobial proteins correlates with an increased occurrence of skin infections, one can speculate that other recurrent skin infections may also be associated with a dysregulation of antimicrobial proteins.

5.1.3 Other Skin Diseases and Infections

In contrast to AD and in addition to psoriasis, increased levels of antimicrobial proteins were also found in other inflammatory skin diseases and skin infections. For example, in acne vulgaris intense hBD-2 immunoreactivity was shown in the lesional and perilesional epithelium, indicating that upregulated beta-defensins may be involved in the pathogenesis of this disease (Chronnell et al. 2001). Induced expression of hBD-2 was also associated with superficial folliculitis (Oono et al. 2003) and in tinea pedum (Kawai et al. 2006). In viral skin infections the expression of LL-37 was found to be increased in keratinocytes of patients with *Condyloma acuminata* and *Verrucae vulgares*, suggesting a role of LL-37 in cutaneous infections caused by papillomavirus (Conner et al. 2002).

5.2 Wound Healing

Chronic wounds are frequently contaminated and colonized by bacteria; and it remains unclear whether there is sufficient expression of inducible antimicrobial peptides in the margin of wounds compared with normal skin. Decreased levels of hBD-2 were demonstrated in full-thickness burns and chronic wounds (Ortega et al. 2000) whereas moderate to strong hBD-2 immunostaining was detected in chronic ulcers (Butmarc et al. 2004). The authors concluded that the constitutively high baseline expression of hBD-2 in chronic wounds might be due to ongoing tissue injury and bacterial colonization. Investigation of the expression of hBD-2 after injury showed a marked upregulation of hBD-2 after wounding (Butmarc et al. 2004).

LL-37 was also shown to be upregulated after cutaneous injury due to synthesis within epidermal keratinocytes and deposition from infiltrated granulocytes (Dorschner et al. 2001). Using blocking-antibodies to LL-37 it was demonstrated that LL-37 promotes the re-epithelialization of human skin wounds (Heilborn et al. 2003). The authors speculated that the reduction of LL-37 in chronic wounds impairs re-epithelialization and may contribute to their failure to heal.

In general, growth factors stimulate the regeneration of tissue after wounding. A recent study showed that growth factors essential in wound healing, like IGF-I and TGF-α, induce the expression of LL-37, hBD-3, NGAL and ALP in human keratinocytes (Sorensen et al. 2003). These findings offer an explanation for the expression of these antimicrobial peptides in wound healing and define a new host defense role for growth factors.

Sterile wounding of human skin resulted in induced hBD-3 expression through activation of the epidermal growth factor receptor (EGFR). After skin wounding, the receptor was activated by heparin-binding EGF that was released by a metalloprotease-dependent mechanism. Activation of the EGFR generated antimicrobial concentrations of hBD-3 and increased the activity of organotypic epidermal cultures against *S. aureus*, indicating that sterile wounding initiates an innate immune response that increases resistance to infection and microbial colonization (Sorensen et al. 2006).

Very recently, it was shown that hBD-2, -3 and -4, but not hBD-1, stimulated human keratinocytes to increase their gene expression and protein production of IL-6 and -10, IP-10, monocyte chemoattractant protein (MCP)-1, macrophage inflammatory protein (MIP)-3alpha and RANTES (Niyonsaba et al. 2007). In addition, hBDs elicited intracellular Ca^{2+}-mobilization, increased keratinocyte migration and proliferation, and induced phosphorylation of EGFR and signal transducer and activator of transcription (STAT)-1 and -3, which are intracellular signaling molecules involved in keratinocyte migration and proliferation. This data provides evidence that human antimicrobial peptides participate in wound healing by stimulating cytokine/chemokine production and promoting keratinocyte migration and proliferation.

A recent publication investigated the role of S100A7 in human wound exudate and granulation tissue (Lee and Eckert 2007). Immunohistological studies suggested that S100A7 is produced by keratinocytes surrounding the wound and is released into the wound exudate.

5.3 Diseases of the Airway Epithelia: Cystic Fibrosis

Cystic fibrosis (CF) is a life-threatening autosomal recessive disorder caused by mutant cystic fibrosis transmembrane conductance regulator (CFTR), a cAMP-regulated chloride channel (Gadsby et al. 2006; Soferman 2006). Patients with CF have a predisposition to subsequent chronic colonization and infection with bacteria, especially *P. aeruginosa*, leading to progressive lung destruction. It was postulated that the activity of antimicrobial molecules is compromised by changes in the composition of the airway surface liquid in lungs of CF patients (Smith et al. 1996). The high salt concentration in CF airway fluid, caused by the lack of functional chloride channels, was shown to inhibit salt-sensitive antimicrobial peptides, such as the defensins, from killing bacteria, and subsequently leading to increased susceptibility to infections (Guggino 2001).

Furthermore, it was demonstrated that hBD-2 and hBD-3 are susceptible to degradation and inactivation by the cysteine proteases cathepsins B, L, and S (Taggart et al. 2003). These three cathepsins are present and active in CF bronchoalveolar lavage and incubation of hBD-2 and -3 with CF lavage leads to the degradation of the antimicrobial molecules. These results show that any potential use of host-derived AMP for the treatment of CF should consider the potential inactivation of these proteins by endogenous and bacterial proteases.

Overexpression of LL-37 in CF xenografts was shown to increase the antimicrobial activity of airway surface fluid and to restore bacterial killing (Bals et al. 1999), providing an alternative genetic approach for the treatment of CF based on enhanced expression of an endogenous antimicrobial peptide. LL-37 was also investigated for its role in the regulation of wound closure of the airway epithelium. LL-37 stimulated the healing of mechanically induced wounds and induced cell proliferation and the migration of airway epithelial cells (Shaykhiev et al. 2005).

5.4 Gastrointestinal Diseases: Inflammatory Bowel Diseases

Crohn's disease (CD) and ulcerative colitis (UC) are the two major entities of chronic inflammatory bowel diseases (IBDs). UC is typically restricted to the colon and CD can be found most commonly in the ileum of the small intestine and in the colon (Podolsky 2002). The expression of HNP 1–3 in the normal intestinal mucosa and in cases of inflammatory bowel disease was studied by Cunliffe et al. (Cunliffe 2003). In the normal intestinal mucosa, HNPs were shown to be only weakly expressed in lamina propria neutrophils and not in Paneth cells. In cases of active CD and UC, scattered surface epithelial cells, as well as numerous lamina propria neutrophils, were seen to express HNPs.

Investigation of mRNA and peptide expression of biopsies from colonic resections under basal and inflammatory conditions revealed that HD-5 mRNA expression was enhanced in inflammatory states of the large bowel, suggesting that antimicrobial peptides in the colon may be of importance in maintaining the mucosal barrier and controlling microbial invasion in IBD (Wehkamp et al. 2002).

The impact of chronic inflammation on the expression of hBD-1 and -2, HD-5 and -6 and lysozyme in epithelial cells of the small and large intestine was systematically characterized by another group (Fahlgren et al. 2003). Colonic epithelial cells from patients with UC displayed a significant increase of hBD-2, HD-5, -6 and lysozyme mRNA as compared with the controls; and HD-5 and lysozyme protein were shown to be located in metaplastic Paneth-like cells in UC colon. Colonic epithelial cells of CD patients showed increased mRNA levels of HD-5 and lysozyme mRNA, whereas ileal epithelial cells of Crohn's patients with ileo-caecal inflammation did not. Chronic inflammation in colon results therefore in the induction of hBD-2 and alpha-defensins and increased lysozyme expression.

Recently, it was shown that, in contrast to UC, CD is characterized by an impaired expression of intestinal defensins (Fellermann et al. 2003). The expression of hBD-2, -3 and -4 was shown to be attenuated in the inflamed tissue of patients with CD (Wehkamp et al. 2003; Fahlgren et al. 2004). CD patients with ileal involvement are characterized by a diminished expression of the ileal Paneth cell alpha-defensins HD-5 and HD-6, which correlates with decreased antimicrobial activity in the ileal mucosa (Wehkamp et al. 2005). The decrease of alpha-defensin expression in Paneth cells is even more pronounced in CD patients with a mutation in the nucleotide-binding oligomerization domain (NOD)-2 gene, coding for an intracellular peptidoglycan receptor, which is associated with CD and ileal involvement (Wehkamp et al. 2004; Wehkamp et al. 2005). A disturbance in antimicrobial defense, as provided by Paneth cells of the small intestine, seems to be a critical factor in the pathogenesis of ileal CD.

In addition to the defensins, it was very recently demonstrated that the antimicrobial serine antiproteases elafin and secretory leukocyte protease inhibitor (SLPI) showed enhanced expression in inflamed versus non-inflamed UC and this phenomenon was significantly less pronounced in CD (Schmid et al. 2007). Elafin and SLPI may be therefore added to the list of antimicrobial peptides with diminished induction comparing CD and UC.

5.5 Diseases Associated with Phagocyte Dysfunction

5.5.1 Specific Granule Deficiency

The important role of neutrophil alpha-defensins in the killing of phagocytosed microbes has been demonstrated in patients with specific granule deficiency (SGD) who suffer from frequent and severe bacterial infections (Ganz et al. 1988). Polymorphonuclear leukocytes (PMN) from these patients display a nearly complete deficiency of defensins. These findings suggest that this profound deficiency of PMN microbicidal defensins may contribute to the clinical manifestations of the disorder (Ganz et al. 1988).

5.5.2 Kostmann Syndrome

Kostmann syndrome (also known as severe congenital neutropenia) is a rare autosomal recessive disorder in which the neutrophilic granulocytes fail to reach a mature and functional state. Infants with Kostmann syndrome often suffer from severe infections, beginning in the first months of life. It was reported that neutrophilic granulocytes from patients with Kostmann syndrome reduced the concentrations of alpha-defensins HNP-1-3 and were deficient in LL-37 (Putsep et al. 2002). Missing LL-37 in the saliva of patients with Kostmann syndrome led to the suggestion that the recurrent oral infections are the result of the deficiency of oral antimicrobials, such as LL-37.

6 General Conclusion and Future Aspects

Findings of the past two decades clearly document that epithelial cells have the capacity to mount a "chemical barrier" apart from the physical defense shield against invading microorganisms. This "chemical barrier" includes preformed compounds present at the uppermost layers of the epithelium as well as newly synthesized antimicrobial proteins that are produced upon stimulation. The stimulus could be contact with pathogenic bacteria or bacterial products, endogenous proinflammatory cytokines and/or disruption of the physical barrier by wounding with subsequently released growth factors.

Exogenous application of antimicrobial peptides could be a promising therapeutic option in the near future for the treatment of patients with epithelial infections and chronic wounds, but a much more promising option would be a promotion of the endogenous expression of antimicrobial peptides.

The focal expression of various antimicrobial peptides in healthy skin without signs of inflammation provides evidence that conditions may exist which cause the induction of antimicrobial peptides in the absence of proinflammatory cytokines or growth factors produced during wound healing. Microbes always present at skin surfaces may therefore facilitate antimicrobial peptide induction without inflammation or wounding. This hypothesis is supported by a recent observation showing that a number of probiotic bacteria, including *E. coli* Nissle 1917, induce the expression of hBD-2 in intestinal epithelial cells (Wehkamp et al. 2004). The beneficial effects of probiotic bacteria may result from their properties to induce antimicrobial peptides. It would be intriguing to speculate about bacterial components inducing antimicrobial peptides without undesirable adverse reactions. Application of such artificial "antimicrobial peptide inducers" could serve as an optimal future therapy to achieve an increased resistance towards infection in various epithelia.

Acknowledgement The authors are supported by the Deutsche Forschungsgemeinschaft (SFB 617) and by the Bundesministerium für Bildung und Forschung (BMBF: SkinStaph).

References

Abiko Y, Nishimura M, Kusano K, Yamazaki M, Arakawa T, Takuma T, Kaku T (2003) Upregulated expression of human beta defensin-1 and -3 mRNA during differentiation of keratinocyte immortalized cell lines, HaCaT and PHK16-0b. J Dermatol Sci 31:225–228

Agerberth B, Gunne H, Odeberg J, Kogner P, Boman HG, Gudmundsson GH (1995) FALL-39, a putative human peptide antibiotic, is cysteine-free and expressed in bone marrow and testis. Proc Natl Acad Sci USA 92:195–199

Ali RS, Falconer A, Ikram M, Bissett CE, Cerio R, Quinn AG (2001) Expression of the peptide antibiotics human beta defensin-1 and human beta defensin-2 in normal human skin. J Invest Dermatol 117:106–111

Alkemade JA, Molhuizen HO, Ponec M, Kempenaar JA, Zeeuwen PL, Jongh GJ de, Vlijmen-Willems IM van, Erp PE van, Kerkhof PC van de, Schalkwijk J (1994) SKALP/elafin is an inducible proteinase inhibitor in human epidermal keratinocytes. J Cell Sci 107:2335–2342

Allaker RP, Kapas S (2003) Adrenomedullin expression by gastric epithelial cells in response to infection. Clin Diagn Lab Immunol 10:546–551

Allaker RP, Zihni C, Kapas S (1999) An investigation into the antimicrobial effects of adrenomedullin on members of the skin, oral, respiratory tract and gut microflora. FEMS Immunol Med Microbiol 23:289–293

Bals R, Wang X, Wu Z, Freeman T, Bafna V, Zasloff M, Wilson JM (1998) Human beta-defensin 2 is a salt-sensitive peptide antibiotic expressed in human lung. J Clin Invest 102:874–880

Bals R, Weiner DJ, Meegalla RL, Wilson JM (1999) Transfer of a cathelicidin peptide antibiotic gene restores bacterial killing in a cystic fibrosis xenograft model. J Clin Invest 103:1113–1117

Bensch KW, Raida M, Magert HJ, Schulz-Knappe P, Forssmann WG (1995) hBD-1: a novel beta-defensin from human plasma. FEBS Lett 368:331–335

Bevins CL (2006) Paneth cell defensins: key effector molecules of innate immunity. Biochem Soc Trans 34:263–266

Braff MH, Di Nardo A, Gallo RL (2005) Keratinocytes store the antimicrobial peptide cathelicidin in lamellar bodies. J Invest Dermatol 124:394–400

Butmarc J, Yufit T, Carson P, Falanga V (2004) Human beta-defensin-2 expression is increased in chronic wounds. Wound Repair Regen 12:439–443

Casewell MW, Desai N (1983) Survival of multiply-resistant *Klebsiella aerogenes* and other gram-negative bacilli on finger-tips. J Hosp Infect 4:350–360

Chronnell CM, Ghali LR, Ali RS, Quinn AG, Holland DB, Bull JJ, Cunliffe WJ, McKay IA, Philpott MP, Muller-Rover S (2001) Human beta defensin-1 and -2 expression in human pilosebaceous units: upregulation in acne vulgaris lesions. J Invest Dermatol 117:1120–1125

Chung WO, Dale BA (2004) Innate immune response of oral and foreskin keratinocytes: utilization of different signaling pathways by various bacterial species. Infect Immun 72:352–358

Clohessy PA, Golden BE (1995) Calprotectin-mediated zinc chelation as a biostatic mechanism in host defence. Scand J Immunol 42:551–556

Cole AM, Kim YH, Tahk S, Hong T, Weis P, Waring AJ, Ganz T (2001) Calcitermin, a novel antimicrobial peptide isolated from human airway secretions. FEBS Lett 504:5–10

Cole AM, Liao HI, Stuchlik O, Tilan J, Pohl J, Ganz T (2002) Cationic polypeptides are required for antibacterial activity of human airway fluid. J Immunol 169:6985–6991

Conner K, Nern K, Rudisill J, O'Grady T, Gallo RL (2002) The antimicrobial peptide LL-37 is expressed by keratinocytes in condyloma acuminatum and verruca vulgaris. J Am Acad Dermatol 47:347–350

Cunliffe RN (2003) Alpha-defensins in the gastrointestinal tract. Mol Immunol 40:463–467

Dale BA, Krisanaprakornkit S (2001) Defensin antimicrobial peptides in the oral cavity. J Oral Pathol Med 30:321–327

De Y, Chen Q, Schmidt AP, Anderson GM, Wang JM, Wooters J, Oppenheim JJ, Chertov O (2000) LL-37, the neutrophil granule- and epithelial cell-derived cathelicidin, utilizes formyl peptide receptor-like 1 (FPRL1) as a receptor to chemoattract human peripheral blood neutrophils, monocytes, and T cells. J Exp Med 192:1069–1074

Domachowske JB, Dyer KD, Adams AG, Leto TL, Rosenberg HF (1998) Eosinophil cationic protein/RNase 3 is another RNase A-family ribonuclease with direct antiviral activity. Nucleic Acids Res 26:3358–3363

Domachowske JB, Dyer KD, Bonville CA, Rosenberg HF (1998) Recombinant human eosinophil-derived neurotoxin/RNase 2 functions as an effective antiviral agent against respiratory syncytial virus. J Infect Dis 177:1458–1464

Dorschner RA, Pestonjamasp VK, Tamakuwala S, Ohtake T, Rudisill J, Nizet V, Agerberth B, Gudmundsson GH, Gallo RL (2001) Cutaneous injury induces the release of cathelicidin anti-microbial peptides active against group A *Streptococcus*. J Invest Dermatol 117:91–97

Eckert RL, Broome AM, Ruse M, Robinson N, Ryan D, Lee K (2004) S100 proteins in the epidermis. J Invest Dermatol 123:23–33

Elias PM (2005) Stratum corneum defensive functions: an integrated view. J Invest Dermatol 125:183–200

Ellison RT 3rd, Giehl TJ (1991) Killing of gram-negative bacteria by lactoferrin and lysozyme. J Clin Invest 88:1080–1091

Ericksen B, Wu Z, Lu W, Lehrer RI (2005) Antibacterial activity and specificity of the six human alpha-defensins. Antimicrob Agents Chemother 49:269–275

Ezoe K, Katsumata M (1990) Immunohistochemical study of lysozyme in human apocrine glands. J Dermatol 17:159–163

Fahlgren A, Hammarstrom S, Danielsson A, Hammarstrom ML (2003) Increased expression of antimicrobial peptides and lysozyme in colonic epithelial cells of patients with ulcerative colitis. Clin Exp Immunol 131:90–101

Fahlgren A, Hammarstrom S, Danielsson A, Hammarstrom ML (2004) beta-Defensin-3 and -4 in intestinal epithelial cells display increased mRNA expression in ulcerative colitis. Clin Exp Immunol 137:379–385

Fellermann K, Wehkamp J, Herrlinger KR, Stange EF (2003) Crohn's disease: a defensin deficiency syndrome? Eur J Gastroenterol Hepatol 15:627–634

Fleming A (1922) On a remarkable bacteriolytic element found in tissues and secretions. Proc R Soc Lond 93:306–310

Fritz H (1988) Human mucus proteinase inhibitor (human MPI). Human seminal inhibitor I (HUSI-I), antileukoprotease (ALP), secretory leukocyte protease inhibitor (SLPI). Biol Chem Hoppe Seyler 369[Suppl]:79–82

Frohm M, Agerberth B, Ahangari G, Stahle-Backdahl M, Liden S, Wigzell H, Gudmundsson GH (1997) The expression of the gene coding for the antibacterial peptide LL-37 is induced in human keratinocytes during inflammatory disorders. J Biol Chem 272:15258–15263

Frye M, Bargon J, Gropp R (2001) Expression of human beta-defensin-1 promotes differentiation of keratinocytes. J Mol Med 79:275–282

Fulton C, Anderson GM, Zasloff M, Bull R, Quinn AG (1997) Expression of natural peptide antibiotics in human skin. Lancet 350:1750–1751

Gadsby DC, Vergani P, Csanady L (2006) The ABC protein turned chloride channel whose failure causes cystic fibrosis. Nature 440:477–483

Ganz T (2004) Antimicrobial polypeptides. J Leukoc Biol 75:34–38

Ganz T, Metcalf JA, Gallin JI, Boxer LA, Lehrer RI (1988) Microbicidal/cytotoxic proteins of neutrophils are deficient in two disorders: Chediak-Higashi syndrome and "specific" granule deficiency. J Clin Invest 82:552–556

Garcia JR, Jaumann F, Schulz S, Krause A, Rodriguez-Jimenez J, Forssmann U, Adermann K, Kluver E, Vogelmeier C, Becker D, Hedrich R, Forssmann WG, Bals R (2001) Identification of a novel, multifunctional beta-defensin (human beta-defensin 3) with specific antimicrobial activity. Its interaction with plasma membranes of Xenopus oocytes and the induction of macrophage chemoattraction. Cell Tissue Res 306:257–264

Garcia JR, Krause A, Schulz S, Rodriguez-Jimenez FJ, Kluver E, Adermann K, Forssmann U, Frimpong-Boateng A, Bals R, Forssmann WG (2001) Human beta-defensin 4: a novel inducible peptide with a specific salt-sensitive spectrum of antimicrobial activity. FASEB J 15:1819–1821

Gasior-Chrzan B, Bostad L, Falk ES (1994) An immunohistochemical study of lysozyme in the skin of psoriatic patients. Acta Derm Venereol 74:344–346

Gearing AJ, Fincham NJ, Bird CR, Wadhwa M, Meager A, Cartwright JE, Camp RD (1990) Cytokines in skin lesions of psoriasis. Cytokine 2:68–75

Ghosh D, Porter E, Shen B, Lee SK, Wilk D, Drazba J, Yadav SP, Crabb JW, Ganz T, Bevins CL (2002) Paneth cell trypsin is the processing enzyme for human defensin-5. Nat Immunol 3:583–590

Gläser R, Harder J, Lange H, Bartels J, Christophers E, Schröder JM (2005) Antimicrobial psoriasin (S100A7) protects human skin from *Escherichia coli* infection. Nat Immunol 6:57–64

Goetz DH, Holmes MA, Borregaard N, Bluhm ME, Raymond KN, Strong RK (2002) The neutrophil lipocalin NGAL is a bacteriostatic agent that interferes with siderophore-mediated iron acquisition. Mol Cell 10:1033–1043

Gombart AF, Borregaard N, Koeffler HP (2005) Human cathelicidin antimicrobial peptide (CAMP) gene is a direct target of the vitamin D receptor and is strongly up-regulated in myeloid cells by 1,25-dihydroxyvitamin D3. FASEB J 19:1067–1077

Gottsch JD, Eisinger SW, Liu SH, Scott AL (1999) Calgranulin C has filariacidal and filariastatic activity. Infect Immun 67:6631–6636

Gudmundsson GH, Agerberth B, Odeberg J, Bergman T, Olsson B, Salcedo R (1996) The human gene FALL39 and processing of the cathelin precursor to the antibacterial peptide LL-37 in granulocytes. Eur J Biochem 238:325–332

Guggino WB (2001) Cystic fibrosis salt/fluid controversy: in the thick of it. Nat Med 7:888–889

Harder J, Bartels J, Christophers E, Schröder JM (1997) A peptide antibiotic from human skin. Nature 387:861

Harder J, Bartels J, Christophers E, Schröder JM (2001) Isolation and characterization of human beta-defensin-3, a novel human inducible peptide antibiotic. J Biol Chem 276:5707–5713

Harder J, Meyer-Hoffert U, Teran LM, Schwichtenberg L, Bartels J, Maune S, Schröder JM (2000) Mucoid *Pseudomonas aeruginosa*, TNF-alpha, and IL-1beta, but not IL-6, induce human beta-defensin-2 in respiratory epithelia. Am J Respir Cell Mol Biol 22:714–721

Harder J, Meyer-Hoffert U, Wehkamp K, Schwichtenberg L, Schröder JM (2004) Differential gene induction of human beta-defensins (hBD-1, -2, -3, and -4) in keratinocytes is inhibited by retinoic acid. J Invest Dermatol 123:522–529

Harder J, Schröder JM (2002) RNase 7, a novel innate immune defense antimicrobial protein of healthy human skin. J Biol Chem 277:46779–46784

Harder J, Schröder JM (2005) Psoriatic scales: a promising source for the isolation of human skin-derived antimicrobial proteins. J Leukoc Biol 77:476–486

Heilborn JD, Nilsson MF, Kratz G, Weber G, Sorensen O, Borregaard N, Stahle-Backdahl M (2003) The cathelicidin anti-microbial peptide LL-37 is involved in re-epithelialization of human skin wounds and is lacking in chronic ulcer epithelium. J Invest Dermatol 120:379–389

Heizmann CW, Fritz G, Schafer BW (2002) S100 proteins: structure, functions and pathology. Front Biosci 7:d1356–d1368

Helmerhorst EJ, Van't Hof W, Veerman EC, Simoons-Smit I, Nieuw Amerongen AV (1997) Synthetic histatin analogues with broad-spectrum antimicrobial activity. Biochem J 326:39–45

Henseler T, Christophers E (1995) Disease concomitance in psoriasis. J Am Acad Dermatol 32:982–986

Hoover DM, Wu Z, Tucker K, Lu W, Lubkowski J (2003) Antimicrobial characterization of human beta-defensin 3 derivatives. Antimicrob Agents Chemother 47:2804–2809

Howell MD, Boguniewicz M, Pastore S, Novak N, Bieber T, Girolomoni G, Leung DY (2006) Mechanism of HBD-3 deficiency in atopic dermatitis. Clin Immunol 121:332–338

Huang GT, Zhang HB, Kim D, Liu L, Ganz T (2002) A model for antimicrobial gene therapy: demonstration of human beta-defensin 2 antimicrobial activities *in vivo*. Hum Gene Ther 13:2017–2025

Huh WK, Oono T, Shirafuji Y, Akiyama H, Arata J, Sakaguchi M, Huh NH, Iwatsuki K (2002) Dynamic alteration of human beta-defensin 2 localization from cytoplasm to intercellular space in psoriatic skin. J Mol Med 80:678–684

Jainkittivong A, Johnson DA, Yeh CK (1998) The relationship between salivary histatin levels and oral yeast carriage. Oral Microbiol Immunol 13:181–187

Jia HP, Schutte BC, Schudy A, Linzmeier R, Guthmiller JM, Johnson GK, Tack BF, Mitros JP, Rosenthal A, Ganz T, McCray PB Jr (2001) Discovery of new human beta-defensins using a genomics-based approach. Gene 263:211–218

Johansson J, Gudmundsson GH, Rottenberg ME, Berndt KD, Agerberth B (1998) Conformation-dependent antibacterial activity of the naturally occurring human peptide LL-37. J Biol Chem 273:3718–3724

Kao CY, Chen Y, Thai P, Wachi S, Huang F, Kim C, Harper RW, Wu R (2004) IL-17 markedly up-regulates beta-defensin-2 expression in human airway epithelium via JAK and NF-kappaB signaling pathways. J Immunol 173:3482–3491

Kapas S, Bansal A, Bhargava V, Maher R, Malli D, Hagi-Pavli E, Allaker RP (2001) Adrenomedullin expression in pathogen-challenged oral epithelial cells. Peptides 22:1485–1489

Kavanagh K, Dowd S (2004) Histatins: antimicrobial peptides with therapeutic potential. J Pharm Pharmacol 56:285–289

Kawai M, Yamazaki M, Tsuboi R, Miyajima H, Ogawa H (2006) Human beta-defensin-2, an antimicrobial peptide, is elevated in scales collected from tinea pedis patients. Int J Dermatol 45:1389–1390

Kern RA, Kingkade MJ, Kern SF, Behrens OK (1951) Characterization of the action of lysozyme on *Staphylococcus aureus* and on *Micrococcus lysodeikticus*. J Bacteriol 61:171–178

Kjeldsen L, Bainton DF, Sengelov H, Borregaard N (1994) Identification of neutrophil gelatinase-associated lipocalin as a novel matrix protein of specific granules in human neutrophils. Blood 83:799–807

Kjeldsen L, Johnsen AH, Sengelov H, Borregaard N (1993) Isolation and primary structure of NGAL, a novel protein associated with human neutrophil gelatinase. J Biol Chem 268:10425–10432

Klenha J, Krs V (1967) Lysozyme in mouse and human skin. J Invest Dermatol 49:396–399

Lee KC, Eckert RL (2007) S100A7 (Psoriasin) – mechanism of antibacterial action in wounds. J Invest Dermatol 127:945–957

Lehrer RI, Szklarek D, Barton A, Ganz T, Hamann KJ, Gleich GJ (1989) Antibacterial properties of eosinophil major basic protein and eosinophil cationic protein. J Immunol 142:4428–4434

Li X, Leeuw E de, Lu W (2005) Total chemical synthesis of human psoriasin by native chemical ligation. Biochemistry 44:14688–14694

Liang SC, Tan XY, Luxenberg DP, Karim R, Dunussi-Joannopoulos K, Collins M, Fouser LA (2006) Interleukin (IL)-22 and IL-17 are coexpressed by Th17 cells and cooperatively enhance expression of antimicrobial peptides. J Exp Med 203:2271–2279

Liu AY, Destoumieux D, Wong AV, Park CH, Valore EV, Liu L, Ganz T (2002) Human beta-defensin-2 production in keratinocytes is regulated by interleukin-1, bacteria, and the state of differentiation. J Invest Dermatol 118:275–281

Liu L, Wang L, Jia HP, Zhao C, Heng HH, Schutte BC, McCray PB Jr, Ganz T (1998) Structure and mapping of the human beta-defensin HBD-2 gene and its expression at sites of inflammation. Gene 222:237–244

Mackewicz CE, Yuan J, Tran P, Diaz L, Mack E, Selsted ME, Levy JA (2003) alpha-Defensins can have anti-HIV activity but are not CD8 cell anti-HIV factors. Aids 17:F23–F32

Madsen P, Rasmussen HH, Leffers H, Honore B, Dejgaard K, Olsen E, Kiil J, Walbum E, Andersen AH, Basse B, et al (1991) Molecular cloning, occurrence, and expression of a novel partially secreted protein "psoriasin" that is highly up-regulated in psoriatic skin. J Invest Dermatol 97:701–712

Maisetta G, Batoni G, Esin S, Luperini F, Pardini M, Bottai D, Florio W, Giuca MR, Gabriele M, Campa M (2003) Activity of human beta-defensin 3 alone or combined with other antimicrobial agents against oral bacteria. Antimicrob Agents Chemother 47:3349–3351

Mallbris L, O'Brien KP, Hulthen A, Sandstedt B, Cowland JB, Borregaard N, Stahle-Backdahl M (2002) Neutrophil gelatinase-associated lipocalin is a marker for dysregulated keratinocyte differentiation in human skin. Exp Dermatol 11:584–591

Martin E, Ganz T, Lehrer RI (1995) Defensins and other endogenous peptide antibiotics of vertebrates. J Leukoc Biol 58:128–136

Martinez A, Elsasser TH, Muro-Cacho C, Moody TW, Miller MJ, Macri CJ, Cuttitta F (1997) Expression of adrenomedullin and its receptor in normal and malignant human skin: a potential pluripotent role in the integument. Endocrinology 138:5597–5604

Mempel M, Voelcker V, Kollisch G, Plank C, Rad R, Gerhard M, Schnopp C, Fraunberger P, Walli AK, Ring J, Abeck D, Ollert M (2003) Toll-like receptor expression in human keratinocytes: nuclear factor kappaB controlled gene activation by *Staphylococcus aureus* is toll-like receptor 2 but not toll-like receptor 4 or platelet activating factor receptor dependent. J Invest Dermatol 121:1389–1396

Meyer-Hoffert U, Wichmann N, Schwichtenberg L, White PC, Wiedow O (2003) Supernatants of *Pseudomonas aeruginosa* induce the *Pseudomonas*-specific antibiotic elafin in human keratinocytes. Exp Dermatol 12:418–425

Midorikawa K, Ouhara K, Komatsuzawa H, Kawai T, Yamada S, Fujiwara T, Yamazaki K, Sayama K, Taubman MA, Kurihara H, Hashimoto K, Sugai M (2003) *Staphylococcus aureus* susceptibility to innate antimicrobial peptides, beta-defensins and CAP18, expressed by human keratinocytes. Infect Immun 71:3730–3739

Murakami M, Lopez-Garcia B, Braff M, Dorschner RA, Gallo RL (2004) Postsecretory processing generates multiple cathelicidins for enhanced topical antimicrobial defense. J Immunol 172:3070–3077

Murakami M, Ohtake T, Dorschner RA, Schittek B, Garbe C, Gallo RL (2002) Cathelicidin anti-microbial peptide expression in sweat, an innate defense system for the skin. J Invest Dermatol 119:1090–1095

Murthy AR, Lehrer RI, Harwig SS, Miyasaki KT (1993) *In vitro* candidastatic properties of the human neutrophil calprotectin complex. J Immunol 151:6291–6301

Niyonsaba F, Someya A, Hirata M, Ogawa H, Nagaoka I (2001) Evaluation of the effects of peptide antibiotics human beta-defensins-1/-2 and LL-37 on histamine release and prostaglandin D(2) production from mast cells. Eur J Immunol 31:1066–1075

Niyonsaba F, Iwabuchi K, Someya A, Hirata M, Matsuda H, Ogawa H, Nagaoka I (2002) A cathelicidin family of human antibacterial peptide LL-37 induces mast cell chemotaxis. Immunology 106:20–26

Niyonsaba F, Ushio H, Nakano N, Ng W, Sayama K, Hashimoto K, Nagaoka I, Okumura K, Ogawa H (2007) Antimicrobial peptides human beta-defensins stimulate epidermal keratinocyte migration, proliferation and production of proinflammatory cytokines and chemokines. J Invest Dermatol 127:594–604

Nizet V, Ohtake T, Lauth X, Trowbridge J, Rudisill J, Dorschner RA, Pestonjamasp V, Piraino J, Huttner K, Gallo RL (2001) Innate antimicrobial peptide protects the skin from invasive bacterial infection. Nature 414:454–457

Noble WC (1992) Other cutaneous bacteria. The skin microflora and microbial disease. Cambridge University Press, Cambridge

Nomura I, Goleva E, Howell MD, Hamid QA, Ong PY, Hall CF, Darst MA, Gao B, Boguniewicz M, Travers JB, Leung DY (2003) Cytokine milieu of atopic dermatitis, as compared to psoriasis, skin prevents induction of innate immune response genes. J Immunol 171:3262–3269

Ogawa H, Miyazaki H, Kimura M (1971) Isolation and characterization of human skin lysozyme. J Invest Dermatol 57:111–116

Ong PY, Ohtake T, Brandt C, Strickland I, Boguniewicz M, Ganz T, Gallo RL, Leung DY (2002) Endogenous antimicrobial peptides and skin infections in atopic dermatitis. N Engl J Med 347:1151–1160

Oono T, Huh WK, Shirafuji Y, Akiyama H, Iwatsuki K (2003) Localization of human beta-defensin-2 and human neutrophil peptides in superficial folliculitis. Br J Dermatol 148:188–191

Oren A, Ganz T, Liu L, Meerloo T (2003) In human epidermis, beta-defensin 2 is packaged in lamellar bodies. Exp Mol Pathol 74:180–182

Ortega MR, Ganz T, Milner SM (2000) Human beta defensin is absent in burn blister fluid. Burns 26:724–726

Ouellette AJ (1997) Paneth cells and innate immunity in the crypt microenvironment. Gastroenterology 113:1779–1784

Papini M, Simonetti S, Franceschini S, Scaringi L, Binazzi M (1982) Lysozyme distribution in healthy human skin. Arch Dermatol Res 272:167–170

Pernet I, Reymermier C, Guezennec A, Branka JE, Guesnet J, Perrier E, Dezutter-Dambuyant C, Schmitt D, Viac J (2003) Calcium triggers beta-defensin (hBD-2 and hBD-3) and chemokine macrophage inflammatory protein-3 alpha (MIP-3alpha/CCL20) expression in monolayers of activated human keratinocytes. Exp Dermatol 12:755–760

Pfundt R, Wingens M, Bergers M, Zweers M, Frenken M, Schalkwijk J (2000) TNF-alpha and serum induce SKALP/elafin gene expression in human keratinocytes by a p38 MAP kinase-dependent pathway. Arch Dermatol Res 292:180–187

Podolsky DK (2002) Inflammatory bowel disease. N Engl J Med 347:417–429

Putsep K, Carlsson G, Boman HG, Andersson M (2002) Deficiency of antibacterial peptides in patients with morbus Kostmann: an observation study. Lancet 360:1144–1149

Ritonja A, Kopitar M, Jerala R, Turk V (1989) Primary structure of a new cysteine proteinase inhibitor from pig leucocytes. FEBS Lett 255:211–214

Rudolph B, Podschun R, Sahly H, Schubert S, Schröder JM, Harder J (2006) Identification of RNase 8 as a novel human antimicrobial protein. Antimicrob Agents Chemother 50:3194–3196

Sahly H, Schubert S, Harder J, Rautenberg P, Ullmann U, Schroder J, Podschun R (2003) Burkholderia is highly resistant to human Beta-defensin 3. Antimicrob Agents Chemother 47:1739–1741

Salzman NH, Ghosh D, Huttner KM, Paterson Y, Bevins CL (2003) Protection against enteric salmonellosis in transgenic mice expressing a human intestinal defensin. Nature 422:522–526

Schittek B, Hipfel R, Sauer B, Bauer J, Kalbacher H, Stevanovic S, Schirle M, Schroeder K, Blin N, Meier F, Rassner G, Garbe C (2001) Dermcidin: a novel human antibiotic peptide secreted by sweat glands. Nat Immunol 2:1133–1137

Schmid M, Fellermann K, Fritz P, Wiedow O, Stange EF, Wehkamp J (2007) Attenuated induction of epithelial and leukocyte serine antiproteases elafin and secretory leukocyte protease inhibitor in Crohn's disease. J Leukoc Biol 81:907–915

Schröder JM, Harder J (1999) Human beta-defensin-2. Int J Biochem Cell Biol 31:645–651

Shaykhiev R, Beisswenger C, Kandler K, Senske J, Puchner A, Damm T, Behr J, Bals R (2005) Human endogenous antibiotic LL-37 stimulates airway epithelial cell proliferation and wound closure. Am J Physiol Lung Cell Mol Physiol 289:L842–L848

Simpson AJ, Maxwell AI, Govan JR, Haslett C, Sallenave JM (1999) Elafin (elastase-specific inhibitor) has anti-microbial activity against gram-positive and gram-negative respiratory pathogens. FEBS Lett 452:309–313

Simpson AJ, Wallace WA, Marsden ME, Govan JR, Porteous DJ, Haslett C, Sallenave JM (2001) Adenoviral augmentation of elafin protects the lung against acute injury mediated by activated neutrophils and bacterial infection. J Immunol 167:1778–1786

Singh PK, Jia HP, Wiles K, Hesselberth J, Liu L, Conway BA, Greenberg EP, Valore EV, Welsh MJ, Ganz T, Tack BF, McCray PB Jr (1998) Production of beta-defensins by human airway epithelia. Proc Natl Acad Sci USA 95:14961–14966

Smith JJ, Travis SM, Greenberg EP, Welsh MJ (1996) Cystic fibrosis airway epithelia fail to kill bacteria because of abnormal airway surface fluid. Cell 85:229–236

Soferman R (2006) Immunopathophysiologic mechanisms of cystic fibrosis lung disease. Isr Med Assoc J 8:44–48

Sohnle PG, Hunter MJ, Hahn B, Chazin WJ (2000) Zinc-reversible antimicrobial activity of recombinant calprotectin (migration inhibitory factor-related proteins 8 and 14). J Infect Dis 182:1272–1275

Sorensen OE, Cowland JB, Theilgaard-Monch K, Liu L, Ganz T, Borregaard N (2003) Wound healing and expression of antimicrobial peptides/polypeptides in human keratinocytes, a consequence of common growth factors. J Immunol 170:5583–5589

Sorensen OE, Follin P, Johnsen AH, Calafat J, Tjabringa GS, Hiemstra PS, Borregaard N (2001) Human cathelicidin, hCAP-18, is processed to the antimicrobial peptide LL-37 by extracellular cleavage with proteinase 3. Blood 97:3951–3959

Sorensen OE, Thapa DR, Roupe KM, Valore EV, Sjobring U, Roberts AA, Schmidtchen A, Ganz T (2006) Injury-induced innate immune response in human skin mediated by transactivation of the epidermal growth factor receptor. J Clin Invest 116:1878–1885

Taggart CC, Greene CM, Smith SG, Levine RL, McCray PB Jr, O'Neill S, McElvaney NG (2003) Inactivation of human beta-defensins 2 and 3 by elastolytic cathepsins. J Immunol 171:931–937

Tomita T, Hitomi S, Nagase T, Matsui H, Matsuse T, Kimura S, Ouchi Y (2000) Effect of ions on antibacterial activity of human beta defensin 2. Microbiol Immunol 44:749–754

Turner J, Cho Y, Dinh NN, Waring AJ, Lehrer RI (1998) Activities of LL-37, a cathelin-associated antimicrobial peptide of human neutrophils. Antimicrob Agents Chemother 42:2206–2214

Valore EV, Park CH, Quayle AJ, Wiles KR, McCray PB Jr, Ganz T (1998) Human beta-defensin-1: an antimicrobial peptide of urogenital tissues. J Clin Invest 101:1633–1642

Wang Y, Agerberth B, Lothgren A, Almstedt A, Johansson J (1998) Apolipoprotein A-I binds and inhibits the human antibacterial/cytotoxic peptide LL-37. J Biol Chem 273:33115–33118

Wehkamp J, Harder J, Wehkamp K, Wehkamp-von Meissner B, Schlee M, Enders C, Sonnenborn U, Nuding S, Bengmark S, Fellermann K, Schröder JM, Stange EF (2004) NF-kappaB- and AP-1-mediated induction of human beta defensin-2 in intestinal epithelial cells by Escherichia coli Nissle 1917: a novel effect of a probiotic bacterium. Infect Immun 72:5750–5758

Wehkamp J, Harder J, Weichenthal M, Mueller O, Herrlinger KR, Fellermann K, Schröder JM, Stange EF (2003) Inducible and constitutive beta-defensins are differentially expressed in Crohn's disease and ulcerative colitis. Inflamm Bowel Dis 9:215–223

Wehkamp J, Harder J, Weichenthal M, Schwab M, Schaffeler E, Schlee M, Herrlinger KR, Stallmach A, Noack F, Fritz P, Schröder JM, Bevins CL, Fellermann K, Stange EF (2004) NOD2 (CARD15) mutations in Crohn's disease are associated with diminished mucosal alpha-defensin expression. Gut 53:1658–1664

Wehkamp J, Salzman NH, Porter E, Nuding S, Weichenthal M, Petras RE, Shen B, Schaeffeler E, Schwab M, Linzmeier R, Feathers RW, Chu H, Lima H Jr, Fellermann K, Ganz T, Stange EF, Bevins CL (2005) Reduced Paneth cell alpha-defensins in ileal Crohn's disease. Proc Natl Acad Sci USA 102:18129–18134

Wehkamp J, Schwind B, Herrlinger KR, Baxmann S, Schmidt K, Duchrow M, Wohlschlager C, Feller AC, Stange EF, Fellermann K (2002) Innate immunity and colonic inflammation: enhanced expression of epithelial alpha-defensins. Dig Dis Sci 47:1349–1355

Weinberg A, Krisanaprakornkit S, Dale BA (1998) Epithelial antimicrobial peptides: review and significance for oral applications. Crit Rev Oral Biol Med 9:399–414

Weinberg ED (2001) Human lactoferrin: a novel therapeutic with broad spectrum potential. J Pharm Pharmacol 53:1303–1310

Wiedow O, Harder J, Bartels J, Streit V, Christophers E (1998) Antileukoprotease in human skin: an antibiotic peptide constitutively produced by keratinocytes. Biochem Biophys Res Commun 248:904–909

Wiedow O, Schröder JM, Gregory H, Young JA, Christophers E (1990) Elafin: an elastase-specific inhibitor of human skin. Purification, characterization, and complete amino acid sequence. J Biol Chem 265:14791–14795

Wilde CG, Griffith JE, Marra MN, Snable JL, Scott RW (1989) Purification and characterization of human neutrophil peptide 4, a novel member of the defensin family. J Biol Chem 264:11200–11203

Wingens M, Bergen BH van, Hiemstra PS, Meis JF, Vlijmen-Willems IM van, Zeeuwen PL, Mulder J, Kramps HA, Ruissen F van, Schalkwijk J (1998) Induction of SLPI (ALP/HUSI-I) in epidermal keratinocytes. J Invest Dermatol 111:996–1002

Wolk K, Kunz S, Witte E, Friedrich M, Asadullah K, Sabat R (2004) IL-22 increases the innate immunity of tissues. Immunity 21:241–254

Wu Z, Hoover DM, Yang D, Boulegue C, Santamaria F, Oppenheim JJ, Lubkowski J, Lu W (2003) Engineering disulfide bridges to dissect antimicrobial and chemotactic activities of human beta-defensin 3. Proc Natl Acad Sci USA 100:8880–8885

Yang D, Chertov O, Bykovskaia SN, Chen Q, Buffo MJ, Shogan J, Anderson M, Schroder JM, Wang JM, Howard OM, Oppenheim JJ (1999) Beta-defensins: linking innate and adaptive immunity through dendritic and T cell CCR6. Science 286:525–528

Zaiou M, Nizet V, Gallo RL (2003) Antimicrobial and protease inhibitory functions of the human cathelicidin (hCAP18/LL-37) prosequence. J Invest Dermatol 120:810–816

Zasloff M (1987) Magainins, a class of antimicrobial peptides from *Xenopus* skin: isolation, characterization of two active forms, and partial cDNA sequence of a precursor. Proc Natl Acad Sci USA 84:5449–5453

Zhang J, Dyer KD, Rosenberg HF (2002) RNase 8, a novel RNase A superfamily ribonuclease expressed uniquely in placenta. Nucleic Acids Res 30:1169–1175

Zhang L, Yu W, He T, Yu J, Caffrey RE, Dalmasso EA, Fu S, Pham T, Mei J, Ho JJ, Zhang W, Lopez P, Ho DD (2002) Contribution of human alpha-defensin 1, 2, and 3 to the anti-HIV-1 activity of CD8 antiviral factor. Science 298:995–1000

Zheng Y, Danilenko DM, Valdez P, Kasman I, Eastham-Anderson J, Wu J, Ouyang W (2006) Interleukin-22, a T(H)17 cytokine, mediates IL-23-induced dermal inflammation and acanthosis. Nature 445: 648–651

Zucht HD, Grabowsky J, Schrader M, Liepke C, Jurgens M, Schulz-Knappe P, Forssmann WG (1998) Human beta-defensin-1: a urinary peptide present in variant molecular forms and its putative functional implication. Eur J Med Res 3:315–323

Zudaire E, Martinez A, Cuttitta F (2003) Adrenomedullin and cancer. Regul Pept 112:175–183

Chapter 10
The Complement System in Innate Immunity

K.R. Mayilyan, Y.H. Kang, A.W. Dodds, and R.B. Sim()

1	The Complement System in Mammals	220
	1.1 Classical Pathway	221
	1.2 The Lectin Pathway	223
	1.3 Alternative Pathway	225
	1.4 Regulation of the Complement System	228
	1.5 Complement Receptors	229
2	The Structure of Complement Proteins	230
3	Complement Across Species	232
References		233

Abstract Complement is an important component of the innate immune defence of animals against infectious agents. The complement system in mammals is well characterised and consists of about 35–40 proteins, present in blood plasma and other body fluids, and also on cell surfaces. The function of complement is to recognise and opsonise particulate materials including invading micro-organisms and "altered-self" cells (dying, infected or damaged host cells). Recognition of a target by large polymeric complement proteins including C1q, MBL and the ficolins results in activation of proteases which cleave complement protein C3, a thiolester-containing protein (TEP) which binds covalently to the target. Target-bound complement proteins opsonise the target by promoting interaction with phagocytic cells which express complement receptors. The complement system appears to be highly conserved in vertebrates, although research on reptiles and amphibians is limited. Only a few invertebrate animals have been studied, but likely orthologues of complement target-recognition proteins, proteases and TEPs have been demonstrated in cephalochordates, urochordates, echinoderms, arthropods and coelenterates. This suggests that complement-like activity has been important in host defence since an early stage in the evolution of multicellular animals.

MRC Immunochemistry Unit, Department of Biochemistry, University of Oxford, South Parks Road, Oxford OX1 3QU, UK, *bob.sim@bioch.ox.ac.uk*

H. Heine (ed.), *Innate Immunity of Plants, Animals, and Humans.*
Nucleic Acids and Molecular Biology 21.
© Springer-Verlag Berlin Heidelberg 2008

1 The Complement System in Mammals

Complement is a central component of the innate immune system which is involved in host defence against infectious agents. The complement system in mammals is well characterised (particularly in mice and humans) and consists of about 35–40 proteins, present in blood plasma and other body fluids, and also on cell surfaces. In human blood plasma, the combined concentration of complement proteins is about 3.0–3.5 mg/ml, or about 4–5% of the total plasma protein. The function of complement is to recognise and then opsonise or lyse particulate materials

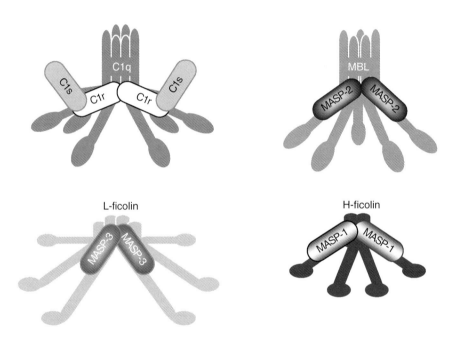

Fig. 1 The recognition proteins of the classical and lectin pathways. C1q, MBL and the ficolins are oligomeric proteins with a "bunch of tulips" shape. The "stalks" are made up of collagen triple helices, and the "heads" are globular domains, each with three lobes. One head and one collagenous stalk is made up of three polypeptide chains. C1q has six heads. It associates with a dimer of the serine protease proenzyme C1r, which in turn binds the serine protease proenzyme C1s. When C1q binds to a target, C1r autoactivates, then activates C1s. MBL has variable polymerisation and is thought to circulate in plasma in forms with four to six heads. The ficolins are probably mainly four-headed structures. MBL and the ficolins associate with one homodimer of MASP-1, or MASP2,or MASP3, or of Map19, a truncated alternative splicing product of the MASP2 gene, which lacks a serine protease domain. When MBL or ficolins bind to a target, MASP1 or MASP2 autoactivate. The activation mechanism for MASP3 is not yet known. MASP1 and MASP3 are alternative splicing products from a single gene and have different serine protease domains. C1r, C1s and the MASPs are homologues, and have the same domain structure (Fig. 3)

including invading micro-organisms and "altered-self" cells (dying, infected or damaged host cells). Opsonisation of micro-organisms with complement components targets them for phagocytosis by cells expressing complement receptors. In the complement system, large polymeric pattern recognition molecules including C1q, MBL and the ficolins (Fig. 1) have the capability to recognise micro-organisms via their highly conserved surface features (or "pathogen-associated molecular patterns"; PAMPS) such as lipopolysaccharides, lipoproteins, peptidoglycan, oligosaccharides and other surface structures (Janeway and Medzhitov 2002). Recent data extends the role of these complement pattern recognition molecules to the recognition and binding to apoptotic and necrotic host cells, and thereby complement contributes to the efficient clearance of cellular debris and apoptotic cells (Mevorach et al. 1998; Nauta et al. 2003). The innate and adaptive immune systems are often regarded as distinct arms of immunity; however, there is increasing data to suggest that innate and adaptive arms of immunity "cross-talk", and that complement has an important role bridging between innate and adaptive immunity.

The complement system can be activated via three routes, the classical pathway, lectin pathway and alternative pathway (Fig. 2). These pathways all converge at the cleavage of C3, the most abundant complement protein. The different pathways produce complex proteases capable of cleaving C3. These C3 convertases (C4b2a, C3bBb; Fig. 2) cleave C3 into two products, C3a and C3b. The major fragment, C3b, can bind covalently onto the surface of the complement-activating "target" particle, and act as an opsonin. C3b can also bind covalently onto the C3 convertases to form C5 convertases (C4b2a3b, C3bBb3b; Fig. 2) and so initiate assembly of the membrane attack complex (MAC), made up of complement proteins C5b, C6, C7, C8 and C9. This MAC can insert into lipid bilayers and cause lysis of a target cell.

1.1 Classical Pathway

The mechanisms of action and activation of complement have been extensively reviewed (McAleer and Sim 1993; Law and Reid 1995; Walport 2001a, b; Sim and Tsiftsoglou 2004). The proteins involved in the classical pathway of complement include C1q, C1r, C1s, and C2-C9 (Fig 2). One molecule of C1q associates with 1 $C1r_2^s$ heterotetramer in the presence of Ca^{2+} (Fig 1) The protein complex is called C1. The classical pathway is activated by the binding of C1q in the C1 complex to the activator. C1q binds via its globular head regions to charge clusters or hydrophobic motifs on a target surface. C1q contains 18 polypeptide chains (6A, 6B, 6C chains) and each globular head contains three domains (one each from A, B, C chains). This provides a total of 18 binding domains of three different types in each molecule, giving great versatility in recognition of targets (Gaboriaud et al. 2003; Kishore et al. 2003). Weak binding interactions of a single binding domain are

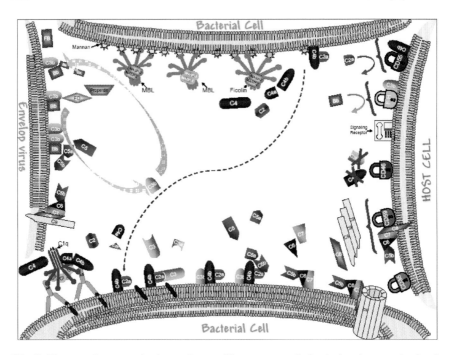

Fig. 2 The complement activation pathways. The sequence of classical pathway activation is shown at the bottom, with IgG antibodies bound to a bacterial surface as an example of a target. C1q binds to the surface (*bottom left*). C1 binding activates C1r and then C1s, which then cleaves C4 and C2. The C4b2a complex (C3 convertase) forms on the bacterial surface and cleaves C3. One C3b binds to C4b2a and forms a binding site for C5. C5 is cleaved, then the C5b6789 complex (the membrane attack complex, MAC) assembles and causes membrane damage. At the *top of the figure*, the lectin pathway is shown. MBL or a ficolin binds directly to a bacterial surface, and MASP2 is activated. This then cleaves C4 and C2, and the pathway follows the same sequence as the classical pathway. *At the left*, the alternative pathway is shown. C3b (derived from the classical or lectin pathway, or by activation by C3(H2O)Bb, as discussed in the text, binds to the surface, then binds Factor B, which is activated by Factor D, forming the C3 convertase, C3bBb. More C3 is cleaved by C3bBb, then C5 activation and MAC assembly follows, as for the classical pathway. *At the right*, the host cell and its mechanisms for protection against complement attack are shown. CD59 binds the C5b678 complex and prevents binding of C9. CD35 (CR1), CD46 (MCP) and CD55 (DAF) all destabilise the C3 convertases or inhibit their formation. Soluble regulators like Factor H (FH) may become transiently bound on the surface and also regulate convertase formation

multiplied by up to six or 18 to provide high avidity binding. The solution of the three-dimensional structure of the globular head region of C1q and the expression of the globular regions of the A, B and C chains have helped to understand why C1q can bind to so many different structures (Gaboriaud et al. 2003; Kishore et al. 2003).

Activation of the classical pathway can be achieved by the binding of C1q to IgG- and IgM-containing immune complexes (antibody–antigen complexes), or to a variety of non-immunoglobulin activators. These include the products of damaged cells, such as nucleic acid, chromatin, cytoplasmic intermediate filaments, mitochondrial membranes, possibly via cardiolipin or via mitochondrial proteins, and also some viruses, gram-positive bacteria via capsular polysaccharide and gram-negative bacteria via the lipid A component of the lipopolysaccharide of the cell wall (Ebenblicher et al. 1991; Alberti et al. 1993; Sim and Malhotra 1994; Butko et al. 1999). There has recently been interest in the role of complement in clearing apoptotic cells, which may occur by direct interaction of C1q with altered anionic phospholipid distributions on these cells (Korb and Ahearn 1997).

The binding of C1q in the C1 complex to the target induces a conformational change in C1q which leads to activation of the serine protease proenzyme C1r, which then activates proenzyme C1s. Activated C1s then cleaves complement component C2 (another serine protease proenzyme) and C4 (Fig. 2), leading to the formation of the protease complex C4b2a, which is the C3 convertase of the classical pathway. C4b binds covalently to the surface of the complement-activating target (see explanation below). C4b2a then cleaves C3 into C3a and C3b. C3 is the most abundant plasma complement component. It plays a central role in the complement system and is homologous to C4 and C5. The major fragment of activated C3, C3b, has an exposed internal thiolester which is extremely reactive with nucleophiles, such as OH or NH_2 groups. This is also the case for C4b. If the activated thiolester of C3b (or C4b) reacts either with a hydroxyl or an amine group on a target surface, C3b becomes covalently bound to the surface by an ester or an amide bond (Sim et al. 1981). C3b can also bind covalently to C4b in the C4b2a complex, forming the C4b2aC3b complex (C5 convertase), in which C4b and C3b form a binding site for C5 and orient it for cleavage by the protease C2a.

Cleavage of C5 initiates the lytic pathway which is common to all three activating pathways (Fig. 2). It is, however, only relevant if the target surface has an accessible lipid bilayer. C5 is cleaved to release C5a, an anaphylotoxin and chemotactic factor, and C5b. C5b has a binding site for C6, which together can bind C7 which alters the conformation of the C5b67 complex. This complex generates a binding site for the phospholipid bilayer of the target cell membrane. C8 binds the C5b67 complex and undergoes a conformational change, allowing insertion of its alpha chain into the membrane. C5b-8 binds a molecule of C9 which can polymerise to form a pore in the target membrane. The C5b-9 complex, or "membrane attack complex" (MAC) can contain between one and 18 C9 molecules (Law and Reid, 1995). The mechanism by which cell damage is initiated is uncertain, but the insertion of C5b-9 in the membrane causes the target to lose its ability to regulate its osmotic pressure causing cell lysis (Bhakdi and Tranum-Jensen 1991; Esser 1991; Fig. 2).

1.2 The Lectin Pathway

The best characterised recognition protein of the lectin pathway is mannan-binding lectin (MBL; Petersen et al. 2001). MBL resembles C1q in that it contains globular heads and triple-helical collagen-like regions (Fig. 1). Its globular heads consist of C-type lectin domains, which bind to neutral sugars such as mannose, N-acetylglucosamine and fucose, in a calcium ion-dependent manner. MBL interacts with any of three serine proteases, called MBL-associated serine proteases (MASPs 1, 2, 3). These

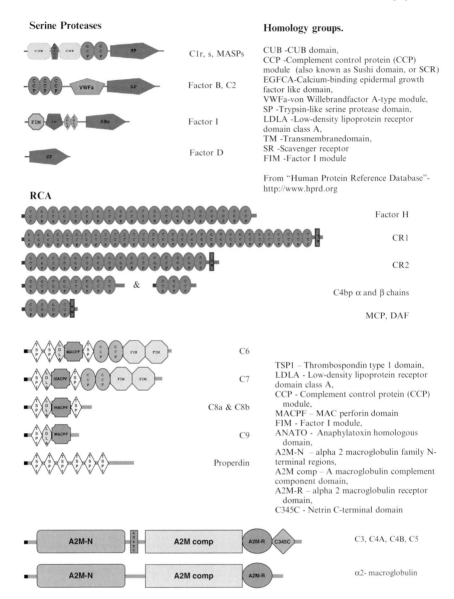

Fig. 3 The domain structures and homologies of complement proteins. Complement proteins are nearly all multi-domain proteins, and can be classified in several homology groups

are close homologues of the C1r, C1s proenzymes which bind to C1q (Fig. 3). MBL forms complexes with dimers of MASP1, or MASP2, or MASP3 (Fig. 1), or of a truncated alternative-splice product of the MASP2 gene, named sMAP or Map19 (Mayilyan et al. 2006). Each complex is similar in overall structure to the C1 complex in the classical pathway (Fig. 1). Thus MBL binds to neutral sugar arrays on targets such as yeasts, fungal spores, bacteria and viruses and, when it binds, the MASP enzymes become activated. Activated MASP2 behaves like C1s and cleaves the complement proteins C4 and C2, and so forms the same C4b2a product as is found in the classical pathway (Fig. 2; Vorup-Jensen et al. 2000). After recognition of a target and activation of MASP2, the cascade of complement activation is the same as the classical pathway (Fig. 2). The other two proteases, MASP1 and MASP3, which are alternative splicing products from a single gene, are of unknown function. Although several papers indicate that they may contribute to complement activation by cleaving C3 directly, this is not conclusively established (Petersen et al. 2001; Hajela et al. 2002). MASP1 does activate coagulation Factor XIII (plasma transglutaminase) and cleaves fibrinogen, so can mediate formation of cross-linked clots, but it is not known if this is a physiologically relevant activity (Hajela et al. 2002). No substrate has so far been found for MASP3. The truncated MASP2 product, Map19, has no proteolytic domain and so is not an enzyme. Its function is also unknown. Thus when MBL-MASP complexes bind to a target, via the lectin domains of MBL, three proteases can be activated, each of which may trigger separate downstream events; but, so far, only the complement activation via MASP2 has been characterised.

Three other, more recently characterised recognition proteins, named H-, L- and M-ficolins, also bind the MASPs and Map19 (Endo et al. 2006) . Each ficolin is likely to form a set of complexes, with each of the MASPs, and when the ficolins bind to a target, they trigger the same set of events as the binding of MBL-MASP complexes to a target. The ficolins, again like C1q and MBL, contain collagenous and globular regions and are multivalent. The globular "head" regions of ficolins are not lectin domains (as in MBL) nor are they charge-recognition domains, as in C1q. Instead they are homologous to fibrinogen domains, and their binding specificity is not yet well understood. L-ficolin binds to some acetylated species (Krarup et al. 2004), including N-acetylglucosamine and N-acetylgalactosamine. The recognition of these sugars led to the ficolins being included in the "lectin" pathway of complement activation, although "lectin" is probably not an appropriate description of ficolin specificity. H and L ficolins have been shown to bind to a range of bacterial species (Krarup et al. 2005).

The lectin pathway, via MBL, has both antibody-dependent and antibody-independent modes of activation. It is activated through binding of the lectin domain of MBL to carbohydrates on micro-organisms. It also interacts with the glycans of the common glycosylation variant of human IgG, named IgG-G0 (Malhotra et al. 1995), and with some IgA forms (Roos et al. 2001) and with glycans on mouse IgM (McMullen et al. 2006). MBL can interact with a subpopulation of human IgM glycoforms, but the interaction does not activate the lectin pathway (Arnold et al. 2005).

1.3 Alternative Pathway

The alternative pathway (Fig. 2) does not have a specific recognition molecule equivalent to C1q or MBL. Activation depends upon the subtle balance between spontaneously deposited C3b on potential activator surfaces and regulatory molecules (Meri and Pangburn, 1990). The alternative pathway is activated by IgG immune complexes and also, in the absence of antibody, by a wide range of bacteria, viruses, yeasts and protozoans (Sim and Malhotra 1994; Law and Reid 1995).

The proteins involved in the alternative pathway are factor B, factor D, factor H, factor I, properdin and C5-C9 (Fig. 2). Factor H, factor I and properdin have major roles in regulating activation of the alternative pathway. In the alternative pathway, the enzyme which cleaves C3 is C3bBb: so that, to form the C3-cleaving enzyme, a fragment of C3 (namely C3b) has to be formed already. The required C3b can be derived in two ways. First, it can arise from C3 cleavage via the classical or lectin pathway. As soon as C3b has been deposited on a complement activator by classical or lectin pathway activation, the C3b can bind Factor B (a homologue of the classical pathway C2) to form a C3bFB complex. This is cleaved by the protease factor D to form the active serine protease C3bBb, which cleaves more C3 (Fig. 2). In this way, the alternative pathway acts as an amplifier for the other pathways and increases the covalent deposition of C3b on the target. Fixation of multiple C3b on the target is essential for opsonisation. Second, the alternative pathway can be activated directly, without the need for prior classical or lectin pathway activation. C3 undergoes spontaneous hydrolysis at a very slow rate to form $C3(H_2O)$, because the internal thiolester of unactivated C3 in the circulation can be attacked by small nucleophiles such as water or ammonia, cleaving the thiolester. $C3(H_2O)$ then forms a complex with the serine protease proenzyme factor B. Factor B is cleaved by factor D to form the soluble C3 convertase $C3(H_2O)Bb$, which is homologous to the classical pathway C3 convertase, C4b2a, and cleaves C3 to form C3b and C3a. The C3b formed binds randomly and covalently to any nearby surfaces including host cells and potential target particles or cells. This can be considered as a surveillance mechanism, as all surfaces in contact with blood receive frequent "hits" from randomly generated C3b molecules. If the hit is amplified, the target surface is opsonised; but if no amplification occurs, the target surface is not damaged. Once a single molecule of C3b is bound to a target, it has two possible fates: it can form the C3 convertase C3bBb, which cleaves more C3, and clusters of C3b are deposited. This amplification of fixation leads to opsonisation of the activator surfaces (i.e. the clusters of C3b engage C3b receptors on phagocytic cells and this promotes adhesion and uptake).

If however the original C3b molecule is deposited on a host cell, it is rapidly inactivated by cell-surface complement regulatory proteins, such as complement receptor type 1 (CR1), decay-accelerating factor (DAF) and membrane cofactor protein (MCP), which are present on host cell membranes. These form a complex with the C3b and either prevent it from interacting with Factor B, or act as cofactors for its proteolysis by factor I which converts C3b to iC3b (Fig. 4). iC3b cannot form a complex with factor B and so does not promote further complement activation.

Other particles or cells which are not host may bind host soluble regulatory proteins such as factor H or C4 binding protein (C4bp) and these down-regulate complement activation on the particle surface. This ensures that C3 convertase formation and amplification of C3b fixation is inhibited. In addition, factor H competes with factor B to bind to C3b deposited on surfaces. It has been shown

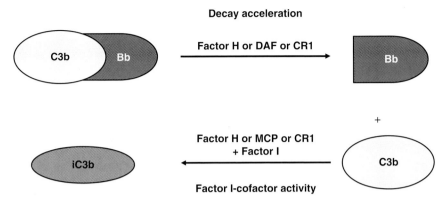

Fig. 4 Decay acceleration, and cofactor activities of factor H and proteolytic activity of factor I. Factor H can dissociate Bb from C3bBb (C3 convertase). This is called "decay-accelerating activity" of factor H. Factor H also has a cofactor activity for inactivation of C3b by factor I. When factor H binds to C3b, C3b is cleaved by factor I. Other proteins which are expressed on host cell surfaces have similar activities to factor H. These include membrane cofactor protein (MCP/CD46), complement receptor type 1 (CR1/CD35) and decay accelerating factor (DAF/CD55). C4bp acts on the classical and lectin pathway C3 convertase, C4b2a, in the same way as Factor H acts on C3bBb

that sites located on factor H interact both with C3b and with polyanions or charge clusters commonly found on the surface of non-activators. Therefore, the interaction of factor H with sites both on a surface and with C3b increases the apparent avidity of factor H for bound C3b (Carreno et al. 1989). Many pathogenic micro-organisms (and other organisms such as multicellular parasites) have evolved mechanisms to bind host Factor H or C4bp to their surface; and this is widely considered to act as a protection against complement attack (Horstmann et al. 1988; Diaz et al. 1997; Hellwage et al. 2001; Schneider et al. 2006). Thus a particle will activate the alternative pathway if:

1. It has surface OH or NH_2 groups to accept covalent binding of randomly-generated C3b *and*:
2. It does not have surface complement regulatory proteins (CR1, DAF, MCP) as host cells do *and*:
3. It does not have surface charge clusters or other binding sites which permit binding of the soluble host complement regulator, Factor H.

Therefore the alternative pathway is regulated by the random fixation of C3b and the presence or absence of complement regulatory proteins or regulatory protein binding sites on potential target surfaces.

Once the alternative pathway has been activated, the C5 convertase (C3bBbC3b) of this pathway is formed by the binding of an activated C3b to the surface bound C3bBb on an activator (Fig. 2). Once C5 convertase is formed, the late stages of complement activation lead to the assembly of the MAC, as described for the classical pathway.

1.4 Regulation of the Complement System

Because activation of each pathway in the complement system involves a high degree of amplification, regulation is required to control proteins that operate at various steps of the pathways. This prevents depletion of complement components and also damage to host cells. Different complement regulators, both fluid phase and membrane-bound, are present to regulate the complement system.

1.4.1 Regulation of Active Serine Proteases

All enzymes participating in the major steps involved with activation and control of the complement pathway belong to the family of mammalian serine proteases. These include C1r, C1s, MASPs, C2, factor B, factor D and factor I, which all have very high substrate specificity. The classical pathway C1r and C1s and the lectin pathway enzymes MASP1 and MASP2 are inhibited by a serpin, named C1-inhibitor (C1-inh), which is abundant in blood plasma (Sim et al. 1979; Ambrus et al. 2002; Presanis et al. 2004). C1-inh is a substrate analogue and forms a covalent 1:1 enzyme–substrate complex with these enzymes. The same serpin, C1-inh, is also the main physiological regulator of the coagulation proteases, Factors XIIa, XIa and kallikrein.

The proteases Factor D and Factor I circulate in active form and do not have an endogenous inhibitor. Control of these enzymes comes from the fact that their substrates are transiently formed complexes: Factor D cleaves Factor B only when it is bound to C3b. Factor I attacks C3b or C4b only when they are bound to a regulatory protein, such as Factor H, CR1, MCP, C4bp.

1.4.2 Control of C3 and C5 Convertases

The C3 and C5 convertases are controlled in three ways. First, the convertases are unstable and decay (i.e. C2a dissociates from C4b, and Bb from C3b) with a half-life of less than 5 min (McAleer and Sim 1993). C2a and Bb, dissociated from C4b and C3b, do not re-bind and are not active as proteases when dissociated. Second, a group of regulatory proteins bind to the C4b2a or C3bBb complexes and accelerate their decay. Proteins with decay-accelerating activity include the soluble proteins factor H, C4b-binding protein (C4bp) and the membrane proteins complement receptor type 1 (CR1) and decay-accelerating factor (DAF). Third, once C2a or Bb have dissociated, the remaining C4b or C3b, in complex with the same regulatory proteins, are inactivated by the protease factor I (Fig. 4). This activity of the regulatory proteins is called "cofactor" activity, as they act as cofactors for Factor I. The regulatory protein C4bp acts only as a cofactor for C4b cleavage by factor I, and Factor H acts only in C3b breakdown. CRI and MCP, however, act for both C3b and C4b.

This group of regulatory proteins, although they have confusing and non-standardised nomenclature, are all close homologues and are encoded in a gene cluster, called the regulation of complement activation (RCA) locus on human chromosome 1 (Rodriguez de Cordoba et al. 1985; Heine-Suner et al. 1997; Fig. 3). The RCA locus contains genes for the complement regulators Factor H, C4bp, MCP, DAF, a group of proteins of unknown function related to Factor H, called FHR1-5 (Diaz-Guillen et al. 1999) and the complement receptors 1 (CR1) and 2 (CR2). CR1 is both a receptor and a regulatory protein. The structure of these proteins is discussed briefly in Section 3.

Properdin is the only positive regulatory protein and acts only on the alternative pathway (Fearon and Austen 1975). Properdin binds to C3b in the activated C3bBb complex and stabilises the complex, thus increasing the life of the C3 and C5 convertases.

1.4.3 Regulation of the Membrane Attack Complex

A number of proteins (soluble and membrane bound) have regulatory roles on the lytic activity of the MAC (McAleer and Sim 1993). There are two soluble inhibitory proteins, S-protein (vitronectin) and SP40, 40 (clusterin). S-protein can bind to the C5b-7 complex, diminishing its binding to membranes. A cell-surface protein called CD59 has an inhibitory role on the MAC by binding to the C5b-8 complex, thereby preventing C9 interaction to form the MAC.

1.5 Complement Receptors

When C1q, MBL or the ficolins bind to targets they themselves act as opsonins and, in addition, they trigger complement activation resulting in the deposition of C3b, also an opsonin. Most molecules of C3b deposited on targets are gradually broken down (over a time-course of minutes to hours) to iC3b, a more effective opsonin. Eventually target-bound iC3b is more slowly degraded by proteases to a fragment called C3dg, which acts as an adjuvant. To interact with cells, these molecules require receptors.

A common receptor for C1q, MBL and another group of proteins related to MBL (the collectins) was described in 1990 (Malhotra et al. 1990) and identified later as calreticulin (Sim et al. 1998). However, calreticulin had been characterised mainly as an intracellular protein; and it took some further research to demonstrate how calreticulin functioned as part of a cell-surface receptor. It was shown by Henson and colleagues (Ogden et al. 2001; Vandivier et al. 2002) that calreticulin binds to the cell-surface receptor CD91 and acts as an adaptor to bind the collagenous region of C1q, MBL and other proteins with similar collagenous structures (the collectins SP-A and SP-D, and probably also the ficolins) to the cell surface via CD91. Phagocytic uptake of apoptotic cells, mediated by MBL or C1q binding

Opsonisation, adjuvant activity and receptors

C3 is the major source of opsonic fragments. These fragments bind to Complement Receptors CR1, CR2, CR3, CR4

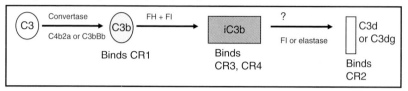

CR1: mainly on RBC: also on neutrophils, monocytes, B cells, dendritic cells.
> CR1 is the IMMUNE ADHERENCE receptor. It mediates adherence of immune complexes to RBC (for transport), contributes to phagocytosis, and also to stimulation of adaptive immune system.

CR3, 4: these are the integrins MAC1 and p150, 95. They are present on phagocytic cells and mediate phagocytosis.

CR2: on B cells, dendritic cells: mediates stimulation of the adaptive immune system by complement activators.

Fig. 5 A summary of C3 receptor functions. *RBC* Red blood cells

followed by interaction with the Calreticulin-CD91 complex was demonstrated (Vandivier et al. 2002).

CD91 has a limited cellular distribution, while C1q receptor activity is very widespread, occurring on most cell types except red blood cells. On cells which do not express CD91, HLA class I heavy chain (Arosa et al. 1999), or possibly CD59 (Ghiran et al. 2003) may act as calreticulin-binding proteins, allowing particles coated with C1q or MBL to adhere to the cells.

Receptors for fragments of C3 (C3b, iC3b, C3dg) are relatively well characterised and are summarised in Fig. 5. Recently, a new C3b and iC3b receptor on phagocytic cells was described, named CRIg (Ig indicating that it is a member of the immunoglobulin superfamily; Helmy et al. 2006).

2 The Structure of Complement Proteins

Complement proteins are nearly all multi-domain glycoproteins and can be classified in several structurally related groups (Fig. 3). These glycoproteins are well characterised and the details of their sequence, gene localisation and orthologues in other species can be found readily from the Expert Protein Analysis System (ExPASy) proteomics server of the Swiss Institute of Bioinformatics (SIB; http://expasy.org/) and its links (Gasteiger et al. 2003).

The proteases of the complement system are all serine proteases, with specificity for cleavage on the carboxyl side of arginyl or lysyl residues (Sim and Tsiftsoglou 2004). Their serine protease (SP) domains are homologous to trypsin, but unlike trypsin they have specificity for a very restricted group of protein substrates and generally have a low catalytic efficiency (compared with trypsin or even with specific proteases like thrombin or plasmin). Several of the complement proteases are likely to have arisen by duplication events, either at the gene level or as a result of larger-scale duplication. C1r, C1s and the MASPs are close homologues or paralogues. C1r and C1s are linked genes on human chromosome 12p13. MASP1 and MASP3 (alternative splicing products from a single gene) are encoded on 3q27, while MASP2 is on 1p36. Similarly, C2 of the classical pathway and Factor B of the alternative pathway are encoded by adjacent genes on 6p21 and are close homologues. Factor I, however, has an unique sequence of domains which is not found in any other human protein.

The regulation of complement activation (RCA) proteins are all encoded in the RCA gene cluster on human chromosome 1q32. Each of these proteins is composed mainly or entirely of CCP domains, each about 60 amino acids long (Fig. 3). Nearly all CCP domains are encoded by single exons, With long relatively repetitive stretches of exons of similar sequence, it is not surprising that there may be considerable gene rearrangement in this region such that, between human and mouse, there are differences in the numbers of functional genes and pseudogenes and in the size of the gene products (Sim et al. 1993a, b; Hourcade et al. 2000).

The proteins involved in MAC formation (C6, 7, 8, 9) have related structures (Fig. 3), including a domain type also present in the protein perforin, which is found in granules of cytolytic T lymphocytes and, like the MAC, induces cell lysis. C6 and C7 are encoded on linked genes on human chromosome 5p13, with C9 on 5p14. C8 is made up of α, β chains, which are homologous to each other, and a γ chain which is unrelated in structure to the α, β chains. The C8A and C8B genes are on chromosome 1p32, while the C8G gene is on 9q34.

The C3, C4, C5 family of proteins (Fig. 3) are related in structure to the major plasma protease inhibitor, $\alpha 2$ Macroglobulin ($\alpha 2M$; Sim and Sim 1981). Each of these is encoded as a single polypeptide chain of about 180–190 kDa, but C3 and C5 are cleaved into two chains before secretion, C4 into three chains and $\alpha 2M$, in contrast, oligomerises in most, but not all mammals to form a tetramer. Each of these except C5 contains an internal thiolester, formed between the Cys and Gln residues in a sequence: Gly-Cys-X-Glu-Gln. Exposure of the thiolester on proteolytic activation allows C3 and C4 to form a covalent bond, by transesterification, with hydroxyl or amino groups on complement-activating surfaces. $\alpha 2M$ can form covalent bonds with the proteases which it inhibits, although this is not essential for inhibition. C5 lacks the thiolester as the Cys and Gln in the sequence above are replaced by Ser and Ala, respectively. A thiolester is a rare post-synthetic modification in proteins; and there has been great interest in detecting thiolester proteins or "TEPs" in non-mammalian species as an indication of whether the species expresses C3- or C4-like proteins (i.e. has a complement system) . Among human proteins, the only other TEP is CD109, a cell-surface protein (Lin et al. 2002). Within this group of complement proteins, a gene duplication has formed two isotypes of C4, C4A and C4B, encoded on human chromosome 6p21, which differ in only a few amino acid residues, but have subtle functional differences.

Activation of C3, C4 and C5 leads to release of similar bioactive peptides, called anaphylotoxins (C3a, C4a, C5a; all about 10 kDa), which have vasoactive properties. C5a is also a chemotactic factor for neutrophils.

The collagen-containing complement proteins, C1q, MBL and the ficolins (not shown in Fig. 3), also show evidence of gene duplication. C1q has three homologous but not identical polypeptides, A, B and C, encoded by closely linked genes on human chromosome 1p36. There are two MBLs in many mammalian species, encoded by distinct genes. However in human, one gene is a pseudogene, so only one MBL is expressed. The MBL gene is on human chromosome 10q11. Among the ficolins, the genes for M and l-ficolin are on chromosome 9q34, but the H-ficolin gene is close to the C1q genes at 1p36.

3 Complement Across Species

The complement system of humans has been extensively explored by functional analysis, exploration of disease-related complement abnormalities, protein and nucleic acid sequencing, and determination of three-dimensional structures of proteins. Mouse complement research has been stimulated by extensive work on knock-out strains lacking various complement proteins. Few other vertebrate species have been examined in detail, but from genomic sequencing and some protein purification and functional studies, it is clear that mammals (e.g. chimpanzee, dog, horse, sheep, guinea-pig, pig, cattle) birds (mainly chicken) and bony fish (extensive work on carp, trout, etc.) have complement systems very similar to mouse and human. There is a striking feature of bony fish complement, namely that several complement components are encoded by multiple genes. In carp, all the members of the thioester-containing protein family are present in multiple isotypes, differing in the primary structures at various functional sites. Three factor B/C2-like gene products identified in carp have distinct expression pattern (sites and inducibility; Nakao et al. 2003).

There has been less work on amphibians and reptiles (mainly *Xenopus* species and cobras) but these are also thought to have similar complement systems, with all three activation pathways. Details of complement component orthologues in other species can be found by exploring the HomoloGene search function of NCBI (http://www.ncbi.nlm.nih.gov/Database/index.html); and the evolution of complement systems has been reviewed in detail (Dodds and Law 1998; Dodds 2002; Fujita et al. 2004; Nonaka and Kimura 2006; Dodds and Matsushita 2007).

In cartilaginous fish, there is quite extensive work on two species of shark, *Triakis scyllium* and *Ginglymostoma cirratum*. These appear to have classical and alternative pathways and lytic activities; and the lectin pathway is also likely to be present, since a MASP-like protease has been identified (Smith 1998). Jawless fish (agnathans, including hagfish, lamprey) have no immunoglobulin-like proteins and no other features of mammalian adaptive immunity. However, as summarised by Dodds and Matsushita (2007), lampreys have a C3/C4-like TEP, a MASP, a C2 or

Factor B-like protease, and proteins which appear related to C1q and MBL. They have no complement-like lytic mechanism. The C1q-like protein has only one type of polypeptide chain (not A, B, C as in higher vertebrates; Matsushita et al. 2004).

Among the cephalochordates and urochordates, *Amphioxus* and two ascidians (*Halocynthia roretzi, Ciona intestinalis*) have been studied, largely by M. Nonaka, T. Fujita, M. Matsushita and colleagues (for summaries, see Fujita et al. 2004; Nonaka and Kimura 2006; Dodds and Matsushita 2007). In *Amphioxus*, cDNA sequencing indicates the presence of MASP-like and C3/C4-like molecules. In the ascidians, more extensive DNA sequencing and functional studies indicate that *H. roretzi* has a complement system activated by a pathway most resembling the mammalian lectin pathway. It has a C3/C4-like protein which is opsonic and a receptor for this opsonin which is an integrin, like CR3 and CR4. Target recognition appears to be by a glucose-binding lectin, or by four ficolin-like molecules, all of which bind a MASP-like protease. *C. intestinalis* has been the subject of extensive genome sequencing (Azumi et al. 2003) and contains two C1q-like sequences, nine ficolin-like, nine MBL-like, three MASP-like, two C3/C4-like, three C2/Factor B-like and ten sequences possibly related to MAC proteins (C6, 7, 8, 9). It also has 15 integrin-like sequences and more than 100 CCP-containing proteins (possibly related to the RCA cluster proteins of mammals).

Among the rest of the animal kingdom, data are limited (for a summary, see Dodds et al. 2007) . C3/C4-like and C2/Factor B-like proteins have been found in echinoderms (sea urchins). Among arthropods, the horseshoe crab *Limulus polyphemus* has a TEP which is like α2M, but does not act like C3/C4. Two insects, the mosquito and the fruitfly have TEPs, but these do not resemble C3/C4 or α2M. However, another horseshoe crab, *Carcinoscorpius rotundicanda*, does have a C3/C4-like TEP and a C2/Factor B-like protein: no functional information is available yet on these proteins. Finally, among the coelenterates there are data on two cnidaria A C3/C4-like TEP has been found by DNA sequencing in a coral, *Swiftia excreta*, and also in the sea anemone *Nematostella vectensis*. The latter also has a C2/FB-like gene. These findings suggest that the origin of complement-like activity is near to the development of multicellular organisms.

References

Alberti S, Marques G, Camprubi S, Merino S, Tomas JM, Vivanco F, Benedi VJ (1993) C1q binding and activation of the complement classical pathway by *Klebsiella pneumoniae* outer membrane proteins. Infect Immun 61:852

Arnold JN, Wormald MR, Suter DM, Radcliffe CM, Harvey DJ, Dwek RA, Rudd PM, Sim RB (2005) Human serum IgM glycosylation: identification of glycoforms that can bind to mannan-binding lectin. J Biol Chem 280:29080

Arosa FA, De Jesus O, Porto G, Carmo AM, De Sousa M (1999) Calreticulin is expressed on the cell surface of activated human peripheral blood T lymphocytes in association with major histocompatibility complex class I molecules. J Biol Chem 274:16917

Azumi K, De Santis R, De Tomaso A, Rigoutsos I, Yoshizaki F, Pinto MR, Marino R, Shida K, Ikeda M, Ikeda M, Arai M, Inoue Y, Shimizu T, Satoh N, Rokhsar DS, Du Pasquier L,

Kasahara M, Satake M, Nonaka M (2003) Genomic analysis of immunity in a Urochordate and the emergence of the vertebrate immune system: "waiting for Godot". Immunogenetics 55:570–581

Bhakdi S, Tranum-Jensen J (1991) Complement lysis: a hole is a hole. Immunol Today 12:318

Butko P, Nicholson-Weller A, Wessels MR (1999) Role of complement component C1q in the IgG-independent opsonophagocytosis of group B streptococcus. J Immunol 163:2761

Carreno MP, Labarre D, Maillet F, Jozefowicz M, Kazatchkine MD (1989) Regulation of the human alternative complement pathway: formation of a ternary complex between factor H, surface-bound C3b and chemical groups on nonactivating surfaces. Eur J Immunol 19:2145

Dodds AW (2002) Which came first, the lectin/classical pathway or the alternative pathway of complement? Immunobiology. 205:340–354

Dodds AW, Law SK (1998) The phylogeny and evolution of the thioester bond-containing proteins C3, C4 and alpha 2-macroglobulin.Immunol Rev. 166:15–26

Dodds AW, Matsushita M (2007) The phylogeny of the complement system and the origin of the complement classical pathway. Immunobiology 212:233–243

Ebenbichler CF, Thielens NM, Vornhagen R, Marschang P, Arlaud GJ, Dierich MP (1991) Human immunodeficiency virus type 1 activates the classical pathway of complement by direct C1 binding through specific sites in the transmembrane glycoprotein gp41. J Exp Med 174:1417

Endo Y, Takahashi M, Fujita T (2006) Lectin complement system and pattern recognition. Immunobiology 211:283

Esser AF (1991) Big MAC attack: complement proteins cause leaky patches. Immunol Today 12:316

Fearon DT, Austen KF (1975) Properdin: binding to C3b and stabilization of the C3b-dependent C3 convertase. J Exp Med 142:856

Fujita T, Matsushita M, Endo Y (2004) The lectin-complement pathway – its role in innate immunity and evolution. Immunol Rev 198:185–202

Gaboriaud C, Juanhuix J, Gruez A, Lacroix M, Darnault C, Pignol D, Verger D, Fontecilla-Camps JC, Arlaud GJ (2003) The crystal structure of the globular head of complement protein C1q provides a basis for its versatile recognition properties. J Biol Chem 278:46974

Gasteiger E, Gattiker A, Hoogland C, Ivanyi I, Appel RD, Bairoch A (2003) ExPASy: the proteomics server for in-depth protein knowledge and analysis. Nucleic Acids Res 31:3784–3788

Ghiran I, Klickstein LB, Nicholson-Weller A (2003) Calreticulin is at the surface of circulating neutrophils and uses CD59 as an adaptor molecule. J Biol Chem. 278:21024–21031

Hajela K, Kojima M, AmbrusG, Wong KH, Moffatt BE, Ferluga J, Hajela S, Gal P, Sim RB (2002) The biological functions of MBL-associated serine proteases (MASPs). Immunobiology 205:467

Heine-Suner D, Diaz-Guillen MA, De Villena FP, Robledo M, Benitez J, Rodriguez de Cordoba S (1997) A high-resolution map of the regulator of the complement activation gene cluster on 1q32 that integrates new genes and markers. Immunogenetics 45:422–427

Hellwage J, Meri T, Heikkila T, Alitalo A, Panelius J, Lahdenne P, Seppala IJ, Meri S (2001) The complement regulator factor H binds to the surface protein OspE of *Borrelia burgdorferi*. J Biol Chem 276:8427

Helmy KY, Katschke KJ, Gorgani NN, Kljavin NM, Elliott JM, Diehl L, Scales SJ, Ghilardi N, Van Lookeren Campagne M (2006) CRIg: a macrophage complement receptor required for phagocytosis of circulating pathogens. Cell 124:915–927

Horstmann RD, Sievertsen HJ, Knobloch J, Fischetti VA (1988) Antiphagocytic activity of streptococcal M protein: selective binding of complement control protein factor H. Proc Natl Acad Sci USA 85:1657

Hourcade D, Liszewski ME, Krych-Goldberg M, Atkinson JP (2000) Functional domains, structural variations and pathogen interactions of MCP, DAF and CR1. Immunopharmacology 49:103–116

Janeway CA Jr, Medzhitov R (2002) Innate immune recognition. Annu Rev Immunol 20:197

Kishore U, Gupta SK, Perdikoulis MV, Kojouharova MS, Urban BC, Reid KB (2003) Modular organization of the carboxyl-terminal, globular head region of human C1q A, B, and C chains. J Immunol 171:812

Korb LC, Ahearn JM (1997) C1q binds directly and specifically to surface blebs of apoptotic human keratinocytes: complement deficiency and systemic lupus erythematosus revisited. J Immunol 158:4525

Krarup A, Thiel S, Hansen A, Fujita T, Jensenius J-C (2004) L-ficolin is a pattern recognition molecule specific for acetyl groups. J Biol Chem 279:47513

Krarup A, Sorensen UB, Matsushita M, Jensenius J-C, Thiel S (2005) Effect of capsulation of opportunistic pathogenic bacteria on binding of the pattern recognition molecules mannan-binding lectin, L-ficolin, and H-ficolin. Infect Immun 73:1052

Law SKA, Reid KBM (1995) Complement, second edition. In: Male D (ed) In focus. Oxford, UK: IRL Press.

Lin M, Sutherland DR, Horsfall W, Totty N, Yeo E, Nayar R, Wu XF, Schuh AC (2002) Cell surface antigen CD109 is a novel member of the alpha(2) macroglobulin/C3, C4, C5 family of thioester-containing proteins. Blood 99:1683–1691

Malhotra R, Wormald MR, Rudd PM, Fischer PB, Dwek RA, Sim RB (1995) Glycosylation changes of IgG associated with rheumatoid arthritis can activate complement via the mannose-binding protein. Nat Med 1:237

Mayilyan KR, Presanis JS, ArnoldJN, Hajela K, Sim RB (2006) Heterogeneity of MBL-MASP complexes. Mol Immunol 43:1286–1292

McAleer MA, Sim RB (1993) The complement system. In: Sim RB (ed) Activators and Inhibitors of Complement. Kluwer, Dordrecht, pp 1–15

McMullen ME, Hart ML, Walsh MC, Buras J, Takahashi K, Stahl GL (2006) Mannose-binding lectin binds IgM to activate the lectin complement pathway in vitro and in vivo. Immunobiology 211:759–766

Meri S, Pangburn MK (1990) Discrimination between activators and nonactivators of the alternative pathway of complement: regulation via a sialic acid/polyanion binding site on factor H. Proc Natl Acad Sci USA 87:3982

Mevorach D, Mascarenhas JO, Gershov D, Elkon KB (1998) Complement-dependent clearance of apoptotic cells by human macrophages. J Exp Med 188:2313

Nakao M, Mutsuro J, Nakahara M, Kato Y, Yano T (2003) Expansion of genes encoding complement components in bony fish: biological implications of the complement diversity. Dev Comp Immunol 27:749–762

Nauta AJ, Daha MR, Van Kooten C, Roos A (2003) Recognition and clearance of apoptotic cells: a role for complement and pentraxins. Trends Immunol 24:148

Nonaka M, Kimura A (2006) Genomic view of the evolution of the complement system. Immunogenetics 58:701–713

Ogden CA, deCathelineau A, Hoffmann PR, Bratton D, Ghebrehiwet B, Fadok VA, Henson PM (2001) C1q and mannose binding lectin engagement of cell surface calreticulin and CD91 initiates macropinocytosis and uptake of apoptotic cells. J Exp Med 194:781

Petersen SV, Thiel S, Jensenius J-C (2001) The mannan-binding lectin pathway of complement activation: biology and disease association. Mol Immunol 38:133–149

Presanis JS, Hajela K, Ambrus G, Gal P, Sim RB (2004) Differential substrate and inhibitor profiles for human MASP-1 and MASP-2. Mol Immunol 40:921

Rodriguez de Cordoba S, Lublin DM, Rubinstein P,Atkinson JP (1985) Human genes for three complement components that regulate the activation of C3 are tightly linked. J Exp Med 161:1189–1195

Roos A, Bouwman LH, Gijlswijk-Janssens DJ van, Faber-Krol MC, Stahl GL, Daha MR (2001) Human IgA activates the complement system via the mannan-binding lectin pathway. J Immunol 167:2861

Schneider MC, Exley RM, Chan H, Feavers I, Kang YH, Sim RB, Tang CM (2006) Functional significance of factor H binding to *Neisseria meningitidis*. J Immunol 176:7566

Sim RB, Malhotra R(1994) Interactions of carbohydrates and lectins with complement. Biochem Soc Trans 22:106

Sim RB, Sim E (1981) Autolytic fragmentation of complement components C3 and C4 under denaturing conditions, a property shared with alpha 2-macroglobulin. Biochem J 193:129–141

Sim RB, Tsiftsoglou SA (2004) Proteases of the complement system. Biochem Soc Trans 32:21
Sim RB, Arlaud GJ, Colomb MG (1979) C1 inhibitor-dependent dissociation of human complement component C1 bound to immune complexes. Biochem J 179:449
Sim RB, Twose TM, Paterson DS, Sim E (1981) The covalent-binding reaction of complement component C3. Biochem J 193:115–127
Sim RB, Day AJ, Moffatt BE, Fontaine M (1993a) Complement factor I and cofactors in control of complement system convertase enzymes. Methods Enzymol 223:13
Sim RB, Kolble K, McAleer MA, Dominguez O, Dee VM (1993b) Genetics and deficiencies of the soluble regulatory proteins of the complement system. Int Rev Immunol 10:65–86
Sim RB, Moestrup SK, Stuart GR, Lynch NJ, Lu J, Schwaeble WJ, Malhotra R (1998) Interaction of C1q and the collectins with the potential receptors calreticulin (cC1qR/collectin receptor) and megalin. Immunobiology 199:208
Smith SL (1998) Shark complement:an assessment. Immunol Rev 166:67–78
Vandivier RW, Ogden CA, Fadok VA, Hoffmann PR, Brown KK, Botto M, Walport MJ, Fisher JH, Henson PM, Greene KE (2002) Role of surfactant proteins A, D, and C1q in the clearance of apoptotic cells in vivo and in vitro: calreticulin and CD91 as a common collectin receptor complex. J Immunol 169:3978
Vorup-Jensen T, Petersen SV, Hansen AG, Poulsen K, Schwaeble W, Sim RB, Reid KB, Davis SJ, Thiel S, Jensenius J-C (2000) Distinct pathways of mannan-binding lectin (MBL)- and C1-complex autoactivation revealed by reconstitution of MBL with recombinant MBL-associated serine protease-2. J Immunol 165:2093
Walport MJ (2001a) Complement. First of two parts. N Engl J Med 344:1058
Walport MJ (2001b) Complement. Second of two parts. N Engl J Med 344:1140

Index

A

Acrorhagi tentacles, 31
Aiptasia, 36, 37
Allorecognition, 27, 31–34, 104–108
Anopheles gambiae, 84
Anthopleura, 31, 36
Antibacterial peptides, 117
Antibody-antigen complex, 222
Antimicrobial activity, 39, 103, 190–196, 198–202, 208, 209
Antimicrobial peptide, 29, 35, 39, 43–45, 48, 50–53, 55, 56, 60, 61, 103–104, 118, 175, 188–210
Apoptosis-associated speck-like protein (ASC), 170, 172, 173, 176, 177
Asobara tabida, 76
Atopic dermatitis, 206
Attacin, 44
Aurelia aurita, 35
Aurelin, 35

B

Bacteria, 34, 44, 46, 47, 56, 61, 75, 83, 85–86, 103, 114, 116–117, 121–122, 124, 126, 176, 188, 191
Baculovirus inhibitor of apoptosis repeat (BIR) domain, 171, 173, 177
Bendless, 58
Bilateria, 27–30, 38
Blau syndrome, 178, 179
Botryllus schlosseri, 104, 105, 108

C

C1-inh (C1-inhibitor), 228
C3 convertase, 106, 221–223, 226, 227
C5 convertase, 221, 223, 227–229
Cactus, 48, 53–55
Caenorhabditis elegans, 8, 27, 29, 76, 118, 146
Candida albicans, 51, 84, 103, 104, 139, 140, 195
Caspase, 49, 58–60, 63
  Caspase-1, 170, 172, 176, 177
Caspase-activating recruitment domain (CARD), 171–174, 176–178
Catch tentacle, 31–32
Calreticulin, 229, 230
Cathelicidin, 200, 201, 205
CCP, 86, 224, 231, 233
CD11b/CD18, 137, 138, 141
CD14, 57, 137, 138, 140, 141, 150
CD46, 106, 222, 227
CD55, DAF (Decay-accelerating factor) 106, 222, 227
CD59, 106, 222, 229, 230
CD91, 229, 230
CD94, 108
Cecropin, 44, 54, 103
Chemical barrier, 2, 187, 188, 202–204, 210
Chemokine, 122, 124, 136, 142, 176, 191–193, 207
Chlorella, 37, 38
Chronic neurologic cutaneous and articular syndrome, 179
Ciona intestinalis, 99, 101, 102, 104, 105, 108, 233
Ciona savignyi, 101
Cnidaria, 27–39, 100, 102, 233
Colony specificity, 104
Complement, 7, 62, 84–86, 101, 106–107, 114, 127, 154, 219–233.

237

Complement proteins, 107, 219–221, 224, 225, 230–232
   C1q, 219–225, 229, 230, 232, 233
   C1r, 220–224, 228, 231
   C1s, 220–225, 228, 231
   C2, 221–226, 228, 231–233
   C3, 84, 101, 106, 219, 221–233
   C3a, 221, 223, 226, 232
   C3b, 221–223, 226–230
   C3dg, 229, 230
   C4, 107, 222, 223, 225, 226, 231–233
   C5, 221–224, 226–229, 231, 232
   C5a, 223, 232
   C5b, 221, 223, 229
   C6, 101, 107, 221, 223, 224, 231, 233
   C7, 107, 221, 223, 224, 231
   C8, 106, 107, 221, 223, 231
   C9, 101, 106, 107, 221–224, 226, 229, 231
   iC3b, 226, 227, 229, 230
Coral, 28–32, 36, 37, 233
CR1, 106, 222, 224, 226–230
CR2, 106, 224, 229, 230
CR3, 13, 106, 230, 233
CR4, 106, 230, 233
Crohn's disease 151, 178, 208, 209
Croquemort, 87
Cross-talk, 155, 221
Crystal cells, 74, 77, 83
Cryopyrin (NALP-3) 172, 177, 179
Cystic fibrosis, 207–208
Cytokines, 114, 116, 118, 121, 122, 124, 135, 142–147, 153, 171, 176, 187, 191, 195, 196, 203–206, 210

D
Death domain (DD), 49, 52, 57, 117, 142, 147
Defensins, 35, 44, 54, 175, 176, 189–194, 200, 206, 208–210
   hBD-1, 189–191, 207, 208
   hBD-2, 190–193, 203, 205–210
   hBD-3, 190–194, 206–208
Dermcidin, 190, 195
γ-D-glutamyl-meso-diaminopimelic acid (iE-DAP), 170, 172, 174
dIAP2, 53, 58
DIF, 48, 49, 52–55
Differentiation, 116, 141, 191, 192, 196
Diptericin, 44, 60, 85, 90, 117
Disease resistance genes, 171
Domeless (DOME), 61, 90

Domino, 73
Dorsal, 48, 49, 52–55, 74, 116, 117
DREDD, 53, 58–60
Drosocin, 44
Drosomycin, 44, 52, 54, 55, 85, 117
Drosophila, 4, 6, 29, 43–55, 58–63, 73–90, 116, 136, 146
Dscam, 63, 84–85

E
Early-onset-sarcoidosis, 178
E.coli, 56, 74, 84–87, 90, 191–192, 195–200, 202, 210
Eater, 86–87
EGF-like domains, 105, 106
Encapsulation, 44, 45, 73–80, 82–83, 90, 93
Erwinia carotovora, 56, 90
Evolution, 1–3, 6, 7, 9, 10, 12, 13, 16, 17, 19, 20, 33, 37–39, 99–101, 105, 107
   coevolution, 1, 3, 10, 12–14, 16–20

F
Factor B, 222, 224, 226, 228, 231–233
Factor D, 222, 224, 226, 228
Factor H, 222, 224, 226–229
Factor I, 201, 224, 226–228, 231
Factor XIa, 228
Factor XIIa, 228
Factor XIII, 225
FADD, 53
Familial cold autoinflammatory syndrome, 179
Fat body, 43, 45, 57, 59, 62, 63, 77, 79, 85, 90
Flagellin, 3, 137, 138, 140, 152, 170, 172, 173, 177
Fu/Hc-receptor, 104, 105
Fungia, 36

G
Gorgonians, 31
Graft-versus-host-disease, 178
Gram-negative bacteria, 35, 43–47, 49, 50, 56, 59, 60, 84–86, 89, 90, 103, 117, 174, 189, 192, 195, 201, 223
Gram-negative binding protein, (GNBP) 49–51
Gram-positive bacteria, 35, 44, 46–52, 55, 85, 86, 103, 174, 189, 191, 193, 195, 199, 200, 223
Grass, 50, 52

H

Halocynthia roretzi, 103–105, 233
Hemocytes, 45, 54, 62, 63, 73–76, 78, 83–85, 90, 102–103, 106
Histocompatibility, 31, 100, 104, 105, 171
Homophilic interactions, 176
HrVC120, 105
HvAPX1, 37, 38
Hydra, 27–29, 31, 34–39
Hydractinia, 27, 28, 31–34
Hydramacin-1, 35
Hydrozoa, 27–29, 31

I

Ice protease-activating factor (IPAF), 137, 170, 172, 176, 177, 179
IKK, 53, 54, 56, 59, 60, 63, 118, 143–145, 147, 174
IL-1 receptor-associated kinase (IRAK-4), 137, 142, 144–148, 152–155
  deficiency, 148, 154
  dissociation, 149
  genes, 136, 152
  knock-out mice, 148
  mRNA, 147, 154
  mutations, 136, 137, 149–155
  protein, 152, 154, 155
  recruitment, 147, 148
IL-1R-associated kinase (IRAK), 52, 53, 122
  IRAK-1, 142–149, 154, 155
  IRAK-2, 147–149
  IRAK-M, 147–149
IMD, 45, 53, 57, 58
  pathway, 43, 45, 47, 50, 52, 53, 55–61
Inflammasome, 176, 179
Immulectin, 76–78
Integrin, 76, 78, 80, 82, 230, 233
Interferon-γ, 193
Interleukin, 4, 5, 116–118, 204
Interstitial stem cell, 28, 32
Intraspecies competition, 27, 31, 38
IRF-3, 142–145, 147
IRF-5, 144, 146
IRF-7, 144–146

J

JAK/STAT, 45, 61, 62, 75, 90
Janeway, Charles v, 100, 115, 136, 221
Jellyfish, 27–29, 31, 34–36, 39
JNK, 53, 58, 59, 143, 192

K

Kallikrein, 228
Keratinocyte, 190–194, 196–199, 201–207, 210

L

Lactoferrin, 190, 202
Lamellar bodies, 192
Lamellocytes, 74, 75, 78
Lectin, 76, 78–79, 106, 108, 114, 125–127, 220–228, 232, 233
Leptopilina boulardi, 75
Leucin-rich repeats (LRR), 1, 2, 4, 5, 30, 35, 137, 170–172
  domain, 1, 6–9, 15
Lipocalin, 202
Lipopolysaccharide (LPS) 3, 55, 57, 76, 77, 90, 116, 117, 136–138, 140–142, 145, 147–150, 152–155, 221, 223
Listeria, 35, 88, 103, 104, 121, 174
Lysozyme, 46, 51, 189, 190, 205, 208, 209

M

Macroglobulin complement-related (MCR), 84
Macrophages, 74, 83, 126, 137, 138, 149, 176, 177
Manduca sexta, 76
Mannan-binding lectin (MBL), 219–225, 229, 230, 232, 233
MAP kinases, 81, 118, 135, 143, 144
MASPs, 106, 220, 223–225, 228, 231
  MASP-1, 220
  MASP-2, 220
  MASP-3, 220
  Map19, 220, 224, 225
MD-2, 137, 138, 140, 142, 150, 154
Membrane attack complex (MAC), 106, 107, 221–224, 227, 229, 231, 233
Membrane cofactor protein (MCP), 106, 207, 222, 224–229
Metchnikoff, 30
Metchnikowin, 44, 54
Millepora, 31
Muckle–Wells syndrome, 179
MurNAc-L-Ala-D-isoGln, 172
Mycobacterium, 87, 121, 126, 138, 141, 151
MyD88, 7, 52, 53, 63, 116–118, 122, 137–140, 142–149, 154, 155

N

Natural killer (NK) cells, 101, 108, 122–125
Necrotic, 31, 32, 50, 52, 140, 221
Nematocysts, 31–33
Neutrophil, 46, 126, 137, 154, 188, 193, 194, 200–202, 209
NF-κB, 48, 49, 52, 53, 55, 57–59, 61, 63, 116–118, 141–146, 148–151, 154, 170, 172, 174, 177, 178, 192
Nitric oxide (NO), 90, 135, 142
NKR-P1, 108
Nod-like receptors (NLRs), 169–180
Nonspecific cytotoxic cells, 122–124
Nucleotide-binding oligomerisation domain (NOD), 6, 7, 46, 136, 169–179, 209
Nucleotide binding site (NBS)-LRR
    genes 6, 7, 9, 10, 13, 20
    proteins 6–9, 11, 13

P

Pacifastacus leniusculus, 76
Paneth cell, 175, 194, 208, 209
Parasitic wasps, 74, 75
Parasitoid parasites, 73, 75, 78
Pathogen-associated molecular patters (PAMPs), 29, 30, 135–138, 140–142, 144, 146–147, 149–152, 170, 174–175, 203, 221
Pattern recognition receptor (PRR), 29, 51, 76, 79, 86, 87, 114, 115, 136, 203
Pelle, 52, 53
Peptidoglycan (PGN), 45–51, 53, 55, 76, 77, 85, 136, 174, 209, 221
Peptidoglycan Recognition receptor (PGRP), 45–51, 53, 55–58, 60–61, 85–87
    PGRP-LC, 47, 48, 50, 53, 55–58, 85–87
    PGRP-LE, 48, 50, 55–58, 85
    PGRP-SA, 46, 49–51, 85
    PGRP-SD, 49–51
    PGRP-SC1a, 85, 86
Perforin, 107, 123, 124, 224, 231
Peroxidasin, 76, 78, 79
Persephone, 50, 51
Peste, 87, 88
Petromyzon marinus, 101
Phagocytes, 30, 83, 126
Phagocytosis, 30, 44, 45, 63, 73, 75, 76, 79–81, 83–90, 106, 114, 142, 221, 230
Physical barrier, 29, 36, 45, 187, 188, 203, 210
Plasmatocytes, 74, 75, 77, 78, 87, 88
Protease, 8, 11, 49–52, 89, 90, 106, 123, 177, 190, 197, 199–209, 219, 220, 223, 226, 228–303

Properdin, 224, 226, 229
Psoriasis, 191, 205, 206
Pyrin domain (PYD), 171–173, 176, 177

R

Rab, 89
Rac, 78–82, 88
Reactive oxygen, 4, 75, 83, 126, 135, 142, 200
Regulation of complement activation (RCA) locus, 224, 229, 231, 233
Relish, 48, 49, 52, 53, 55, 58–60, 62, 63
RHIM domain, 57
Rho, 78–82, 88
Rick, 172, 174, 175
RNAi, 52, 58, 60, 74, 83–91
RNase
    RNase 7, 190, 195, 203
    RNase 8, 190, 196

S

S100, 190, 196–198, 201
Scavenger receptor protein (dSR-CI), 86, 87
Sea anemones 27–32, 35
Self/non-self discrimination, 104, 105
Self-sterility, 104, 105
Serpin, 52, 228
Signalosome, 175
Skin, 39, 46, 103, 178, 188–210
Spatial competition, 31
Spätzle, 49–53, 63
Spätzle processing enzyme (SPE), 50, 52
Spirit, 50, 52
Staphylococcus aureus, 51, 84, 103, 104, 138, 189, 194, 195,
Styela clava, 103
Styelin, 103
Stylophora, 31
Symbiosis, 36–38, 47

T

TAB2, 53, 58, 118
TAK1, 53, 58–60, 118, 175
Thioester proteins (TEPs), 62, 84, 219, 231, 233
TIR domain-containing adapter inducing IFN-β (TRIF), 53, 57, 63, 116, 118, 144–147
TIR domain-containing adapter protein/ MyD88 adapter-like protein (TIRAP/MAL), 116, 118, 143–149

Toll, 2, 4–7, 30, 49–55, 58, 61–63, 75, 85, 86, 114–117, 135–137
    pathway 43, 45, 47, 49–55, 61, 63
Toll-like Receptors (TLRs), 7, 35, 49, 50, 63, 104, 117–118, 121, 122, 125, 127, 136–155, 203
    activation, 141
    expression, 137
    oligomerization, 141, 144
    polymorphisms,149–152, 155
    recognition, 139
    signal transduction, 142
    specificity, 146, 147
    tyrosine phosphorylation, 141, 142
Toll/IL-1R (TIR) domain, 52
Tracheal cytotoxin (TCT), 47–50, 56, 57
TRAF2, 52, 58
TRAF6, 142–146, 148
TRIF-related adapter molecule (TRAM), 53, 116, 118, 144, 145, 147
Tube, 52, 53
Tunicates, 100
Turandot (Tot) genes, 61, 62

U
Ubiquitin, 53, 54, 58–60, 117–118, 143, 174
UEV1a, 53, 58, 117, 143
Ulcerative colitis, 208, 209
Unpaired (UPD), 61, 90
Urochordates, 30, 34, 99–107, 219, 233

V
vCRL1, 107
Virus, 2, 3, 8, 34, 44, 62, 103, 114, 116, 121, 123, 125, 136, 137, 139, 140, 200, 202, 206
Vitelline coat, 105–106
Vitronectin, 229

W
Wounds, 188, 203, 206–208, 210

Z
Zebrafish, 113–115, 117–122, 124–127
Zooxanthella, 36, 37
ZP domain, 105, 106

Printing: Krips bv, Meppel, The Netherlands
Binding: Stürtz, Würzburg, Germany